The Composite Materials Handbook-MIL 17

VOLUME 3

Materials Usage, Design, and Analysis

CRC Press
Taylor & Francis Group
Boca Raton London New York

CRC Press is an imprint of the
Taylor & Francis Group, an **informa** business

The Composite Materials Handbook-MIL 17
Release 3E, 1997

Distributed by:
Technomic Publishing Company, Inc.
851 New Holland Avenue, Box 3535
Lancaster, Pennsylvania 17604 U.S.A.

This edition is published jointly by Technomic Publishing Company, Inc. and Materials Sciences Corporation in cooperation with ASTM. Distributed by Technomic Publishing Co., Inc.

Printed in the United States of America
10 9 8 7 6 5 4 3 2 1

Main entry under title:
The Composite Materials Handbook-MIL 17: Materials Usage, Design, and Analysis, Volume 3

ISBN No. 1-56676-828-4 (Volume 3)
ISBN No. 1-56676-829-2 (3-Volume Set)
ISBN No. 1-56676-831-4 (3-Volume Set and CD-ROM)

ACKNOWLEDGEMENT

The services necessary for the development and maintenance of the Composite Materials Handbook (MIL-HDBK-17) are provided by the handbook Secretariat, Materials Sciences Corporation. This work is performed under contract with the US Army Research Laboratory (Contract Number DAAL01-97-C-0140).

The primary source of funding for the current contract is the Federal Aviation Administration. Other sources include NASA, Army, Department of Energy, and Air Force. Volunteer committee members from government, industry, and academia coordinate and review all the information provided in this handbook. The time and effort of the volunteers and the support of their respective departments, companies, and universities make it possible to insure completeness, accuracy, and state-of-the-art composite technology.

FOREWORD

1. This handbook is approved for use by all Departments and Agencies of the Department of Defense.

2. This handbook is for guidance only. This handbook cannot be cited as a requirement. If it is, the contractor does not have to comply. This mandate is a DoD requirement only; it is not applicable to the Federal Aviation Administration (FAA) or other government agencies.

3. Every effort has been made to reflect the latest information on polymeric composites. The handbook is continually reviewed and revised to ensure its completeness and currentness. Documentation for the secretariat should be directed to: Materials Sciences Corporation, MIL-HDBK-17 Secretariat, 500 Office Center Drive, Suite 250, Fort Washington, PA 19034.

4. MIL-HDBK-17 provides guidelines and material properties for polymer (organic) matrix composite materials. The first three volumes of this handbook currently focus on, but are not limited to, polymeric composites intended for aircraft and aerospace vehicles. Metal matrix composites (MMC), ceramic matrix composites (CMC), and carbon/carbon composites (C/C) will be covered in separate volumes as developments occur.

5. This standardization handbook has been developed and is being maintained as a joint effort of the Department of Defense and the Federal Aviation Administration.

6. The information contained in this handbook was obtained from materials producers, industry, reports on Government sponsored research, the open literature, and by contact with research laboratories and those who participate in the MIL-HDBK-17 coordination activity.

7. All information and data contained in this handbook have been coordinated with industry and the US Army, Navy, Air Force, NASA, and Federal Aviation Administration prior to publication.

8. Copies of this document and revisions thereto may be obtained from the Standardization Document Order Desk, Bldg. 4D, 700 Robbins Avenue, Philadelphia, PA 19111-5094.

9. Beneficial comments (recommendations, additions, deletions) and any pertinent data which may be of use in improving this document should be addressed to: Director, U.S. Army Research Laboratory, Weapons and Materials Research Directorate, Attn: AMSRL-WM-M, Aberdeen Proving Ground, MD 21005-5069, by using the Standardization Document Improvement Proposal (DD Form 1426) appearing at the end of this document or by letter.

CONTENTS

SUMMARY OF CHANGES IN MIL-HDBK-17-3E

Chapter 1

This chapter has been reorganized.

Chapter 2

This chapter has had a major reorganization. Several sections have been revised. New sections on shipping and storage, process control, and specifications have been added.

Chapter 3

The statistical methods formerly in this chapter have been moved to Volume 1, Section 8.4.2.

Chapter 4

The section on buckling has been revised.

The section on critical fluids, sensitivity and evaluation (formerly 4.12) was moved to Volume 1, Section 2.3.1.3.

Chapter 8

This chapter has been reorganized.

Chapter 9

The outline and sections of this chapter were added.

Editorial corrections and conversions to S.I. units have been made as needed.

CHAPTER 1 GENERAL INFORMATION

Page

1.1 INTRODUCTION

The standardization of a statistically-based mechanical property data base, procedures used, and overall material guidelines for characterization of composite material systems is recognized as being beneficial to both manufacturers and government agencies. A complete characterization of the capabilities of any engineering material system depends on the inherent material physical and chemical composition which are independent of specific applications. Therefore, at the material system characterization level, the data and guidelines contained in this handbook apply to military and commercial products and provide the technical basis for establishing statistically valid design values acceptable to certificating or procuring agencies.

This standardization handbook has been developed and is maintained as a joint effort of the Department of Defense and the Federal Aviation Administration. It is oriented toward the standardization of methods used to develop and analyze mechanical property data on current and emerging composite materials.

1.2 PURPOSE, SCOPE, AND ORGANIZATION OF VOLUME 3

This handbook is for guidance only. This handbook cannot be cited as a requirement. If it is, the contractor does not have to comply. This mandate is a DoD requirement only; it is not applicable to the Federal Aviation Administration (FAA) or other government agencies.

Volume 3 of MIL-HDBK-17 provides methodologies and lessons learned for the design, manufacture, and analysis of composite structures and for utilization of the material data provided in Volume II consistent with the guidance provided in Volume I. The information provided is included primarily as background and as a basis for consistent use of terminology, notation, and methodology and is not offered for regulatory purposes. The volume represents a compilation of the relevant composites design, manufacture, and analysis experience of engineers and scientists from industry, government, and academia.

The scope of Volume 3 is limited to the introduction of concepts, methodologies, and potential pitfalls in the manufacture and analysis of composites. **Chapter 2, Materials and Processes**, defines major material systems and processing methods. Effects of various processing parameters on final composite product performance are emphasized. **Chapter 3, Quality Control of Production Materials**, reviews important issues related to quality control in the production of composite materials. It reviews recommended manufacturing inspection procedures and techniques for material property verification and statistical quality control. **Chapter 4, Design and Analysis**, addresses the design and analysis of various composite systems. It provides an overview of the current techniques and describes how the various constituent properties reported in MIL-HDBK-17 are used in the design and analysis of a composite system. From micromechanics to simple laminate constructions, it presents standard analyses which provide a common nomenclature and methodology basis for users of MIL-HDBK-17. **Chapter 5, Design and Analysis of Structural Joints**, describes accepted design procedures and analytical methods for determining stresses and deformations in structural joints for composite structures. **Chapter 6, Structural Reliability**, discusses some of the important factors affecting composite structure reliability including static strength, environmental effects, fatigue, and damage tolerance. **Chapter 7, Thick-Section Composites**, details methods of thick-section laminate analysis, thick-section structural analysis techniques, physical property requirements for three-dimensional analysis, experimental property determination techniques, and process simulation techniques and models. **Chapter 8, Supportability**, considers the design for and the design of repairs in composite structures based on maintainability and reliability. It provides guidelines to the designer of new structures with supportability/maintainability in mind, and it provides the information relevant to cost-effective repair procedures. Appropriately concluding Volume 3 is **Chapter 9, Lessons Learned**, which addresses a variety of issues related to earlier topics providing a depository of knowledge gained from a number of involved contractors, agencies, and businesses.

1.3 SYMBOLS, ABBREVIATIONS, AND SYSTEMS OF UNITS

This section defines the symbols and abbreviations which are used within MIL-HDBK-17 and describes the system of units which is maintained. Common usage is maintained where possible. References 1.3(a), 1.3(b), and 1.3(c) served as primary sources for this information.

1.3.1 Symbols and abbreviations

The symbols and abbreviations used in this document are defined in this section with the exception of statistical symbols. These latter symbols are defined in Chapter 8. The lamina/laminate coordinate axes used for all properties and a summary of the mechanical property notation are shown in Figure 1.3.1.

- The symbols f and m, when used as either subscripts or superscripts, always denote fiber and matrix, respectively.
- The type of stress (for example, cy - compression yield) is always used in the superscript position.
- Direction indicators (for example, x, y, z, 1, 2, 3, etc.) are always used in the subscript position.
- Ordinal indicators of laminae sequence (e.g., 1, 2, 3, etc.) are used in the superscript position and must be parenthesized to distinguish them from mathematical exponents.
- Other indicators may be used in either subscript or superscript position, as appropriate for clarity.
- Compound symbols (such as, basic symbols plus indicators) which deviate from these rules are shown in their specific form in the following list.

The following general symbols and abbreviations are considered standard for use in MIL-HDBK-17. Where exceptions are made, they are noted in the text and tables.

A - (1) area (m^2,in^2)
- (2) ratio of alternating stress to mean stress
- (3) A-basis for mechanical property values

a - (1) length dimension (mm,in)
- (2) acceleration (m/sec^2,ft/sec^2)
- (3) amplitude
- (4) crack or flaw dimension (mm,in)

B - (1) B-basis for mechanical property values
- (2) biaxial ratio

Btu - British thermal unit(s)

b - width dimension (mm,in), e.g., the width of a bearing or compression panel normal to load, or breadth of beam cross-section

C - (1) specific heat (kJ/kg °C,Btu/lb °F)
- (2) Celsius

CF - centrifugal force (N,lbf)

CPF - crossply factor

CPT - cured ply thickness (mm, in.)

CG - (1) center of mass, "center of gravity"
- (2) area or volume centroid

℄ - centerline

c - column buckling end-fixity coefficient

$\bar{c}$ - honeycomb sandwich core depth (mm,in)

cpm - cycles per minute

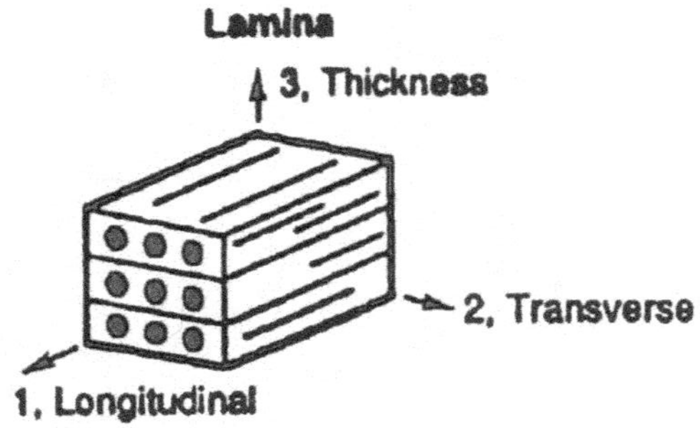

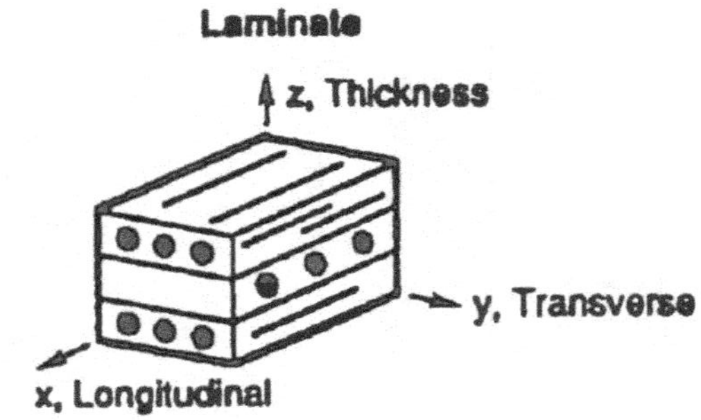

Notation = H_i^{jk}

Where,

$$H = \begin{cases} \sigma, \tau; \text{ Applied Normal, Shear Stress} \\ F; \text{ Allowable Stress} \\ \varepsilon, \gamma; \text{ Extensional, Shear Strain} \\ E, G; \text{ Young's, Shear Modulus} \\ \nu; \text{ Poisson's Ratio} \end{cases}$$

Note: ν_{12}^{t} = Major Poisson's Ratio = $-\dfrac{\varepsilon_2}{\varepsilon_1^{t}}$

ν_{21}^{t} = Minor Poisson's Ratio = $-\dfrac{\varepsilon_1}{\varepsilon_2^{t}}$

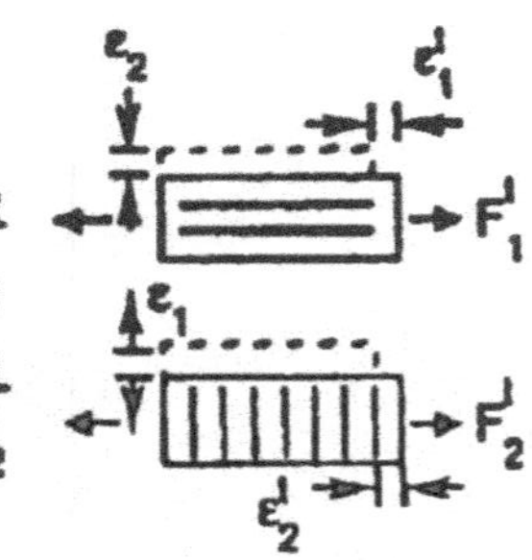

$$i = \begin{cases} \left.\begin{array}{l} 1; \text{ Longitudinal} \\ 2; \text{ Transverse} \\ 3; \text{ Thickness} \\ 12, 13, 32; \text{ Shear, Poisson's} \end{array}\right\} \text{Lamina} \\ \left.\begin{array}{l} x; \text{ Longitudinal} \\ y; \text{ Transverse} \\ z; \text{ Thickness} \\ xy, xz, zy; \text{ Shear, Poisson's} \end{array}\right\} \text{Laminate} \end{cases}$$

$$j = \begin{cases} c; \text{ Compression} \\ t; \text{ Tension} \\ s; \text{ Shear} \end{cases}$$

$$k = \begin{cases} y; \text{ Yield} \\ u; \text{ Ultimate, Not Used for Stiffness} \end{cases}$$

Examples, F_2^{tu} = Lamina Ultimate Transverse Tensile Allowable Stress

E_z^{c} = Laminate Compressive Young's Modulus, Thickness Direction

FIGURE 1.3.1 *Mechanical property notation.*

D - (1) diameter (mm,in)
- (2) hole or fastener diameter (mm,in)
- (3) plate stiffness (N-m,lbf-in)

d - mathematical operator denoting differential

E - modulus of elasticity in tension, average ratio of stress to strain for stress below proportional limit (Gpa,Msi)

E' - storage modulus (GPa,Msi)

E" - loss modulus (GPa,Msi)

E_c - modulus of elasticity in compression, average ratio of stress to strain for stress below proportional limit (GPa,Msi)

E_c' - modulus of elasticity of honeycomb core normal to sandwich plane (GPa,Msi)

E^{sec} - secant modulus (GPa,Msi)

E^{tan} - tangent modulus (GPa,Msi)

e - minimum distance from a hole center to the edge of the sheet (mm,in)

e/D - ratio of edge distance to hole diameter (bearing strength)

F - (1) stress (MPa,ksi)
- (2) Fahrenheit

F^b - bending stress (MPa,ksi)

F^{ccr} - crushing or crippling stress (upper limit of column stress for failure) (MPa,ksi)

F^{su} - ultimate stress in pure shear (this value represents the average shear stress over the cross-section) (MPa,ksi)

FAW - fiber areal weight (g/m^2, lb/in^2)

FV - fiber volume (%)

f - (1) internal (or calculated) stress (MPa,ksi)
- (2) stress applied to the gross flawed section (MPa,ksi)
- (3) creep stress (MPa,ksi)

f^c - internal (or calculated) compressive stress (MPa,ksi)

f_c - (1) maximum stress at fracture (MPa,ksi)
- (2) gross stress limit (for screening elastic fracture data (MPa,ksi)

ft - foot, feet

G - modulus of rigidity (shear modulus) (GPa,Msi)

GPa - gigapascal(s)

g - (1) gram(s)
- (2) acceleration due to gravity (m/s^2,ft/s^2)

H/C - honeycomb (sandwich)

h - height dimension (mm,in) e.g. the height of a beam cross-section

hr - hour(s)

I - area moment of inertia (mm^4,in^4)

i - slope (due to bending) of neutral plane in a beam, in radians

in. - inch(es)

J - (1) torsion constant (= I_p for round tubes) (m^4,in^4)
- (2) Joule

K - (1) Kelvin
- (2) stress intensity factor ($MPa\sqrt{m}$,$ksi\sqrt{in}$)
- (3) coefficient of thermal conductivity (W/m °C, Btu/ft²/hr/in/°F)
- (4) correction factor
- (5) dielectric constant

K_{app} - apparent plane strain fracture toughness or residual strength ($MPa\sqrt{m}$,$ksi\sqrt{in}$)

K_c - critical plane strain fracture toughness, a measure of fracture toughness at point of crack growth instability ($MPa\sqrt{m}$,$ksi\sqrt{in}$)

K_{Ic} - plane strain fracture toughness ($MPa\sqrt{m}$,$ksi\sqrt{in}$)

K_N - empirically calculated fatigue notch factor

K_s - plate or cylinder shear buckling coefficient

K_t - (1) theoretical elastic stress concentration factor

- (2) t_w/c ratio in H/C sandwich

Kv - dielectric strength (KV/mm, V/mil)
K_x,K_y - plate or cylinder compression buckling coefficient
k - strain at unit stress (m/m,in/in)
L - cylinder, beam, or column length (mm,in)
L' - effective column length (mm,in)
lb - pound
M - applied moment or couple (N-m,in-lbf)
Mg - megagram(s)
MPa - megapascal(s)
MS - military standard
M.S. - margin of safety
MW - molecular weight
MWD - molecular weight distribution
m - (1) mass (kg,lb)
- (2) number of half wave lengths
- (3) metre
- (4) slope

N - (1) number of fatigue cycles to failure
- (2) number of laminae in a laminate
- (3) distributed in-plane forces on a panel (lbf/in)
- (4) Newton
- (5) normalized

NA - neutral axis
n - (1) number of times in a set
- (2) number of half or total wavelengths
- (3) number of fatigue cycles endured

P - (1) applied load (N,lbf)
- (2) exposure parameter
- (3) probability
- (4) specific resistance (Ω)

P^u - test ultimate load, (N,lb per fastener)
P^y - test yield load, (N,lb per fastener)
p - normal pressure (Pa,psi)
psi - pounds per square inch
Q - area static moment of a cross-section (mm^3,in^3)
q - shear flow (N/m,lbf/in)
R - (1) algebraic ratio of minimum load to maximum load in cyclic loading
- (2) reduced ratio

RA - reduction of area
R.H. - relative humidity
RMS - root-mean-square
RT - room temperature
r - (1) radius (mm,in)
- (2) root radius (mm,in)
- (3) reduced ratio (regression analysis)

S - (1) shear force (N,lbf)
- (2) nominal stress in fatigue (MPa,ksi)
- (3) S-basis for mechanical property values

S_a - stress amplitude in fatigue (MPa,ksi)
S_e - fatigue limit (MPa,ksi)
S_m - mean stress in fatigue (MPa,ksi)
S_{max} - highest algebraic value of stress in the stress cycle (MPa,ksi)
S_{min} - lowest algebraic value of stress in the stress cycle (MPa,ksi)

S_R - algebraic difference between the minimum and maximum stresses in one cycle (MPa,ksi)
S.F. - safety factor
s - (1) arc length (mm,in)
- (2) H/C sandwich cell size (mm,in)
T - (1) temperature (°C,°F)
- (2) applied torsional moment (N-m,in-lbf)
T_d - thermal decomposition temperature (°C,°F)
T_F - exposure temperature (°C,°F)
T_g - glass transition temperature (°C,°F)
T_m - melting temperature (°C,°F)
t - (1) thickness (mm,in)
- (2) exposure time (s)
- (3) elapsed time (s)
V - (1) volume (mm^3,in^3)
- (2) shear force (N,lbf)
W - (1) weight (N,lbf)
- (2) width (mm,in)
- (3) Watt
x - distance along a coordinate axis
Y - nondimensional factor relating component geometry and flaw size
y - (1) deflection (due to bending) of elastic curve of a beam (mm,in)
- (2) distance from neutral axis to given point
- (3) distance along a coordinate axis
Z - section modulus, I/y (mm^3,in^3)
α - coefficient of thermal expansion (m/m/°C,in/in/°F)
γ - shear strain (m/m,in/in)
Δ - difference (used as prefix to quantitative symbols)
δ - elongation or deflection (mm,in)
ϵ - strain (m/m,in/in)
ϵ - elastic strain (m/m,in/in)
ϵ - plastic strain (m/m,in/in)
μ - permeability
η - plasticity reduction factor
$[\eta]$ - intrinsic viscosity
η^* - dynamic complex viscosity
ν - Poisson's ratio
ρ - (1) density (kg/m3,lb/in3)
- (2) radius of gyration (mm,in)
ρ_c' - H/C sandwich core density (kg/m^3,lb/in^3)
Σ - total, summation
σ - standard deviation
σ_{ij}, τ_{ij} - stress in j direction on surface whose outer normal is in i direction (i, j = 1, 2, 3 or x, y, z) (MPa,ksi)
T - applied shear stress (MPa,ksi)
ω - angular velocity (radians/s)
∞ - infinity

1.3.1.1 Constituent properties

The following symbols apply specifically to the constituent properties of a typical composite material.

E^f - Young's modulus of filament material (MPa,ksi)
E^m - Young's modulus of matrix material (MPa,ksi)

E_x^g - Young's modulus of impregnated glass scrim cloth in the filament direction or in the warp direction of a fabric (MPa,ksi)

E_y^g - Young's modulus of impregnated glass scrim cloth transverse to the filament direction or to the warp direction in a fabric (MPa,ksi)

G^f - shear modulus of filament material (MPa,ksi)

G^m - shear modulus of matrix (MPa,ksi)

G_{xy}^g - shear modulus of impregnated glass scrim cloth (MPa,ksi)

$G_{cx}^{'}$ - shear modulus of sandwich core along X-axis (MPa,ksi)

$G_{cy}^{'}$ - shear modulus of sandwich core along Y-axis (MPa,ksi)

ℓ - filament length (mm,in)

α^f - coefficient of thermal expansion for filament material (m/m/°C,in/in/°F)

α^m - coefficient of thermal expansion for matrix material (m/m/°C,in/in/°F)

α_x^g - coefficient of thermal expansion of impregnated glass scrim cloth in the filament direction or in the warp direction of a fabric (m/m/°C,in/in/°F)

α_y^g - coefficient of thermal expansion of impregnated glass scrim cloth transverse to the filament direction or to the warp direction in a fabric (m/m/°C,in/in/°F)

ν^f - Poisson's ratio of filament material

ν^m - Poisson's ratio of matrix material

ν_{xy}^g - glass scrim cloth Poisson's ratio relating to contraction in the transverse (or fill) direction as a result of extension in the longitudinal (or warp) direction

ν_{yx}^g - glass scrim cloth Poisson's ratio relating to contraction in the longitudinal (or warp) direction as a result of extension in the transverse (or fill) direction

σ - applied axial stress at a point, as used in micromechanics analysis (MPa,ksi)

τ - applied shear stress at a point, as used in micromechanics analysis (MPa,ksi)

1.3.1.2 Laminae and laminates

The following symbols, abbreviations, and notations apply to composite laminae and laminates. At the present time the focus in MIL-HDBK-17 is on laminae properties. However, commonly used nomenclature for both laminae and laminates are included here to avoid potential confusion.

A_{ij} $(i,j = 1,2,6)$ - extensional rigidities (N/m,lbf/in)

B_{ij} $(i,j = 1,2,6)$ - coupling matrix (N,lbf)

C_{ij} $(i,j = 1,2,6)$ - elements of stiffness matrix (Pa,psi)

D_x, D_y - flexural rigidities (N-m,lbf-in)

D_{xy} - twisting rigidity (N-m,lbf-in)

D_{ij} $(i,j = 1,2,6)$ - flexural rigidities (N-m,lbf-in)

E_1 - Young's modulus of lamina parallel to filament or warp direction (GPa,Msi)

E_2 - Young's modulus of lamina transverse to filament or warp direction (GPa,Msi)

E_x - Young's modulus of laminate along x reference axis (GPa,Msi)

E_y - Young's modulus of laminate along y reference axis (GPa,Msi)

G_{12} - shear modulus of lamina in 12 plane (GPa,Msi)

G_{xy} - shear modulus of laminate in xy reference plane (GPa,Msi)

h_i - thickness of ith ply or lamina (mm,in)

M_x, M_y, M_{xy} - bending and twisting moment components (N-m/m, in-lbf/in in plate and shell analysis)

n_f - number of filaments per unit length per lamina

Q_x, Q_y - shear force parallel to z axis of sections of a plate perpendicular to x and y axes, respectively (N/m,lbf/in)

Q_{ij} $(i,j = 1,2,6)$ - reduced stiffness matrix (Pa,psi)

u_x, u_y, u_z	- components of the displacement vector (mm,in)
u_x^o, u_y^o, u_z^o	- components of the displacement vector at the laminate's midsurface (mm,in)
V_v	- void content (% by volume)
V_f	- filament content or fiber volume (% by volume)
V_g	- glass scrim cloth content (% by volume)
V_m	- matrix content (% by volume)
V_x, V_y	- edge or support shear force (N/m,lbf/in)
W_f	- filament content (% by weight)
W_g	- glass scrim cloth content (% by weight)
W_m	- matrix content (% by weight)
W_s	- weight of laminate per unit surface area (N/m^2,lbf/in^2)
α_1	- lamina coefficient of thermal expansion along 1 axis (m/m/°C,in/in/°F)
α_2	- lamina coefficient of thermal expansion along 2 axis (m/m/°C,in/in/°F)
α_x	- laminate coefficient of thermal expansion along general reference x axis (m/m/°C,in/in/°F)
α_y	- laminate coefficient of thermal expansion along general reference y axis (m/m/°C,in/in/°F)
α_{xy}	- laminate shear distortion coefficient of thermal expansion (m/m/°C,in/in/°F)
θ	- angular orientation of a lamina in a laminate, i.e., angle between 1 and x axes (°)
λ_{xy}	- product of ν_{xy} and ν_{yx}
ν_{12}	- Poisson's ratio relating contraction in the 2 direction as a result of extension in the 1 direction[1]
ν_{21}	- Poisson's ratio relating contraction in the 1 direction as a result of extension in the 2 direction[1]
ν_{xy}	- Poisson's ratio relating contraction in the y direction as a result of extension in the x direction[1]
ν_{yx}	- Poisson's ratio relating contraction in the x direction as a result of extension in the y direction[1]
ρ_c	- density of a single lamina (kg/m^3,lb/in^3)
$\bar{\rho}_c$	- density of a laminate (kg/m^3,lb/in^3)
ϕ	- (1) general angular coordinate, (°)
	- (2) angle between x and load axes in off-axis loading (°)

1.3.1.3 Subscripts

The following subscript notations are considered standard in MIL-HDBK-17.

1, 2, 3	- laminae natural orthogonal coordinates (1 is filament or warp direction)
A	- axial
a	- (1) adhesive
	- (2) alternating
app	- apparent
byp	- bypass
c	- composite system, specific filament/matrix composition. Composite as a whole, contrasted to individual constituents. Also, sandwich core when used in conjunction with prime (')
	- (4) critical
cf	- centrifugal force
e	- fatigue or endurance
eff	- effective

[1]The convention for Poisson's ratio should be checked before comparing different sources as different conventions are used.

eq	- equivalent
f	- filament
g	- glass scrim cloth
H	- hoop
i	- i^{th} position in a sequence
L	- lateral
m	- (1) matrix
	- (2) mean
max	- maximum
min	- minimum
n	- (1) n^{th} (last) position in a sequence
	- (2) normal
p	- polar
s	- symmetric
st	- stiffener
T	- transverse
t	- value of parameter at time t
x, y, z	- general coordinate system
Σ	- total, or summation
o	- initial or reference datum
( )	- format for indicating specific, temperature associated with term in parentheses. RT - room temperature (21°C,70°F); all other temperatures in °F unless specified.

1.3.1.4 Superscripts

The following superscript notations are considered standard in MIL-HDBK-17.

b	- bending
br	- bearing
c	- (1) compression
	- (2) creep
cc	- compression crippling
cr	- compression buckling
e	- elastic
f	- filament
g	- glass scrim cloth
is	- interlaminar shear
(i)	- i^{th} ply or lamina
lim	- limit, used to indicate limit loading
m	- matrix
ohc	- open hole compression
oht	- open hole tension
p	- plastic
pl	- proportional limit
rup	- rupture
s	- shear
scr	- shear buckling
sec	- secant (modulus)
so	- offset shear
T	- temperature or thermal
t	- tension
tan	- tangent (modulus)
u	- ultimate
y	- yield

- secondary (modulus), or denotes properties of H/C core when used with subscript c

CAI - compression after impact

1.3.1.5 Acronyms

The following acronyms are used in MIL-HDBK-17.

AA - atomic absorption
AES - Auger electron spectroscopy
AIA - Aerospace Industries Association
ANOVA - analysis of variance
ARL - US Army Research Laboratory - Materials Directorate
ASTM - American Society for Testing and Materials
BMI - bismaleimide
BVID - barely visible impact damage
CAI - compression after impact
CCA - composite cylinder assemblage
CLS - crack lap shear
CMCS - Composite Motorcase Subcommittee (JANNAF)
CPT - cured ply thickness
CTA - cold temperature ambient
CTD - cold temperature dry
CTE - coefficient of thermal expansion
CV - coefficient of variation
CVD - chemical vapor deposition!
DCB - double cantilever beam
DDA - dynamic dielectric analysis
DLL - design limit load
DMA - dynamic mechanical analysis
DOD - Department of Defense
DSC - differential scanning calorimetry
DTA - differential thermal analysis
DTRC - David Taylor Research Center
ENF - end notched flexure
EOL - end-of-life
ESCA - electron spectroscopy for chemical analysis
ESR - electron spin resonance
ETW - elevated temperature wet
FAA - Federal Aviation Administration
FFF - field flow fractionation
FMECA - Failure Modes Effects Criticality Analysis
FOD - foreign object damage
FTIR - Fourier transform infrared spectroscopy
FWC - finite width correction factor
GC - gas chromatography
GSCS - Generalized Self Consistent Scheme
HDT - heat distortion temperature
HPLC - high performance liquid chromatography
ICAP - inductively coupled plasma emission
IITRI - Illinois Institute of Technology Research Institute
IR - infrared spectroscopy
ISS - ion scattering spectroscopy
JANNAF - Joint Army, Navy, NASA, and Air Force
LC - liquid chromatography
LPT - laminate plate theory

LSS	- laminate stacking sequence
MMB	- mixed mode bending
MOL	- material operational limit
MS	- mass spectroscopy
MSDS	- material safety data sheet
MTBF	- Mean Time Between Failure
NAS	- National Aerospace Standard
NASA	- National Aeronautics and Space Administration
NDI	- nondestructive inspection
NMR	- nuclear magnetic resonance
PEEK	- polyether ether ketone
RDS	- rheological dynamic spectroscopy
RH	- relative humidity
RT	- room temperature
RTA	- room temperature ambient
RTD	- room temperature dry
RTM	- resin transfer molding
SACMA	- Suppliers of Advanced Composite Materials Association
SAE	- Society of Automotive Engineers
SANS	- small-angle neutron scattering spectroscopy
SEC	- size-exclusion chromatography
SEM	- scanning electron microscopy
SFC	- supercritical fluid chromatography
SI	- International System of Units (Le Système Interational d'Unités)
SIMS	- secondary ion mass spectroscopy
TBA	- torsional braid analysis
TEM	- transmission electron microscopy
TGA	- thermogravimetric analysis
TLC	- thin-layer chromatography
TMA	- thermal mechanical analysis
TOS	- thermal oxidative stability
TVM	- transverse microcrack
UDC	- unidirectional fiber composite
VNB	- V-notched beam
XPS	- X-ray photoelectron spectroscopy

1.3.2 System of units

To comply with Department of Defense Instructive 5000.2, Part 6, Section M, "Use of the Metric System," dated February 23, 1991, the data in MIL-HDBK-17 are generally presented in both the International System of Units (SI units) and the U. S. Customary (English) system of units. ASTM E 380, Standard for Metric Practice, provides guidance for the application for SI units which are intended as a basis for worldwide standardization of measurement units (Reference 1.3.2(a)). Further guidelines on the use of the SI system of units and conversion factors are contained in the following publications (References 1.3.2(b) - (e)):

(1) DARCOM P 706-470, *Engineering Design Handbook: Metric Conversion Guide*, July 1976.

(2) NBS Special Publication 330, "The International System of Units (SI)," National Bureau of Standards, 1986 edition.

(3) NBS Letter Circular LC 1035, "Units and Systems of Weights and Measures, Their Origin, Development, and Present Status," National Bureau of Standards, November 1985.

(4) NASA Special Publication 7012, "The International System of Units Physical Constants and Conversion Factors", 1964.

English to SI conversion factors pertinent to MIL-HDBK-17 data are contained in Table 1.3.2.

TABLE 1.3.2 *English to SI conversion factors.*

To convert from	to	Multiply by
Btu (thermochemical)/in^2-s	watt/meter^2 (W/m^2)	1.634 246 E+06
Btu-in/(s-ft^2-°F)	W/(m K)	5.192 204 E+02
degree Fahrenheit	degree Celsius (°C)	T = (T - 32)/1.8
degree Fahrenheit	kelvin (K)	T = (T + 459.67)/1.8
foot	meter (m)	3.048 000 E-01
ft^2	m^2	9.290 304 E-02
foot/second	meter/second (m/s)	3.048 000 E-01
ft/s^2	m/s^2	3.048 000 E-01
inch	meter (m)	2.540 000 E-02
in.^{2}	meter^2 (m^2)	6.451 600 E-04
in.^{3}	m^3	1.638 706 E-05
kilogram-force (kgf)	newton (N)	9.806 650 E+00
kgf/m^2	pascal (Pa)	9.806 650 E+00
kip (1000 lbf)	newton (N)	4.448 222 E+03
ksi (kip/in^2)	MPa	6.894 757 E+00
lbf-in	N-m	1.129 848 E-01
lbf-ft	N-m	1.355 818 E+00
lbf/in^2 (psi)	pascal (Pa)	6.894 757 E+03
lb/in^2	gm/m^2	7.030 696 E+05
lb/in^3	kg/m^3	2.767 990 E+04
Msi (10^6 psi)	GPa	6.894 757 E+00
pound-force (lbf)	newton (N)	4.488 222 E+00
pound-mass (lb avoirdupois)	kilogram (kg)	4.535 924 E-01
torr	pascal (Pa)	1.333 22 E+02

*The letter "E" following the conversion factor stands for exponent and the two digits after the letter "E" indicate the power of 10 by which the number is to be multiplied.

1.4 DEFINITIONS

The following definitions are used within MIL-HDBK-17. This glossary of terms is not totally comprehensive but it does represent nearly all commonly used terms. Where exceptions are made, they are noted in the text and tables. For ease of identification the definitions have been organized alphabetically.

A-Basis (or A-Value) -- A statistically-based material property; a 95% lower confidence bound on the first percentile of a specified population of measurements. Also a 95% lower tolerance bound for the upper 99% of a specified population.

A-Stage -- An early stage in the reaction of thermosetting resins in which the material is still soluble in certain liquids and may be liquid or capable of becoming liquid upon heating. (Sometimes referred to as **resol.**)

Absorption -- A process in which one material (the absorbent) takes in or absorbs another (the absorbate).

Accelerator -- A material which, when mixed with a catalyzed resin, will speed up the chemical reaction between the catalyst and the resin.

Accuracy -- The degree of conformity of a measured or calculated value to some recognized standard or specified value. Accuracy involves the systematic error of an operation.

Addition Polymerization -- Polymerization by a repeated addition process in which monomers are linked together to form a polymer without splitting off of water or other simple molecules.

Adhesion -- The state in which two surfaces are held together at an interface by forces or interlocking action or both.

Adhesive -- A substance capable of holding two materials together by surface attachment. In the handbook, the term is used specifically to designate structural adhesives, those which produce attachments capable of transmitting significant structural loads.

ADK -- Notation used for the k-sample Anderson-Darling statistic, which is used to test the hypothesis that k batches have the same distribution.

Aliquot -- A small, representative portion of a larger sample.

Aging -- The effect, on materials, of exposure to an environment for a period of time; the process of exposing materials to an environment for an interval of time.

Ambient -- The surrounding environmental conditions such as pressure or temperature.

Anelasticity -- A characteristic exhibited by certain materials in which strain is a function of both stress and time, such that, while no permanent deformations are involved, a finite time is required to establish equilibrium between stress and strain in both the loading and unloading directions.

Angleply -- Same as **Crossply**.

Anisotropic -- Not isotropic; having mechanical and/or physical properties which vary with direction relative to natural reference axes inherent in the material.

Aramid -- A manufactured fiber in which the fiber-forming substance consisting of a long-chain synthetic aromatic polyamide in which at least 85% of the amide (-CONH-) linkages are attached directly to two aromatic rings.

Areal Weight of Fiber -- The weight of fiber per unit area of prepreg. This is often expressed as grams per square meter. See Table 1.6.2 for conversion factors.

Artificial Weathering -- Exposure to laboratory conditions which may be cyclic, involving changes in temperature, relative humidity, radiant energy and any other elements found in the atmosphere in various geographical areas.

Aspect Ratio -- In an essentially two-dimensional rectangular structure (e.g., a panel), the ratio of the long dimension to the short dimension. However, in compression loading, it is sometimes considered to be the ratio of the load direction dimension to the transverse dimension. Also, in fiber micro-mechanics, it is referred to as the ratio of length to diameter.

Autoclave -- A closed vessel for producing an environment of fluid pressure, with or without heat, to an enclosed object which is undergoing a chemical reaction or other operation.

Autoclave Molding -- A process similar to the pressure bag technique. The lay-up is covered by a pressure bag, and the entire assembly is placed in an autoclave capable of providing heat and pressure for curing the part. The pressure bag is normally vented to the outside.

Axis of Braiding -- The direction in which the braided form progresses.

B-Basis (or B-Value) -- A statistically-based material property; a 95% lower confidence bound on the tenth percentile of a specified population of measurements. Also a 95% lower tolerance bound for the upper 90% of a specified population. (See Volume 1, Section 8.1.4)

B-Stage -- An intermediate stage in the reaction of a thermosetting resin in which the material softens when heated and swells when in contact with certain liquids but does not entirely fuse or dissolve. Materials are usually precured to this stage to facilitate handling and processing prior to final cure. (Sometimes referred to as **resitol.**)

Bag Molding -- A method of molding or laminating which involves the application of fluid pressure to a flexible material which transmits the pressure to the material being molded or bonded. Fluid pressure usually is applied by means of air, steam, water or vacuum.

Balanced Laminate -- A composite laminate in which all laminae at angles other than 0 degrees and 90 degrees occur only in ± pairs (not necessarily adjacent).

Batch (or Lot) -- For fibers and resins, a quantity of material formed during the same process and having identical characteristics throughout. For prepregs, laminae, and laminates, material made from one batch of fiber and one batch of resin.

Bearing Area -- The product of the pin diameter and the specimen thickness.

Bearing Load -- A compressive load on an interface.

Bearing Yield Strength -- The bearing stress at which a material exhibits a specified limiting deviation from the proportionality of bearing stress to bearing strain.

Bend Test -- A test of ductility by bending or folding, usually with steadily applied forces. In some instances the test may involve blows to a specimen having a cross section that is essentially uniform over a length several times as great as the largest dimension of the cross section.

Binder -- A bonding resin used to hold strands together in a mat or preform during manufacture of a molded object.

Binomial Random Variable -- The number of successes in independent trials where the probability of success is the same for each trial.

Birefringence -- The difference between the two principal refractive indices (of a fiber) or the ratio between the retardation and thickness of a material at a given point.

Bleeder Cloth -- A nonstructural layer of material used in the manufacture of composite parts to allow the escape of excess gas and resin during cure. The bleeder cloth is removed after the curing process and is not part of the final composite.

Bobbin-- A cylinder or slightly tapered barrel, with or without flanges, for holding tows, rovings, or yarns.

Bond -- The adhesion of one surface to another, with or without the use of an adhesive as a bonding agent.

Braid -- A system of three or more yarns which are interwoven in such a way that no two yarns are twisted around each other.

Braid Angle -- The acute angle measured from the axis of braiding.

Braid, Biaxial -- Braided fabric with two-yarn systems, one running in the $+\theta$ direction, the other in the $-\theta$ direction as measured from the axis of braiding.

Braid Count -- The number of braiding yarn crossings per inch measured along the axis of a braided fabric.

Braid, Diamond -- Braided fabric with an over one, under one weave pattern, (1 x 1).

Braid, Flat -- A narrow bias woven tape wherein each yarn is continuous and is intertwined with every other yarn in the system without being intertwined with itself.

Braid, Hercules -- A braided fabric with an over three, under three weave pattern, (3 x 3).

Braid, Jacquard -- A braided design made with the aid of a jacquard machine, which is a shedding mechanism by means of which a large number of ends may be controlled independently and complicated patterns produced.

Braid, Regular -- A braided fabric with an over two, under two weave pattern (2 x 2).

Braid, Square -- A braided pattern in which the yarns are formed into a square pattern.

Braid, Two-Dimensional -- Braided fabric with no braiding yarns in the through thickness direction.

Braid, Three-Dimensional -- Braided fabric with one or more braiding yarns in the through thickness direction.

Braid, Triaxial -- A biaxial braided fabric with laid in yarns running in the axis of braiding.

Braiding-- A textile process where two or more strands, yarns or tapes are intertwined in the bias direction to form an integrated structure.

Broadgoods -- A term loosely applied to prepreg material greater than about 12 inches in width, usually furnished by suppliers in continuous rolls. The term is currently used to designate both collimated uniaxial tape and woven fabric prepregs.

Buckling (Composite) -- A mode of structural response characterized by an out-of-plane material deflection due to compressive action on the structural element involved. In advanced composites, buckling may take the form not only of conventional general instability and local instability but also a micro-instability of individual fibers.

Bundle -- A general term for a collection of essentially parallel filaments or fibers.

C-Stage -- The final stage of the curing reaction of a thermosetting resin in which the material has become practically infusable and insoluble. (Normally considered fully cured and sometimes referred to as **resite**.)

Capstan -- A friction type take-up device which moves braided fabric away from the fell. The speed of which determines the braid angle.

Carbon Fibers -- Fibers produced by the pyrolysis of organic precursor fibers such as rayon, polyacrylonitrile (PAN), and pitch in an inert atmosphere. The term is often used interchangeably with "graphite"; however, carbon fibers and graphite fibers differ in the temperature at which the fibers are made and heat-treated, and the amount of carbon produced. Carbon fibers typically are carbonized at about 2400°F (1300°C) and assay at 93 to 95% carbon, while graphite fibers are graphitized at 3450 to 5450°F (1900 to 3000°C) and assay at more than 99% elemental carbon.

Carrier-- A mechanism for carrying a package of yarn through the braid weaving motion. A typical carrier consists of a bobbin spindle, a track follower, and a tensioning device.

Caul Plates -- Smooth metal plates, free of surface defects, the same size and shape as a composite lay-up, used immediately in contact with the lay-up during the curing process to transmit normal pressure and to provide a smooth surface on the finished laminate.

Censoring -- Data is right (left) censored at M, if, whenever an observation is less than or equal to M (greater than or equal to M), the actual value of the observation is recorded. If the observation exceeds (is less than) M, the observation is recorded as M.

Chain-Growth Polymerization -- One of the two principal polymerization mechanisms. In chain-growth polymerization, the reactive groups are continuously regenerated during the growth process. Once started, the polymer molecule grows rapidly by a chain of reactions emanating from a particular reactive initiator which may be a free radical, cation or anion.

Chromatogram-- A plot of detector response against peak volume of solution (eluate) emerging from the system for each of the constituents which have been separated.

Circuit -- One complete traverse of the fiber feed mechanism of a winding machine; one complete traverse of a winding band from one arbitrary point along the winding path to another point on a plane through the starting point and perpendicular to the axis.

Cocuring -- The act of curing a composite laminate and simultaneously bonding it to some other prepared surface during the same cure cycle (see **Secondary Bonding**).

Coefficient of Linear Thermal Expansion -- The change in length per unit length resulting from a one-degree rise in temperature.

Coefficient of Variation -- The ratio of the population (or sample) standard deviation to the population (or sample) mean.

Collimated -- Rendered parallel.

Compatible -- The ability of different resin systems to be processed in contact with each other without degradation of end product properties. (See **Compatible**, Volume 1, Section 8.1.4)

Composite Class -- As used in the handbook, a major subdivision of composite construction in which the class is defined by the fiber system and the matrix class, e.g., organic-matrix filamentary laminate.

Composite Material -- Composites are considered to be combinations of materials differing in composition or form on a macroscale. The constituents retain their identities in the composite; that is, they do not dissolve or otherwise merge completely into each other although they act in concert. Normally, the components can be physically identified and exhibit an interface between one another.

Compound-- An intimate mixture of polymer or polymers with all the materials necessary for the finished product.

Condensation Polymerization -- This is a special type of step-growth polymerization characterized by the formation of water or other simple molecules during the stepwise addition of reactive groups.

Confidence Coefficient -- See **Confidence Interval**.

Confidence Interval -- A confidence interval is defined by a statement of one of the following forms:

(1) $P\{a<\theta\} \leq 1-\alpha$
(2) $P\{\theta<b\} \leq 1-\alpha$
(3) $P\{a<\theta<b\} \leq 1-\alpha$

where $1-\alpha$ is called the confidence coefficient. A statement of type (1) or (2) is called a one-sided confidence interval and a statement of type (3) is called a two-sided confidence interval. In (1) a is a lower confidence limit and in (2) b is an upper confidence limit. With probability at least $1-\alpha$, the confidence interval will contain the parameter θ.

Constituent -- In general, an element of a larger grouping. In advanced composites, the principal constituents are the fibers and the matrix.

Continuous Filament -- A yarn or strand in which the individual filaments are substantially the same length as the strand.

Coupling Agent -- Any chemical substance designed to react with both the reinforcement and matrix phases of a composite material to form or promote a stronger bond at the interface. Coupling agents are applied to the reinforcement phase from an aqueous or organic solution or from a gas phase, or added to the matrix as an integral blend.

Coverage -- The measure of the fraction of surface area covered by the braid.

Crazing -- Apparent fine cracks at or under the surface of an organic matrix.

Creel -- A framework arranged to hold tows, rovings, or yarns so that many ends can be withdrawn smoothly and evenly without tangling.

Creep -- The time dependent part of strain resulting from an applied stress.

Creep, Rate Of -- The slope of the creep-time curve at a given time.

Crimp -- The undulations induced into a braided fabric via the braiding process.

Crimp Angle -- The maximum acute angle of a single braided yarn's direction measured from the average axis of tow.

Crimp Exchange -- The process by which a system of braided yarns reaches equilibrium when put under tension or compression.

Critical Value(s) -- When testing a one-sided statistical hypothesis, a critical value is the value such that, if the test statistic is greater than (less than) the critical value, the hypothesis is rejected. When testing a two-sided statistical hypothesis, two critical values are determined. If the test statistic is either less than the smaller critical value or greater than the larger critical value, then the hypothesis is rejected. In both cases, the critical value chosen depends on the desired risk (often 0.05) of rejecting the hypothesis when it is true.

Crossply -- Any filamentary laminate which is not uniaxial. Same as Angleply. In some references, the term crossply is used to designate only those laminates in which the laminae are at right angles to one another, while the term angleply is used for all others. In the handbook, the two terms are used synonymously. The reservation of a separate terminology for only one of several basic orientations is unwarranted because a laminate orientation code is used.

Cumulative Distribution Function -- See Volume 1, Section 8.1.4.

Cure -- To change the properties of a thermosetting resin irreversibly by chemical reaction, i.e., condensation, ring closure, or addition. Cure may be accomplished by addition of curing (cross-linking) agents, with or without catalyst, and with or without heat. Cure may occur also by addition, such as occurs with anhydride cures for epoxy resin systems.

Cure Cycle -- The schedule of time periods at specified conditions to which a reacting thermosetting material is subjected in order to reach a specified property level.

Cure Stress -- A residual internal stress produced during the curing cycle of composite structures. Normally, these stresses originate when different components of a lay-up have different thermal coefficients of expansion.

Debond -- A deliberate separation of a bonded joint or interface, usually for repair or rework purposes. (See **Disbond, Unbond**).

Deformation -- The change in shape of a specimen caused by the application of a load or force.

Degradation -- A deleterious change in chemical structure, physical properties or appearance.

Delamination -- The separation of the layers of material in a laminate. This may be local or may cover a large area of the laminate. It may occur at any time in the cure or subsequent life of the laminate and may arise from a wide variety of causes.

Denier -- A direct numbering system for expressing linear density, equal to the mass in grams per 9000 meters of yarn, filament, fiber, or other textile strand.

Density -- The mass per unit volume.

Desorption -- A process in which an absorbed or adsorbed material is released from another material. Desorption is the reverse of absorption, adsorption, or both.

Deviation -- Variation from a specified dimension or requirement, usually defining the upper and lower limits.

Dielectric Constant -- The ratio of the capacity of a condenser having a dielectric constant between the plates to that of the same condenser when the dielectric is replaced by a vacuum; a measure of the electrical charge stored per unit volume at unit potential.

Dielectric Strength -- The average potential per unit thickness at which failure of the dielectric material occurs.

Disbond -- An area within a bonded interface between two adherends in which an adhesion failure or separation has occurred. It may occur at any time during the life of the structure and may arise from a wide variety of causes. Also, colloquially, an area of separation between two laminae in the finished laminate (in this case the term "delamination" is normally preferred.) (See **Debond, Unbond, Delamination**.)

Distribution -- A formula which gives the probability that a value will fall within prescribed limits. (See **Normal, Weibull**, and **Lognormal Distributions**, also Volume 1, Section 8.1.4).

Dry-- a material condition of moisture equilibrium with a surrounding environment at 5% or lower relative humidity.

Dry Fiber Area -- Area of fiber not totally encapsulated by resin.

Ductility -- The ability of a material to deform plastically before fracturing.

Elasticity -- The property of a material which allows it to recover its original size and shape immediately after removal of the force causing deformation.

Elongation -- The increase in gage length or extension of a specimen during a tension test, usually expressed as a percentage of the original gage length.

Eluate -- The liquid emerging from a column (in liquid chromatography).

Eluent -- The mobile phase used to sweep or elute the sample (solute) components into, through, and out of the column.

End -- A single fiber, strand, roving or yarn being or already incorporated into a product. An end may be an individual warp yarn or cord in a woven fabric. In referring to aramid and glass fibers, an end is usually an untwisted bundle of continuous filaments.

Epoxy Equivalent Weight -- The number of grams of resin which contain one chemical equivalent of the epoxy group.

Epoxy Resin -- Resins which may be of widely different structures but are characterized by the presence of the epoxy group. (The epoxy or epoxide group is usually present as a glycidyl ether, glycidyl amine, or as part of an aliphatic ring system. The aromatic type epoxy resins are normally used in composites.)

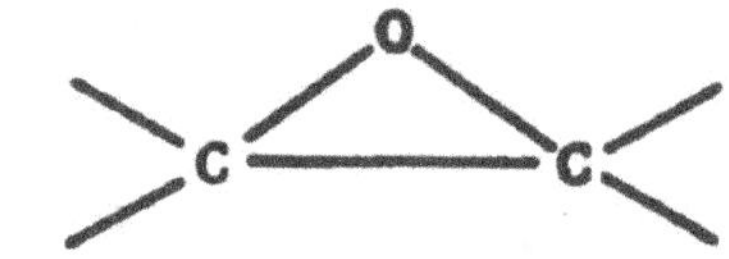

Epoxy group

Extensometer -- A device for measuring linear strain.

F-Distribution -- See Volume 1, Section 8.1.4.

Fabric, Nonwoven -- A textile structure produced by bonding or interlocking of fibers, or both, accomplished by mechanical, chemical, thermal, or solvent means, and combinations thereof.

Fabric, Woven -- A generic material construction consisting of interlaced yarns or fibers, usually a planar structure. Specifically, as used in this handbook, a cloth woven in an established weave pattern from advanced fiber yarns and used as the fibrous constituent in an advanced composite lamina. In a fabric lamina, the warp direction is considered the longitudinal direction, analogous to the filament direction in a filamentary lamina.

Fell -- The point of braid formation, which is defined as the point at which the yarns in a braid system cease movement relative to each other.

Fiber -- A general term used to refer to filamentary materials. Often, fiber is used synonymously with filament. It is a general term for a filament of finite length. A unit of matter, either natural or manmade, which forms the basic element of fabrics and other textile structures.

Fiber Content -- The amount of fiber present in a composite. This is usually expressed as a percentage volume fraction or weight fraction of the composite.

Fiber Count -- The number of fibers per unit width of ply present in a specified section of a composite.

Fiber Direction -- The orientation or alignment of the longitudinal axis of the fiber with respect to a stated reference axis.

Fiber System -- The type and arrangement of fibrous material which comprises the fiber constituent of an advanced composite. Examples of fiber systems are collimated filaments or filament yarns, woven fabric, randomly oriented short-fiber ribbons, random fiber mats, whiskers, etc.

Filament -- The smallest unit of a fibrous material. The basic units formed during spinning and which are gathered into strands of fiber, (for use in composites). Filaments usually are of extreme length and of very small diameter. Filaments normally are not used individually. Some textile filaments can function as a yarn when they are of sufficient strength and flexibility.

Filamentary Composites -- A major form of advanced composites in which the fiber constituent consists of continuous filaments. Specifically, a filamentary composite is a laminate comprised of a number of laminae, each of which consists of a nonwoven, parallel, uniaxial, planar array of filaments (or filament yarns) embedded in the selected matrix material. Individual laminae are directionally oriented and combined into specific multiaxial laminates for application to specific envelopes of strength and stiffness requirements.

Filament Winding -- A reinforced-plastics process that employs a series of continuous, resin-impregnated fibers applied to a mandrel in a predetermined geometrical relationship under controlled tension.

Filament Wound -- Pertaining to an object created by the filament winding method of fabrication.

Fill (Filling) -- In a woven fabric, the yarn running from selvage to selvage at right angles to the warp.

Filler -- A relatively inert substance added to a material to alter its physical, mechanical, thermal, electrical, and other properties or to lower cost. Sometimes the term is used specifically to mean particulate additives.

Finish (or Size System) -- A material, with which filaments are treated, which contains a coupling agent to improve the bond between the filament surface and the resin matrix in a composite material. In addition, finishes often contain ingredients which provide lubricity to the filament surface, preventing abrasive damage during handling, and a binder which promotes strand integrity and facilitates packing of the filaments.

Fixed Effect -- A systematic shift in a measured quantity due to a particular level change of a treatment or condition. (See Volume 1, Section 8.1.4.)

Flash -- Excess material which forms at the parting line of a mold or die, or which is extruded from a closed mold.

Former Plate -- A die attached to a braiding machine which helps to locate the fell.

Fracture Ductility -- The true plastic strain at fracture.

Gage Length -- the original length of that portion of the specimen over which strain or change of length is determined.

Gel-- The initial jelly-like solid phase that develops during formation of a resin from a liquid. Also, a semi-solid system consisting of a network of solid aggregates in which liquid is held.

Gel Coat -- A quick-setting resin used in molding processes to provide an improved surface for the composite; it is the first resin applied to the mold after the mold-release agent.

Gel Point -- The stage at which a liquid begins to exhibit pseudo-elastic properties. (This can be seen from the inflection point on a viscosity-time plot.)

Gel Time -- The period of time from a pre-determined starting point to the onset of gelation (gel point) as defined by a specific test method.

Glass-- An inorganic product of fusion which has cooled to a rigid condition without crystallizing. In the handbook, all reference to glass will be to the fibrous form as used in filaments, woven fabric, yarns, mats, chopped fibers, etc.

Glass Cloth -- Conventionally-woven glass fiber material (see **Scrim**).

Glass Fibers -- A fiber spun from an inorganic product of fusion which has cooled to a rigid condition without crystallizing.

Glass Transition -- The reversible change in an amorphous polymer or in amorphous regions of a partially crystalline polymer from (or to) a viscous or rubbery condition to (or from) a hard and relatively brittle one.

Glass Transition Temperature -- The approximate midpoint of the temperature range over which the glass transition takes place.

Graphite Fibers -- See **Carbon Fibers**.

Greige -- Fabric that has received no finish.

Hand Lay-up -- A process in which components are applied either to a mold or a working surface, and the successive plies are built up and worked by hand.

Hardness -- Resistance to deformation; usually measured by indention. Types of standard tests include Brinell, Rockwell, Knoop, and Vickers.

Heat Cleaned -- Glass or other fibers which have been exposed to elevated temperatures to remove preliminary sizings or binders which are not compatible with the resin system to be applied.

Heterogeneous -- Descriptive term for a material consisting of dissimilar constituents separately identifiable; a medium consisting of regions of unlike properties separated by internal boundaries. (Note that all nonhomogeneous materials are not necessarily heterogeneous).

Homogeneous-- Descriptive term for a material of uniform composition throughout; a medium which has no internal physical boundaries; a material whose properties are constant at every point, in other words, constant with respect to spatial coordinates (but not necessarily with respect to directional coordinates).

Horizontal Shear -- Sometimes used to indicate interlaminar shear. This is not an approved term for use in this handbook.

Humidity, Relative -- The ratio of the pressure of water vapor present to the pressure of saturated water vapor at the same temperature.

Hybrid -- A composite laminate comprised of laminae of two or more composite material systems. Or, a combination of two or more different fibers such as carbon and glass or carbon and aramid into a structure (tapes, fabrics and other forms may be combined).

Hygroscopic -- Capable of absorbing and retaining atmospheric moisture.

Hysteresis -- The energy absorbed in a complete cycle of loading and unloading.

Inclusion-- A physical and mechanical discontinuity occurring within a material or part, usually consisting of solid, encapsulated foreign material. Inclusions are often capable of transmitting some structural stresses and energy fields, but in a noticeably different manner from the parent material.

Integral Composite Structure -- Composite structure in which several structural elements, which would conventionally be assembled by bonding or with mechanical fasteners after separate fabrication, are instead laid up and cured as a single, complex, continuous structure; e.g., spars, ribs, and one stiffened cover of a wing box fabricated as a single integral part. The term is sometimes applied more loosely to any composite structure not assembled by mechanical fasteners.

Interface -- The boundary between the individual, physically distinguishable constituents of a composite.

Interlaminar -- Descriptive term pertaining to some object (e.g., voids), event (e.g., fracture), or potential field (e.g., shear stress) referenced as existing or occurring between two or more adjacent laminae.

Interlaminar Shear -- Shearing force tending to produce a relative displacement between two laminae in a laminate along the plane of their interface.

Intermediate Bearing Stress -- The bearing stress at the point on the bearing load-deformation curve where the tangent is equal to the bearing stress divided by a designated percentage (usually 4%) of the original hole diameter.

Intralaminar -- Descriptive term pertaining to some object (e.g., voids), event (e.g., fracture), or potential field (e.g., temperature gradient) existing entirely within a single lamina without reference to any adjacent laminae.

Isotropic -- Having uniform properties in all directions. The measured properties of an isotropic material are independent of the axis of testing.

Jammed State -- The state of a braided fabric under tension or compression where the deformation of the fabric is dominated by the deformation properties of the yarn.

Knitting -- A method of constructing fabric by interlocking series of loops of one or more yarns.

Knuckle Area -- The area of transition between sections of different geometry in a filament wound part.

k-Sample Data -- A collection of data consisting of values observed when sampling from k batches.

Laid-In Yarns -- A system of longitudinal yarns in a triaxial braid which are inserted between the bias yarns.

Lamina -- A single ply or layer in a laminate made up of a series of layers.

Laminae -- Plural of lamina.

Laminate -- A product made by bonding together two or more layers or laminae of material or materials.

Laminate Orientation -- The configuration of a crossplied composite laminate with regard to the angles of crossplying, the number of laminae at each angle, and the exact sequence of the lamina lay-up.

Lattice Pattern -- A pattern of filament winding with a fixed arrangement of open voids.

Lay-up -- A process of fabrication involving the assembly of successive layers of resin-impregnated material.

Lognormal Distribution -- A probability distribution for which the probability that an observation selected at random from this population falls between a and b $(0 < a < b < B)$ is given by the area under the normal distribution between $\log a$ and $\log b$. The common (base 10) or the natural (base e) logarithm may be used. (See Volume 1, Section 8.1.4.)

Lower Confidence Bound -- See **Confidence Interval**.

Macro -- In relation to composites, denotes the gross properties of a composite as a structural element but does not consider the individual properties or identity of the constituents.

Macrostrain -- The mean strain over any finite gage length of measurement which is large in comparison to the material's interatomic distance.

Mandrel -- A form fixture or male mold used for the base in the production of a part by lay-up, filament winding or braiding.

Mat -- A fibrous material consisting of randomly oriented chopped or swirled filaments loosely held together with a binder.

Material Acceptance -- The testing of incoming material to ensure that it meets requirements.

Material Qualification -- The procedures used to accept a material by a company or organization for production use.

Material System -- A specific composite material made from specifically identified constituents in specific geometric proportions and arrangements and possessed of numerically defined properties.

Material System Class -- As used in this handbook, a group consisting of material systems categorized by the same generic constituent materials, but without defining the constituents uniquely; e.g., the carbon/epoxy class.

Material Variability -- A source of variability due to the spatial and consistency variations of the material itself and due to variation in its processing. (See Volume 1, Section 8.1.4.)

Matrix-- The essentially homogeneous material in which the fiber system of a composite is embedded.

Mean -- See **Sample Mean** and **Population Mean**.

Mechanical Properties -- The properties of a material that are associated with elastic and inelastic reaction when force is applied, or the properties involving the relationship between stress and strain.

Median -- See **Sample Median** and **Population Median**.

Micro -- In relation to composites, denotes the properties of the constituents, i.e., matrix and reinforcement and interface only, as well as their effects on the composite properties.

Microstrain -- The strain over a gage length comparable to the material's interatomic distance.

Modulus, Chord -- The slope of the chord drawn between any two specified points on the stress-strain curve.

Modulus, Initial -- The slope of the initial straight portion of a stress-strain curve.

Modulus, Secant -- The slope of the secant drawn from the origin to any specified point on the stress-strain curve.

Modulus, Tangent -- The ratio of change in stress to change in strain derived from the tangent to any point on a stress-strain curve.

Modulus, Young's -- The ratio of change in stress to change in strain below the elastic limit of a material. (Applicable to tension and compression).

Modulus of Rigidity (also Shear Modulus or Torsional Modulus) -- The ratio of stress to strain below the proportional limit for shear or torsional stress.

Modulus of Rupture, in Bending -- The maximum tensile or compressive stress (whichever causes failure) value in the extreme fiber of a beam loaded to failure in bending. The value is computed from the flexure equation:

$$F^b = \frac{Mc}{I} \qquad 1.4(a)$$

where M = maximum bending moment computed from the maximum load and the original moment arm,
c = initial distance from the neutral axis to the extreme fiber where failure occurs,
I = the initial moment of inertia of the cross section about its neutral axis.

Modulus of Rupture, in Torsion -- The maximum shear stress in the extreme fiber of a member of circular cross section loaded to failure in torsion calculated from the equation:

$$F^s = \frac{Tr}{J} \qquad 1.4(b)$$

where T = maximum twisting moment,
r = original outer radius,
J = polar moment of inertia of the original cross section.

Moisture Content -- The amount of moisture in a material determined under prescribed condition and expressed as a percentage of the mass of the moist specimen, i.e., the mass of the dry substance plus the moisture present.

Moisture Equilibrium -- The condition reached by a sample when it no longer takes up moisture from, or gives up moisture to, the surrounding environment.

Mold Release Agent -- A lubricant applied to mold surfaces to facilitate release of the molded article.

Molded Edge -- An edge which is not physically altered after molding for use in final form and particularly one which does not have fiber ends along its length.

Molding -- The forming of a polymer or composite into a solid mass of prescribed shape and size by the application of pressure and heat.

Monolayer -- The basic laminate unit from which crossplied or other laminates are constructed.

Monomer -- A compound consisting of molecules each of which can provide one or more constitutional units.

NDE -- Nondestructive evaluation. Broadly considered synonymous with NDI.

NDI -- Nondestructive inspection. A process or procedure for determining the quality or characteristics of a material, part, or assembly without permanently altering the subject or its properties.

NDT -- Nondestructive testing. Broadly considered synonymous with NDI.

Necking -- A localized reduction in cross-sectional area which may occur in a material under tensile stress.

Negatively Skewed -- A distribution is said to be negatively skewed if the distribution is not symmetric and the longest tail is on the left.

Nominal Specimen Thickness -- The nominal ply thickness multiplied by the number of plies.

Nominal Value -- A value assigned for the purpose of a convenient designation. A nominal value exists in name only.

Normal Distribution -- A two parameter (μ,σ) family of probability distributions for which the probability that an observation will fall between a and b is given by the area under the curve

$$f(x) = \frac{1}{\sigma\sqrt{2\pi}} \exp\left[-\frac{(x-\mu)^2}{2\sigma^2}\right] \qquad 1.4(c)$$

between a and b. (See Volume 1, Section 8.1.4.)

Normalization -- A mathematical procedure for adjusting raw test values for fiber-dominated properties to a single (specified) fiber volume content.

Normalized Stress -- Stress value adjusted to a specified fiber volume content by multiplying the measured stress value by the ratio of specimen fiber volume to the specified fiber volume. This ratio may be obtained directly by experimentally measuring fiber volume, or indirectly by calculation using specimen thickness and fiber areal weight.

Observed Significance Level (OSL) -- The probability of observing a more extreme value of the test statistic when the null hypotheses is true.

Offset Shear Strength --- (from valid execution of a material property shear response test) the value of shear stress at the intersection between a line parallel to the shear chord modulus of elasticity and the shear stress/strain curve, where the line has been offset along the shear strain axis from the origin by a specified strain offset value.

Oligomer -- A polymer consisting of only a few monomer units such as a dimer, trimer, etc., or their mixtures.

One-Sided Tolerance Limit Factor -- See **Tolerance Limit Factor**.

Orthotropic -- Having three mutually perpendicular planes of elastic symmetry.

Oven Dry -- The condition of a material that has been heated under prescribed conditions of temperature and humidity until there is no further significant change in its mass.

PAN Fibers -- Reinforcement fiber derived from the controlled pyrolysis of poly(acrylonitrile) fiber.

Parallel Laminate -- A laminate of woven fabric in which the plies are aligned in the same position as originally aligned in the fabric roll.

Parallel Wound -- A term used to describe yarn or other material wound into a flanged spool.

Peel Ply -- A layer of resin free material used to protect a laminate for later secondary bonding.

pH -- A measure of acidity or alkalinity of a solution, with neutrality represented by a value of 7, with increasing acidity corresponding to progressively smaller values, and increasing alkalinity corresponding to progressively higher values.

Pick Count -- The number of filling yarns per inch or per centimeter of woven fabric.

Pitch Fibers -- Reinforcement fiber derived from petroleum or coal tar pitch.

Plastic-- A material that contains one or more organic polymers of large molecular weight, is solid in its finished state, and, at some state in its manufacture or processing into finished articles, can be shaped by flow.

Plasticizer -- A material of lower molecular weight added to a polymer to separate the molecular chains. This results in a depression of the glass transition temperature, reduced stiffness and brittleness, and improved processability. (Note, many polymeric materials do not need a plasticizer.)

Plied Yarn -- A yarn formed by twisting together two or more single yarns in one operation.

Poisson's Ratio -- The absolute value of the ratio of transverse strain to the corresponding axial strain resulting from uniformly distributed axial stress below the proportional limit of the material.

Polymer -- An organic material composed of molecules characterized by the repetition of one or more types of monomeric units.

Polymerization -- A chemical reaction in which the molecules of monomers are linked together to form polymers via two principal reaction mechanisms. Addition polymerizations proceed by chain growth and most condensation polymerizations through step growth.

Population -- The set of measurements about which inferences are to be made or the totality of possible measurements which might be obtained in a given testing situation. For example, "all possible ultimate tensile strength measurements for carbon/epoxy system A, conditioned at 95% relative humidity and room temperature". In order to make inferences about a population, it is often necessary to make assumptions about its distributional form. The assumed distributional form may also be referred to as the population. (See Volume 1, Section 8.1.4.)

Population Mean -- The average of all potential measurements in a given population weighted by their relative frequencies in the population. (See Volume 1, Section 8.1.4.)

Population Median -- That value in the population such that the probability of exceeding it is 0.5 and the probability of being less than it is 0.5. (See Volume 1, Section 8.1.4.)

Population Variance -- A measure of dispersion in the population.

Porosity -- A condition of trapped pockets of air, gas, or vacuum within a solid material, usually expressed as a percentage of the total nonsolid volume to the total volume (solid plus nonsolid) of a unit quantity of material.

Positively Skewed -- A distribution is said to be positively skewed if the distribution is not symmetric and the longest tail is on the right.

Postcure -- Additional elevated temperature cure, usually without pressure, to increase the glass transition temperature, to improve final properties, or to complete the cure.

Pot Life -- The period of time during which a reacting thermosetting composition remains suitable for its intended processing after mixing with a reaction initiating agent.

Precision -- The degree of agreement within a set of observations or test results obtained. Precision involves repeatability and reproducibility.

Precursor (for Carbon or Graphite Fiber) -- Either the PAN or pitch fibers from which carbon and graphite fibers are derived.

Preform -- An assembly of dry fabric and fibers which has been prepared for one of several different wet resin injection processes. A preform may be stitched or stabilized in some other way to hold its A shape. A commingled preform may contain thermoplastic fibers and may be consolidated by elevated temperature and pressure without resin injection.

Preply -- Layers of prepreg material, which have been assembled according to a user specified stacking sequence.

Prepreg -- Ready to mold or cure material in sheet form which may be tow, tape, cloth, or mat impregnated with resin. It may be stored before use.

Pressure -- The force or load per unit area.

Probability Density Function -- See Volume 1, Section 8.1.4.

Proportional Limit -- The maximum stress that a material is capable of sustaining without any deviation from the proportionality of stress to strain (also known as Hooke's law).

Quasi-Isotropic Laminate -- A laminate approximating isotropy by orientation of plies in several or more directions.

Random Effect -- A shift in a measured quantity due to a particular level change of an external, usually uncontrollable, factor. (See Volume 1, Section 8.1.4.)

Random Error -- That part of the data variation that is due to unknown or uncontrolled factors and that affects each observation independently and unpredictably. (See Volume 1, Section 8.1.4.)

Reduction of Area -- The difference between the original cross sectional area of a tension test specimen and the area of its smallest cross section, usually expressed as a percentage of the original area.

Refractive Index - The ratio of the velocity of light (of specified wavelength) in air to its velocity in the substance under examination. Also defined as the sine of the angle of incidence divided by the sine of the angle of refraction as light passes from air into the substance.

Reinforced Plastic -- A plastic with relatively high stiffness or very high strength fibers embedded in the composition. This improves some mechanical properties over that of the base resin.

Release Agent -- See **Mold Release Agent**.

Resilience -- A property of a material which is able to do work against restraining forces during return from a deformed condition.

Resin -- An organic polymer or prepolymer used as a matrix to contain the fibrous reinforcement in a composite material or as an adhesive. This organic matrix may be a thermoset or a thermoplastic, and may contain a wide variety of components or additives to influence; handleability, processing behavior and ultimate properties.

Resin Content -- The amount of matrix present in a composite either by percent weight or percent volume.

Resin Starved Area -- Area of composite part where the resin has a non-continuous smooth coverage of the fiber.

Resin System -- A mixture of resin, with ingredients such as catalyst, initiator, diluents, etc. required for the intended processing and final product.

Room Temperature Ambient (RTA) -- 1) an environmental condition of 73±5°F (23±3°C) at ambient laboratory relative humidity; 2) a material condition where, immediately following consolidation/cure, the material is stored at 73±5°F (23±3°C) and at a maximum relative humidity of 60%.

Roving -- A number of strands, tows, or ends collected into a parallel bundle with little or no twist. In spun yarn production, an intermediate state between sliver and yarn.

S-Basis (or S-Value) -- The mechanical property value which is usually the specified minimum value of the appropriate government specification or SAE Aerospace Material Specification for this material.

Sample -- A small portion of a material or product intended to be representative of the whole. Statistically, a sample is the collection of measurements taken from a specified population. (See Volume 1, Section 8.1.4.)

Sample Mean -- The arithmetic average of the measurements in a sample. The sample mean is an estimator of the population mean. (See Volume 1, Section 8.1.4.)

Sample Median -- Order the observation from smallest to largest. Then the sample median is the value of the middle observation if the sample size is odd; the average of the two central observations if n is even. If the population is symmetric about its mean, the sample median is also an estimator of the population mean. (See Volume 1, Section 8.1.4.)

Sample Standard Deviation -- The square root of the sample variance. (See Volume 1, Section 8.1.4.)

Sample Variance -- The sum of the squared deviations from the sample mean, divided by n-1. (See Volume 1, Section 8.1.4.)

Sandwich Construction -- A structural panel concept consisting in its simplest form of two relatively thin, parallel sheets of structural material bonded to, and separated by, a relatively thick, light-weight core.

Saturation -- An equilibrium condition in which the net rate of absorption under prescribed conditions falls essentially to zero.

Scrim (also called **Glass Cloth, Carrier**) -- A low cost fabric woven into an open mesh construction, used in the processing of tape or other B-stage material to facilitate handling.

Secondary Bonding -- The joining together, by the process of adhesive bonding, of two or more already-cured composite parts, during which the only chemical or thermal reaction occurring is the curing of the adhesive itself.

Selvage or Selvedge -- The woven edge portion of a fabric parallel to the warp.

Set -- The strain remaining after complete release of the force producing the deformation.

Shear Fracture (for crystalline type materials) -- A mode of fracture resulting from translation along slip planes which are preferentially oriented in the direction of the shearing stress.

Shelf Life -- The length of time a material, substance, product, or reagent can be stored under specified environmental conditions and continue to meet all applicable specification requirements and/or remain suitable for its intended function.

Short Beam Strength (SBS) -- a test result from valid execution of ASTM test method D 2344.

Significant -- Statistically, the value of a test statistic is significant if the probability of a value at least as extreme is less than or equal to a predetermined number called the significance level of the test.

Significant Digit -- Any digit that is necessary to define a value or quantity.

Size System -- See **Finish**.

Sizing -- A generic term for compounds which are applied to yarns to bind the fiber together and stiffen the yarn to provide abrasion-resistance during weaving. Starch, gelatin, oil, wax, and man-made polymers such as polyvinyl alcohol, polystyrene, polyacrylic acid, and polyacetatates are employed.

Skewness -- See **Positively Skewed, Negatively Skewed**.

Sleeving -- A common name for tubular braided fabric.

Slenderness Ratio -- The unsupported effective length of a uniform column divided by the least radius of gyration of the cross-sectional area.

Sliver -- A continuous strand of loosely assembled fiber that is approximately uniform in cross-sectional area and has no twist.

Solute -- The dissolved material.

Specific Gravity -- The ratio of the weight of any volume of a substance to the weight of an equal volume of another substance taken as standard at a constant or stated temperature. Solids and liquids are usually compared with water at 39°F (4°C).

Specific Heat -- The quantity of heat required to raise the temperature of a unit mass of a substance one degree under specified conditions.

Specimen-- A piece or portion of a sample or other material taken to be tested. Specimens normally are prepared to conform with the applicable test method.

Spindle -- A slender upright rotation rod on a spinning frame, roving frame, twister or similar machine.

Standard Deviation -- See **Sample Standard Deviation**.

Staple -- Either naturally occurring fibers or lengths cut from filaments.

Step-Growth Polymerization -- One of the two principal polymerization mechanisms. In sep-growth polymerization, the reaction grows by combination of monomer, oligomer, or polymer molecules through the consumption of reactive groups. Since average molecular weight increases with monomer consumption, high molecular weight polymers are formed only at high degrees of conversion.

Strain -- the per unit change, due to force, in the size or shape of a body referred to its original size or shape. Strain is a nondimensional quantity, but it is frequently expressed in inches per inch, meters per meter, or percent.

Strand -- Normally an untwisted bundle or assembly of continuous filaments used as a unit, including slivers, tow, ends, yarn, etc. Sometimes a single fiber or filament is called a strand.

Strength -- the maximum stress which a material is capable of sustaining.

Stress-- The intensity at a point in a body of the forces or components of forces that act on a given plane through the point. Stress is expressed in force per unit area (pounds-force per square inch, megapascals, etc.).

Stress Relaxation -- The time dependent decrease in stress in a solid under given constraint conditions.

Stress-Strain Curve (Diagram) -- A graphical representation showing the relationship between the change in dimension of the specimen in the direction of the externally applied stress and the magnitude of the applied stress. Values of stress usually are plotted as ordinates (vertically) and strain values as abscissa (horizontally).

Structural Element -- a generic element of a more complex structural member (for example, skin, stringer, shear panels, sandwich panels, joints, or splices).

Structured Data -- See Volume 1, Section 8.1.4.

Surfacing Mat -- A thin mat of fine fibers used primarily to produce a smooth surface on an organic matrix composite.

Symmetrical Laminate -- A composite laminate in which the sequence of plies below the laminate midplane is a mirror image of the stacking sequence above the midplane.

Tack -- Stickiness of the prepreg.

Tape -- Prepreg fabricated in widths up to 12 inches wide for carbon and 3 inches for boron. Cross stitched carbon tapes up to 60 inches wide are available commercially in some cases.

Tenacity -- The tensile stress expressed as force per unit linear density of the unstrained specimen i.e., grams-force per denier or grams-force per tex.

Tex -- A unit for expressing linear density equal to the mass or weight in grams of 1000 meters of filament, fiber, yarn or other textile strand.

Thermal Conductivity -- Ability of a material to conduct heat. The physical constant for quantity of heat that passes through unit cube of a substance in unit time when the difference in temperature of two faces is one degree.

Thermoplastic -- A plastic that repeatedly can be softened by heating and hardened by cooling through a temperature range characteristic of the plastic, and when in the softened stage, can be shaped by flow into articles by molding or extrusion.

Thermoset -- A plastic that is substantially infusible and insoluble after having been cured by heat or other means.

Tolerance -- The total amount by which a quantity is allowed to vary.

Tolerance Limit -- A lower (upper) confidence limit on a specified percentile of a distribution. For example, the B-basis value is a 95% lower confidence limit on the tenth percentile of a distribution.

Tolerance Limit Factor -- The factor which is multiplied by the estimate of variability in computing the tolerance limit.

Toughness-- A measure of a material's ability to absorb work, or the actual work per unit volume or unit mass of material that is required to rupture it. Toughness is proportional to the area under the load-elongation curve from the origin to the breaking point.

Tow -- An untwisted bundle of continuous filaments. Commonly used in referring to man-made fibers, particularly carbon and graphite fibers, in the composites industry.

Transformation -- A transformation of data values is a change in the units of measurement accomplished by applying a mathematical function to all data values. For example, if the data is given by x, then $y = x + 1$, x , $1/x$, $\log x$, and $\cos x$ are transformations.

Transition, First Order -- A change of state associated with crystallization or melting in a polymer.

Transversely Isotropic -- Descriptive term for a material exhibiting a special case of orthotropy in which properties are identical in two orthotropic dimensions, but not the third; having identical properties in both transverse directions but not the longitudinal direction.

Traveller -- A small piece of the same product (panel, tube, etc.) as the test specimen, used for example to measure moisture content as a result of conditioning.

Twist -- The number of turns about its axis per unit of length in a yarn or other textile strand. It may be expressed as turns per inch (tpi) or turns per centimeter (tpcm).

Twist, Direction of -- The direction of twist in yarns and other textile strands is indicated by the capital letters S and Z. Yarn has S twist if, when held in a vertical position, the visible spirals or helices around its central axis are in the direction of slope of the central portion of the letter S, and Z twist is in the other direction.

Twist multiplier -- The ratio of turns per inch to the square root of the cotton count.

Typical Basis -- A typical property value is a sample mean. Note that the typical value is defined as the simple arithmetic mean which has a statistical connotation of 50% reliability with a 50% confidence.

Unbond -- An area within a bonded interface between two adherends in which the intended bonding action failed to take place. Also used to denote specific areas deliberately prevented from bonding in order to simulate a defective bond, such as in the generation of quality standards specimens. (See **Disbond, Debond**).

Unidirectional Laminate -- A laminate with nonwoven reinforcements and all layers laid up in the same direction.

Unit Cell -- The term applied to the path of a yarn in a braided fabric representing a unit cell of a repeating geometric pattern. The smallest element representative of the braided structure.

Unstructured Data -- See Volume 1, Section 8.1.4.

Upper Confidence Limit -- See **Confidence Interval**.

Vacuum Bag Molding -- A process in which the lay-up is cured under pressure generated by drawing a vacuum in the space between the lay-up and a flexible sheet placed over it and sealed at the edges.

Variance -- See **Sample Variance**.

Viscosity -- The property of resistance to flow exhibited within the body of a material.

Void -- A physical and mechanical discontinuity occurring within a material or part which may be two-dimensional (e.g., disbonds, delaminations) or three-dimensional (e.g., vacuum-, air-, or gas-filled pockets). Porosity is an aggregation of micro-voids. Voids are essentially incapable of transmitting structural stresses or nonradiative energy fields. (See **Inclusion**.)

Warp -- The longitudinally oriented yarn in a woven fabric (see **Fill**); a group of yarns in long lengths and approximately parallel.

Wet Lay-up -- A method of making a reinforced product by applying a liquid resin system while or after the reinforcement is put in place.

Weibull Distribution (Two - Parameter) -- A probability distribution for which the probability that a randomly selected observation from this population lies between a and b ($0 < a < b < \infty$) is given by Equation 1.4(d) where α is called the scale parameter and β is called the shape parameter. (See Volume 1, Section 8.1.4.)

$$\exp\left[-\left(\frac{a}{\alpha}\right)^{\beta}\right]-\exp\left[-\left(\frac{b}{\alpha}\right)^{\beta}\right] \qquad 1.4(d)$$

Wet Lay-up -- A method of making a reinforced product by applying a liquid resin system while the reinforcement is put in place.

Wet Strength -- The strength of an organic matrix composite when the matrix resin is saturated with absorbed moisture. (See **Saturation**).

Wet Winding -- A method of filament winding in which the fiber reinforcement is coated with the resin system as a liquid just prior to wrapping on a mandrel.

Whisker -- A short single crystal fiber or filament. Whisker diameters range from 1 to 25 microns, with aspect ratios between 100 and 15,000.

Work Life -- The period during which a compound, after mixing with a catalyst, solvent, or other compounding ingredient, remains suitable for its intended use.

Woven Fabric Composite -- A major form of advanced composites in which the fiber constituent consists of woven fabric. A woven fabric composite normally is a laminate comprised of a number of laminae, each of which consists of one layer of fabric embedded in the selected matrix material. Individual fabric laminae are directionally oriented and combined into specific multiaxial laminates for application to specific envelopes of strength and stiffness requirements.

Yarn -- A generic term for strands or bundles of continuous filaments or fibers, usually twisted and suitable for making textile fabric.

Yarn, Plied -- Yarns made by collecting two or more single yarns together. Normally, the yarns are twisted together though sometimes they are collected without twist.

Yield Strength -- The stress at which a material exhibits a specified limiting deviation from the proportionality of stress to strain. (The deviation is expressed in terms of strain such as 0.2 percent for the Offset Method or 0.5 percent for the Total Extension Under Load Method.)

X-Axis -- In composite laminates, an axis in the plane of the laminate which is used as the 0 degree reference for designating the angle of a lamina.

X-Y Plane -- In composite laminates, the reference plane parallel to the plane of the laminate.

Y-Axis -- In composite laminates, the axis in the plane of the laminate which is perpendicular to the x-axis.

Z-Axis -- In composite laminates, the reference axis normal to the plane of the laminate.

REFERENCES

1.3(a) Military Standardization Handbook, *"Metallic Materials and Elements for Aerospace Vehicle Structures,"* MIL-HDBK-5F, 1 November 1990.

1.3(b) *DOD/NASA Advanced Composites Design Guide*, Air Force Wright Aeronautical Laboratories, Dayton, OH, prepared by Rockwell International Corporation, 1983 (distribution limited).

1.3(c) ASTM E 206, "Definitions of Terms Relating to Fatigue Testing and the Statistical Analysis of Fatigue Data," *1984 Annual Book of ASTM Standards*, Vol 3.01, ASTM, Philadelphia, PA, 1984.

1.3.2(a) ASTM E 380, "Standard for Metric Practice," *1984 Annual Book of ASTM Standards*, Vol 14.01, ASTM, Philadelphia, PA, 1984.

1.3.2(b) *Engineering Design Handbook: Metric Conversion Guide*, DARCOM P 706-470, July 1976.

1.3.2(c) *The International System of Units (SI)*, NBS Special Publication 330, National Bureau of Standards, 1986 edition.

1.3.2(d) *Units and Systems of Weights and Measures, Their Origin, Development, and Present Status*, NBS Letter Circular LC 1035, National Bureau of Standards, November 1985.

1.3.2(e) *The International System of Units Physical Constants and Conversion Factors*, NASA Special Publication 7012, 1964.

CHAPTER 2 MATERIALS AND PROCESSES -

THE EFFECT OF VARIABILITY ON COMPOSITE PROPERTIES

Page

2.1 INTRODUCTION

The properties of organic matrix composites are, in general, cure and process dependent. This may result in variations of glass transition (service temperature), corrosion stability, susceptibility to microcracking, general strength, or fatigue and service life. In addition, in most cases these materials or structural elements constructed from them are the products of complex multi-step materials processes. Figures 2.1(a) and (b) illustrate the nature of the processing pipeline from raw materials to composite end item. Each rectangle in Figure 2.1(b) represents a process during which additional variability may be introduced into the material. Utilization of a standard composite material property database necessitates an understanding of the dependency of the measured material properties on the characteristics and variability associated with the constituent materials and the sequence of processes used to combine these materials into end products. As a result, development and application of processing controls are essential to achieve the desired mechanical and physical properties for composite structures.

2.2 PURPOSE

The purpose of this chapter is to provide an understanding of the origins and nature of process-induced variability in these materials in the context of an overview of types of composite materials and the associated material processing methodologies. It also seeks to addresses various approaches to minimizing variability, including implementation of process control, and the use of materials and processing specifications.

2.3 SCOPE

This chapter includes descriptions of composite materials from the perspective of their introduction into the material pipeline as the constituent raw material, subsequent conversion of raw materials into intermediate product forms such as prepregs, and finally the utilization of these intermediate product forms by fabricators to process the materials further to form completed composite structures. Emphasis is placed on the cumulative effects that each processing phase in the pipeline contributes to the final products general quality as well as physical, chemical, and mechanical properties. Finally it includes an overview of common process control schemes and discusses preparation of materials and processing specifications.

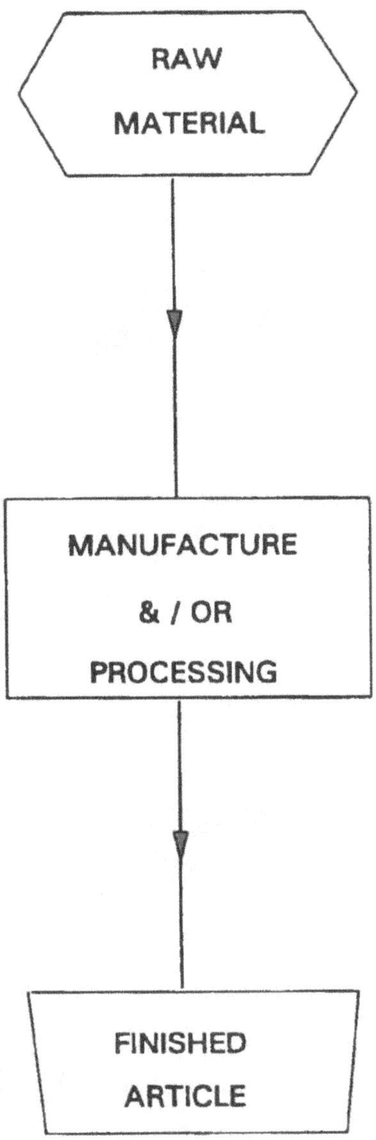

ALL PRODUCTS, REGARDLESS TO STAGE OF WORK, ARE CONSIDERED AS RAW MATERIALS. THESE MAY BE CHEMICALS FOR RESINS, OR SAND TO PROCESS INTO GLASS PRODUCTS, OR PRECURSORS FOR FILAMENTS, OR WOVEN GOODS, OR FLAT PROCESSED SHEETS AND/OR MANY OTHER ARTICLES WHICH HAVE YET TO BE PROCESSED TO AN END ITEM.

EACH RAW PRODUCT THEN IS PROCESSED, OR MIXED WITH OTHER RAW PRODUCTS, OR ALTERED TO BECOME STILL ANOTHER RAW ARTICLE TO EXPERIENCE YET ADDITIONAL PROCESS STEPS THROUGH THE PIPELINE. EACH AS RECEIVED RAW ARTICLE MUST BE PROCESSED IN SUCH A MANNER DURING ITS PROCESSING STEP(S) THAT VARIABILITY IS MINIMIZED AT THE NEXT PIPELINE FUNCTION. PROCESSING FUNCTIONS MAY BE COMPLEX; SUCH AS MATRIX IMPREGNATION, OR THEY MAY BE RELATIVELY SIMPLE; SUCH AS SHIPPING. REGARDLESS, EACH STEP MUST BE EFFECTIVE IN THAT IT DOES NOT INTRODUCE UNCONTROLLED CHANGES THAT ALTER THE PRODUCT FOR SUBSEQUENT USE OR END ITEM PERFORMANCE.

FINISHED ARTICLES LEAVING ONE PROCESSING FUNCTION IS USUALLY STILL CONSIDERED A RAW PRODUCT WHEN DELIVERED TO THE NEXT. THE ONUS FOR CONSISTENCY OF THIS PRODUCT MUST BE RECOGNIZED AND ATTENDED TO DURING THE JUST COMPLETED PROCESSING STEP(S). THE MATERIALS PIPELINE IS NOT COMPLETE UNTIL THE END ARTICLE IS FULLY FUNCTIONAL AS IS.

FIGURE 2.1(a) *Composite materials and processing, basic pipeline common to all materials and processes.*

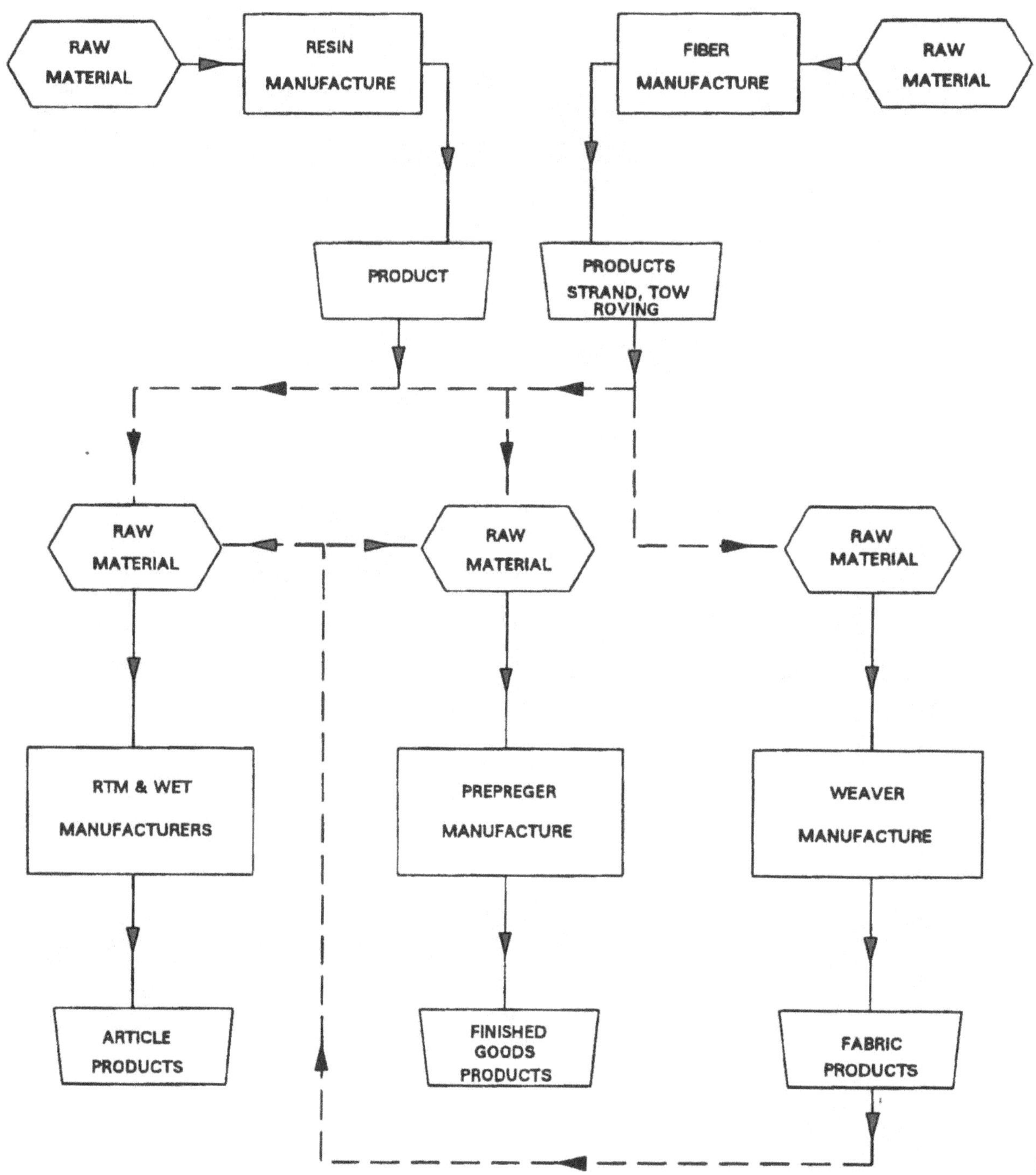

FIGURE 2.1(b) *Raw materials pipeline (example).*

2.4 CONSTITUENT MATERIALS

2.4.1 Reinforcement fibers

2.4.1.1 Carbon

2.4.1.2 Aramid

In the early 1970's, Du Pont Company introduced Kevlar™ aramid, an organic fiber with high specific tensile modulus and strength. This was the first organic fiber to be used as a reinforcement in advanced composites. Today this fiber is used in various structural parts including reinforced plastics, ballistics, tires, ropes, cables, asbestos replacement, coated fabrics, and protective apparel. Aramid fiber is manufactured by extruding a polymer solution through a spinneret. Major forms available from Du Pont are continuous filament yarns, rovings, chopped fiber, pulp, spun-laced sheet, wet-laid papers, thermoplastic-impregnated tows, and thermoformable composite sheets.

Important generic properties of aramid fibers are: low density, high tensile strength, high tensile stiffness, low compressive properties (nonlinear), and exceptional toughness characteristics. The density of aramid is 0.052 lb/in^3 (1.44 gm/cm^3). This is about 40% lower than glass and about 20% lower than commonly used carbon. Aramids do not melt and they decompose at about 900°F (500°C). Tensile strength of yarn, measured in twisted configuration, can be varied from 500 - 600 ksi (3.4 - 4.1 GPa) by choosing different types of aramids. The nominal coefficient of thermal expansion is 3x10^{-6} in/in/F° (-5x10^{6} m/m/C°) in the axial direction. Aramid fibers, being aromatic polyamide polymers, have high thermal stability and dielectric and chemical properties. Excellent ballistic performance and general damage tolerance is derived from fiber toughness. Aramid is used, in fabric or composite form, to achieve ballistic protection for humans, armored tanks, military aircraft, and so on.

Composite systems, reinforced with aramid, have excellent vibration-damping characteristics. They resist shattering upon impact. Temperature of use, in composite form with polymer matrix, range from -33 to 390°F (-36 - 200°C), The nominal tensile properties of composites reinforced with aramid are listed in Table 2.4.1.2(a) - in thermoset (Reference 2.4.1.2(a)) and thermoplastic (Reference 2.4.1.2(b)) resin matrix. At 60% fiber volume fraction, composites of epoxy reinforced with aramid fibers have nominal tensile strength (room temperature) of 200 ksi (1.4 GPa) and nominal tensile modulus of 11 Msi (76 GPa). These composites are ductile under compression and flexure. Ultimate strength, under compression or flexure, is lower than glass or carbon composites. Composite systems, reinforced with aramid, are resistant to fatigue and stress rupture. In the system of epoxy reinforced with aramid, under tension/tension fatigue, unidirectional specimens (V_f ~ 60%) survive 3,000,000 cycles at 50% of their ultimate stress (Reference 2.4.1.2(a)). Recently, thermoplastic resin composites reinforced with aramid have been developed. These thermoplastic composite systems have

TABLE 2.4.1.2(a) *Nominal composite properties reinforced with aramid fiber (V_f ~ 60%)*

		Thermoset (epoxy)		Thermoplastic (J2)	
Tensile Property	Units	unidirectional	fabric[1]	unidirectional	fabric[1]
Modulus	Msi (GPa)	11 (68.5)	6 (41)	10.5-11.5 (73-79)	5.1-5.8 (35-40)
Strength	ksi (GPa)	200 (1.4)	82 (0.56)	180-200 (1.2-1.4)	77-83 (0.53-0.57)

[1] Normalized from V_f = 40%; fabric style S285

exhibited equivalent mechanical properties compared to similar thermoset systems (References 2.4.1.2(b)). In addition, thermoplastic systems provide potential advantages in economical processing (Reference 2.4.1.2(c)), bonding, and repair. A unique thermoformable sheet product, in thermoplastic matrix reinforced with aramid fibers, is available (Reference 2.4.1.2(d)). These composite systems are also used to achieve low coefficient of thermal expansion or high wear resistance. They are non-conductive and exhibit no galvanic reaction with metals. Aramid fibers are available in several forms with different fiber modulus (Table 2.4.1.2(b)). Kevlar™29 has the lowest modulus and highest toughness (strain to failure ~ 4%). These fibers are used mostly in ballistics and other soft composite systems such as cut- and slash- resistance protective apparel, ropes, coated fabric, asbestos replacement, pneumatic tires, etc. These are also used for composites where maximum impact and damage tolerance is critical and stiffness is less important. Kevlar™49 is predominantly used in reinforced plastics - both in thermoplastic and thermoset resin systems. It is also used in soft composites like core of fiber optic cable and mechanical rubber good systems (e.g., high pressure flexible hose, radiator hose, power transmission belts, conveyor belts, etc.). An ultra-high modulus Type 149 has been made available recently. It has 40% higher modulus than Kevlar™49. Kevlar™29 is available in fiber yarn sizes and two rovings sizes. Kevlar™49 is available in six yarn and two rovings sizes. Kevlar™149 is available in three yarn sizes. Yarn sizes range from the very fine 55 denier (30 filaments) to 3000 denier (1300 filaments). Rovings are 4560 denier (3072 filaments) and 7100 denier (5000 filaments). Composite thermoplastic tows, several types of melt-impregnated thermoplastic reinforced with different Kevlar™ yarns and deniers, are also available.

TABLE 2.4.1.2(b) *Nominal properties of aramid fiber*

Tensile Property	Units	Type of Kevlar™ 29	49	149
Modulus	Msi (GPa)	12 (83)	18 (124)	25 (173)
Strength	ksi (GPa)	525 (3.6)	525-600 (3.6-4.1)	500 (3.4)

Aramid composites were first adopted in applications where weight savings were critical - for example, aircraft components, helicopters, space vehicles, and missiles. Armor applications resulted from the superior ballistic and structural performance. In marine recreational industries, light weight, stiffness, vibration damping, and damage tolerance are valued. Composites reinforced with aramids are used in the hulls of canoes, kayaks, and sail and power boats. These same composite attributes have led to use in sports equipment. Composite applications of aramid continue to grow as systems are developed to capitalize on other properties. The stability and frictional properties of aramids at high temperatures have led to brake, clutch, and gasket uses; low coefficient of thermal expansion is being used in printed wiring boards; and exceptional wear resistance is being engineered into injection-molded thermoplastic industrial parts. Melt-impregnated thermoplastic composites, reinforced with aramids, offer unique processing advantages - e.g., in-situ consolidation of filament-wound parts. These can be used for manufacturing thick parts where processing is otherwise very difficult (Reference 2.4.1.2(e)).

Aramid fiber is relatively flexible and tough. Thus it can be combined with resins and processed into composites by most of the methods established for glass. Yarns and rovings are used in filament winding, prepreg tape, and in pultrusion. Woven fabric prepreg is the major form used in thermoset composites.

Aramid fiber is available in various weights, weave patterns, and constructions; from very thin (0.0002 in., 0.005mm) lightweight (275 gm/m^2) to thick (0.026 in., 0.66 mm) heavy (2.8 gm/m^2)woven roving. Thermoplastic-impregnated tows can be woven into various types of fabrics to form prepregs. These composites demonstrate good property retention under hot and humid conditions (Reference 2.4.1.2(f)). Chopped aramid fiber is available in lengths from 6 mm to 100 mm. The shorter lengths are used to reinforce thermoset, thermoplastic, and elastomeric resins in automotive brake and clutch linings, gaskets, and electrical parts. Needle-punched felts and spun yarns for asbestos replacement applications are made from longer fiber staple. A unique very short fiber (0.08 - 0.16 in., 2 - 4 mm) with many attached fibrils is available (aramid pulp). It can provide efficient reinforcement in asbestos replacement uses. Aramid short fibers can be processed into spun-laced and wet-laid papers. These are useful for surfacing veil, thin-printed wiring boards, and gasket material. Uniform dispersion of aramid short fiber in resin formulations is achieved through special mixing methods and equipment. Inherent fiber toughness necessitates special types of tools for cutting fabrics and machining aramid composites.

2.4.1.3 Glass

Glass in the forms used in commerce has been produced by many cultures since the early Etruscan civilization. Glass as a structural material was introduced early in the seventeenth century and became widely used during the twentieth century as the technology for flat pane was perfected. Glass fibrous usage for reinforcement was pioneered in replacement of metals and used for both commercial and military uses with the advent of formulation control and molten material which is die or bushing pulled into continuous filaments. These events lead to a wide range of aerospace and commercial high performance structural applications still in use today.

2.4.1.3.1 Chemical description

Glass is derived from one of our most abundant natural resources--sand. Other than for, possibly, transport and the melting process, it is not petro-chemical dependent. For purposes of this handbook the typical glass compositions are for electrical/Grade "E" glass, a calcium aluminoborosilica composition with an alkali content of less than 2%, chemical resistant "C" glass composed of soda-lime-borsilicates and high strength S-2 glass which is a low-alkali magnesi-alumina-silicate composition (See Table 2.4.1.3.1). Surface treatments (binders/sizing) can be applied directly to the filaments during the pulling step. Organic binders, such as starch oil, are applied to provide optimum weaving and strand protection during weaving of fabrics or "greige goods". These type binders are then washed and heat cleaned off the fabrics for finishing or sizing at the weaver with coupling agents to improve compatibility with resins. (See Figure 2.4.1.3.1) The exception to this process for fabrics is when they are heat treated or "caramelized", which converts the starch to carbon (0.2 - 0.5%). Glass roving products (untwisted) type yarns are most often directly finished with the final coupling agents during the filament manufacturing step. Therefore, the products will be identified with the glass manufacturer's product codes and the desizing step is not necessary as common with fabric "greige goods" forms. Heat cleaned products are also available where the product is essentially pure glass. These products, which are subject to damage, are commonly utilized for silicone laminates. Another finish designation is applicable to the heat cleaned product when it is followed with a demineralized water wash (neutral pH). More common for structural applications are the coupling agents which are applied for use with standard organic polymers. During the 1940's Volan[1] finishes were introduced. Since then, many variations/improvements identified with various company designations have appeared. Perhaps the most recognized is Volan A. This finish provides good wet and dry strength properties in use with polyester, epoxy, and phenolic resins. Prior to the application of this finish the clean(ed) glass is saturated with methacrylate chromic chloride so that the chrome content of the finish is between 0.03% and 0.06%. This addition enhances wet-out of the resin during cure. Perhaps more typically called out for use, but not limited to, with epoxy are the silane finishes. Most all are formulated to enhance laminate wet-out. Some also produce high laminate clarity or good composite properties in aqueous environments. Others improve high-pressure laminating, or resist adverse environment or chemical exposures. Although other finishes are used in

---

[1]E. I. DuPont de Nemours

combination with matrix materials other than epoxy, finishes may have proprietary formulations or varied designations relative to the particular glass manufacturer or weaver, it is believed the compositions are readily available to the resin compounders (prepreggers) to determine compatibility and end use purposes. Note that, non-compatible finishes are purposely applied for ornament applications.

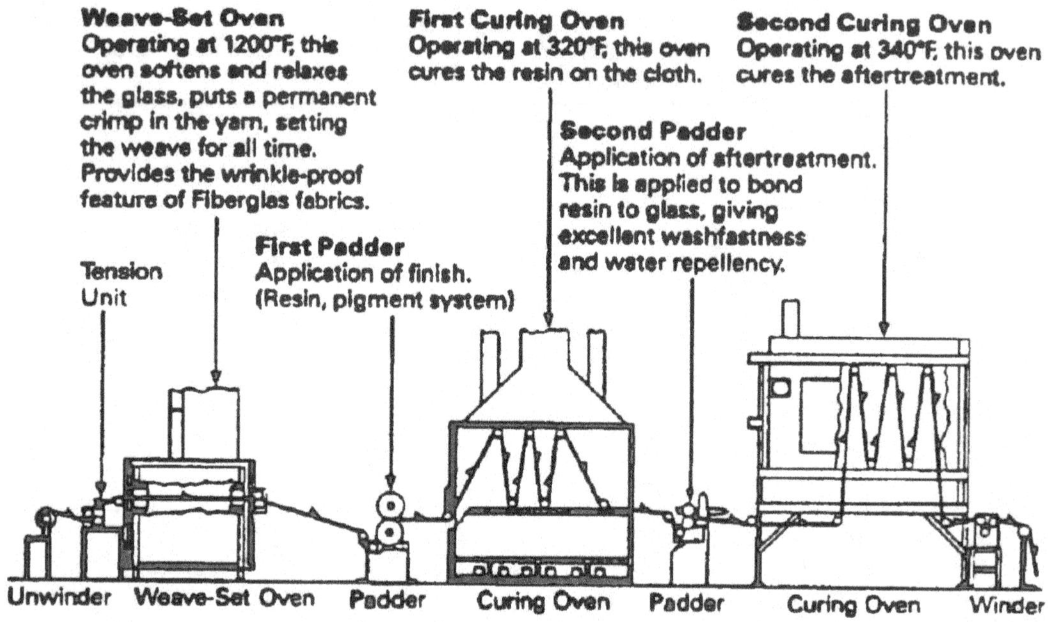

FIGURE 2.4.1.3.1 *Fabric finishing (Reference 2.4.1.3.1(c))*

2.4.1.3.2 Physical forms available

Due to the high quantity of commercial applications for glass products, there are many product forms available. For purposes of this publication glass forms will be limited to continuous filament product forms. These forms fall into four major categories. They are continuous rovings, yarn for fabrics or braiding, mats, and chopped strand. (See Figure 2.4.1.3.2 and ASTM Specification D 579, Reference 2.4.1.3.2(a) for information on glass fabrics. Further discussion of fabrics may be found in Section 2.5.1 on fabrics and preforms.) They are available with a variety of physical surface treatments and finishes. Most structural applications utilize fabric, roving, or rovings converted to unidirectional tapes. Perhaps the most versatile fiber type to produce glass product forms is "E" glass. "E" glass is identified as such for electrical applications. This type or grade of glass has eight or more standardized filament diameters available. These range from 1.4 to 5.1 mils (3.5 to 13 micrometers). (See Table I, ASTM Specification D 578, Reference 2.4.1.3.2(b).) This facilitates very thin product forms. The "S" glasses are identified as such to signify high strength. The S-2 type glasses are available with but one filament diameter. This does not limit the availability of basic structural fabric styles for S-2 glass however. Although there are more "E" roving products, as to yields, available, this has not noticeably restricted the use of S-2 type roving products or roving for unidirectional tape. S-2 type rovings are available in yields of 250, 750, and 1250 yards per lb (500, 1500, and 2500 m/kg).

Although woven rovings may be considered a fabric product form it should be noted for its importance for military applications. Also, there are glass product forms which could be considered as complimentary products for advanced structures. These would include milled fibers and chopped strand.

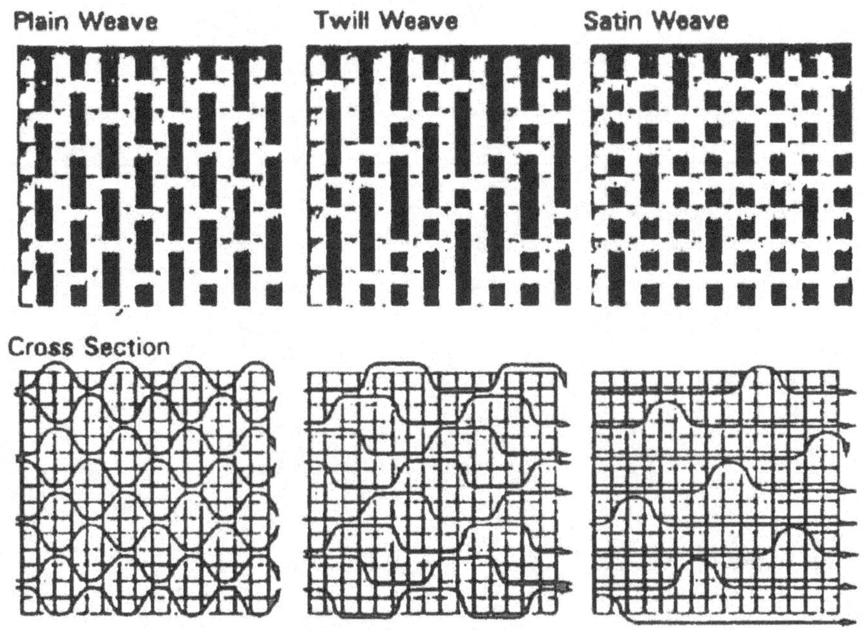

FIGURE 2.4.1.3.2 *Common types of weaves for glass fabrics (Reference 2.4.1.3.1(c)).*

2.4.1.3.3 Advantages and disadvantages

For many years glass composites have had a distinct strength to weight advantage. Although the rapid evolution of carbon and aramid fibers have gained advantages, glass composite products have still prevailed in certain applications. Cost per weight or volume, certain armament applications, chemical or galvanic corrosion resistance, electrical properties, and availability of many product forms remain as examples of advantage. Coefficient of thermal expansion and modulus properties compared to carbon composites may be considered as typical disadvantages. When compared to aramid composites, glass has a disadvantage as to tensile properties but an advantage as to ultimate compression, shear properties, and moisture pick-up.

Commercial uses for glass products are many-fold. These include filtration devices, thermal and electrical insulation, pressure and fluid vessels, and structural products for automotive and recreation vehicles. Many uses are applicable to military and aerospace products as well. A partial listing would include: asbestos replacement, circuitry, optical devices, radomes, helicopter rotor blades, and ballistic applications. Because of the many product forms, structural applications are limitless to fabricate. If there are limitations, compared to other fibers, they may include low thermal and electrical conductivity or perhaps melting temperatures when compared to carbon fibers.

Typical properties for glass fibers and composite materials reinforced with continuous glass fibers are shown in Tables 2.4.1.3.3(a)-(d).

Unburdened costs vary pending product forms and glass types. Typical yield certified "E" glass rovings cost $1.40 per lb., whereas certified S-2 type 750 yield rovings average $6.30 per lb. Lower costing for rovings are experienced with rail car purchases. Typical unburdened fabric costs also vary by weave and fiber type. "E" glass 120 style averages $13.10 per lb., 7781 averages $4.35 per lb., S-2 type 6781 style is $8.40 per lb.

TABLE 2.4.1.3.3(a) *Typical glass fiber electrical properties.*

	E	S-2	H_R
Density			
lb/in^3	0.094	0.089	0.090
g/cm^3	2.59	2.46	2.49
Tensile Strength			
ksi	500	665	665
MPa	34,450	45,818	45,818
Modulus of Elasticity			
Msi	10.5	12.6	12.6
GPa	72.35	86.81	86.81
% Ult. Elongation	4.8	5.4	5.4
Dielectric Constant			
73°F (23°C) @ 1 MHZ	6.3-6.7	4.9-5.3	NA

TABLE 2.4.1.3.3(b) *Typical glass fiber thermal properties.*

	E	S-2	S_B
Coeff. Thermal Expan. 10^6			
in/in/F°	2.8	1.3	
m/m/C°	5.1	2.6	
Softening Point °F (°C)	1530 (832)	1810 (988)	1778 (970.)
Annealing Point °F (°C)	1210 (654)	1510 (821)	1490 (810.)

TABLE 2.4.1.3.3(c) *Typical corrosion resistance of glass fibers (Wt. Loss %).*

Fluid	E	S-2	S_B
10% H_2SO_4	42	6.8	NA
10% HCL	43	4.4	NA
10% HNO_3	43	3.8	NA
H_2O (Distilled)	0.7	0.7	NA
10% Na OH	29	66	NA
10% KOH	23	66	NA

Conditions: 200°F (96°C) - one week immersion

TABLE 2.4.1.3.3(d) *Typical cured epoxy/glass mechanical properties.*

E Glass, Woven 7781 Style	Standard Structural	Dual Purpose Structural/Adhesive
Tensile Strength, ksi (MPa)	63 (430)	48 (330)
Tensile Modulus, Msi (GPa)	3.8 (36)	2.8 (19)
Compressive Strength, ksi (MPa)	60. (410)	50. (340)
Compressive Modulus, Msi (GPa)	3.6 (25)	3.2 (22)
Flexural Strength ksi, (MPa)	80. (550)	65 (450)
Flexural Modulus Msi, (GPa)	3.7 (26)	3.3 (23)
Interlaminar Shear ksi, (MPa)	2.6 (18)	3.8 (26)
Sandwich Peel, lb/in width (N/m width)	N.A.	30. (3.4)
Metal-to-Metal Peel, lb/lin. in. (N/lin. m)	N.A.	55 (6.3)
Specific Gravity gm/cm^3 (lb/in^3)	1.8 (0.065)	1.6 (0.058)
Cured Resin Content % Wt.	33	48

Reference: Fabric MIL-C-9084, VIII B
Resin MIL-R-9300, Ty I MIL-A-25463, Ty I, C1 2

2.4.1.3.4 Common manufacture methods and variable

Most often raw products (and/additives) are mixed and are premelted into marbles. This form facilitates sampling for analysis but, more important, presents a raw product form for automated feeding to the individual melt furnaces. Another method is to feed, via hoppers, dried raw products directly to batch cans. Regardless of the raw form, the material is fed into furnaces to become molten at approximately 2800°F (1500°C). The molten mass flows onto plates which contain many bushings with small orifices from which the individual filaments are drawn. In some cases the individual bushings are heat controlled within <1F° (0.6C°). The diameter of the filaments is controlled by the viscosity of the glass melt and the rate of extrusion. Cooling or solidification occurs rapidly as the glass leaves the bushings in filament form under ambient conditions. Cooling is often added by water spray and/or application of the binders. The individual untwisted filaments are gathered and high speed wound on tubes or "cakes". Sometimes finishes are applied after the strands are wound on the tubes then conditioned (dried). For products common to this document the strands are "C" (continuous) filaments--not "S" (staple) filament. To produce rovings the strands are then creeled, unwound and gathered again to form ends or multiple untwisted strands. (See Table 2.4.1.3.4(a).) This process of gathering or combining is again repeated to form rovings of desired yields (yards per pound). For weaving of fabrics and braiding, the strands are twisted to form yarns. (See Table 2.4.1.3.4(b).) Single yarns are composed of single strands twisted by itself. Two (etc.) strand construction is two strands twisted to produce a single yarn. Plied yarns are made from twisting two or more yarns together. Twisting and plying is often referred to as "throwing". A variable in processing "C" filament products is the repeated tensioning required during the numerous product forms fabrication. Tensioning devices are used--such as: disc-type or "whirls", gate-type, tension bars or "S" bars, and compensating rolls in the delivery from the creels. Humidity is another controlled variable in the twisting, plying, braiding, warping, slashing, gulling and weaving areas. These operations are facilitated to maintain a relative humidity of 60 to 70 percent range. During the glass processing operations surface abrasion is a factor which must be monitored. The many devices such as: guide eyes, spacer bars, rollers and such are subject to wear and must be maintained. Wear could also affect tensioning. These contact devices are manufactured from materials including: stainless steel, chromium plating, and ceramics.

Additional information can be found in References 2.4.1.3.4(a) - (c).

TABLE 2.4.1.3.4(a) *Basic strand fiber designations and strand counts (Reference 2.4.1.3.1(c)).*

Filament Diameter Designation		Strand Count (Number)		
SI (μm)	U.S. Customary (Letter)	TEX g/km	U.S. Customary 100 Yd. Cuts/Lb.	U.S. Customary Yds./Lb.
5	D	11	450	45,000
7	E	22	225	22,500
9	G	33	150	15,000
10	H	45	110	11,000
13	K	66	75	7,500

2.4.1.4 Boron

Elemental boron fiber is formed as a deposition reaction on a hot tungsten wire which is continuously drawn through a reactor containing BCl_3 and H_3. The tungsten wire substrate also reacts to form tungsten boride in the core. The crystalline structure of the deposited boron is considered amorphous due to its small size (20Å). Boron is available as a cylindrical fiber in two nominal diameters, 4- and 5.6-mil (0.10 and 0.14 mm), which have a density of 2.57 and 2.49 g/cm^3 (0.0929 and 0.0900 lb/in^3), respectively. Chemical etching of the fiber surface produces a higher strength, but the process is not used commercially.

Boron fiber is unmatched for its combination of strength, stiffness, and density. The tensile modulus and strength of boron fiber are 60 x 10^6 psi and 0.52 x 10^6 psi (40 GPa and 3600 MPa). Thermal conductivity and thermal expansion are both low, with a coefficient of thermal expansion of 2.5-3.0 x 10^{-6}/F° (4.5-5.4 x 10^{-6}/C°). Typical end-use properties are shown in Table 2.4.1.4. Currently, the cost of boron fiber is approximately an order of magnitude higher than standard carbon fiber.

TABLE 2.4.1.3.4(b) *Typical yarn nomenclature (Reference 2.4.1.3.1(c)).*

Filament Designation	Nominal Filament Diameter, inches (mm)	Strand Count (x100 = yds/lb) (g/km)	Approximate Number of Filaments
D	0.00021 (0.053)	1800 (2.8)	51
D	0.00021 (0.053)	900 (5.5)	102
B	0.00015 (0.0038)	450 (11)	408
D	0.00021 (0.053)	450 (11)	204
D	0.00021 (0.053)	225 (22)	408
E	0.00029 (0.0074)	225 (22)	204
B	0.00015 (0.0038)	150 (33)	1224
C	0.00019 (0.0048)	150 (33)	750
DE	0.00025 (0.0064)	150 (33)	408
G	0.00036(0.0091)	150 (33)	204
H	0.00043 (0.011)	110 (45)	204
C	0.00019 (0.0048)	75 (66)	1500
DE	0.00025 (0.0064)	75 (66)	816
G	0.00036 (0.0091)	75 (66)	408
K	0.00053 (0.014)	75 (66)	204
H	0.00043 (0.011)	55 (90)	408
DE	0.00025 (0.0064)	37 (130)	1632
G	0.00036 (0.0091)	37 (130)	816
K	0.00053 (0.014)	37 (130)	408
H	0.00043 (0.011)	25 (200)	816
K	0.00053 (0.014)	18 (275)	816
G	0.00036 (0.0091)	15 (330)	2052

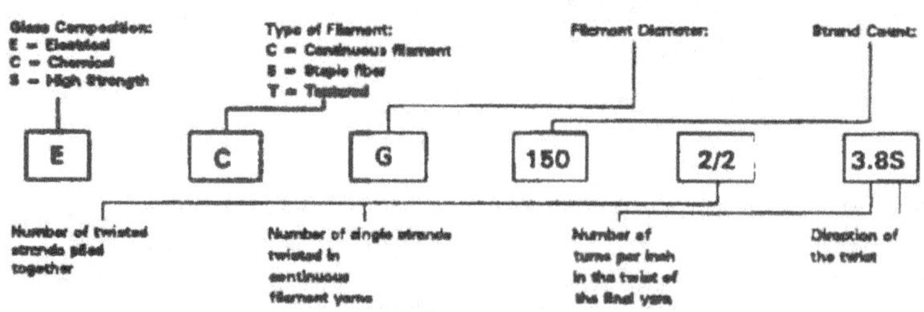

Available almost exclusively in filament or epoxy matrix prepreg form, boron fiber has been used for aerospace applications requiring high strength and/or stiffness, and for selective reinforcement in sporting goods. The most notable use of this fiber is the stabilizer sections of the F-14 and F-15 military aircraft, dorsal longeron of the B-1B bomber, and the repair of metallic airframe structures. High modulus (HM) or high strength (HS) carbon/epoxy composites can match either the tensile modulus or strength of boron composites at a more economical price, but boron/epoxy composites offer twice the composite strength. Additional information can be found in References 2.4.1.4(a) through (g).

2.4.1.5 Alumina

Continuous polycrystalline alumina fiber is ideally suited for the reinforcement of a variety of materials including plastics, metals, and ceramics. Alumina is prepared in the form of continuous yarn containing a nominal 200 filaments. It is supplied in bobbins containing continuous filament yarn, and alumina/aluminum and alumina/magnesium plates. Alumina staple is also available for short fiber reinforcement.

Fibers that are more than 99% purity α alumina have excellent chemical resistance, and have higher modulus and temperature capabilities than ceramic fibers containing silica. The high modulus of 55 Msi (380 GPa) is comparable to that of boron and carbon. The average filament tensile strength is 200 ksi (1.4 GPa) minimum. Since alumina is a good insulator, it can be used in applications where conducting fibers cannot. Nominal properties of alumina are listed in Table 2.4.1.5(a). Cost projections for alumina are competitive with carbon.

Alumina, in continuous form, offers many advantages for composite fabrication including ease of handling, the ability to align fibers in desired directions, and filament winding capability. The fact that alumina is an electrical insulator combined with its high modulus and compressive strength make it of interest for polymer matrix composite applications. For example, alumina/epoxy and aramid/epoxy hybrid composites reinforced with alumina and aramid fibers have been fabricated and are of potential interest for radar transparent structures, circuit boards, and antenna supports. Typical properties of unidirectional composites are listed in Table 2.4.1.5(b).

TABLE 2.4.1.4 *Typical end-use properties of a unidirectional boron/epoxy laminate (V_f = 0.5).*

	Value, ksi (MPa)
Moduli	
Tensile, longitudinal	30 (207)
Tensile, transverse	2.7 (19)
Strength	
Tensile, longitudinal	192 (1323)
Tensile, transverse	10.4 (72)
Compressive, longitudinal	353 (2432)

TABLE 2.4.1.5(a) *Nominal properties of alumina*

Composition	> 99% α-Al_2O_3	Filaments/yarn	200, nominal
Melting Point	3713°F (2045°C)	Tensile Modulus	55 Msi (385 GPa)
Filament Diameter	0.8×10^{-3} in. (20μm)	Tensile Strength	200 ksi (1.4 GPa) minimum
Length/Weight	(~4.7 m/gm)	Density	0.14 lb/in^3 (3.9 gm/cc)

TABLE 2.4.1.5(b) *Nominal properties of alumina composite (V_f ~ 50-55%)*

Moduli		
Tensile, axial		30-32 Msi (210-220 GPa)
Tensile, transverse		20-22 Msi (140-150 GPa)
Shear		7 Msi (50 GPa)
Strength		
Tensile, axial		80 ksi (600 MPa)
Tensile, transverse		26-30 ksi (130-210 MPa)
Shear		12-17 ksi (85-120 GPa)
Fatigue - Axial Endurance Limit		10^7 cycles at 75% of static ultimate (tension-tension, R=0.1, and rotating-bending)
Average Thermal Expansion		68-750° (20-400°C)
Axial		4.0 μin/in/F° (7.2 μm/m/C°)
Transverse		11 μin/in/F° (20 μm/m/C°)
Thermal Conductivity	68-750° (20-400°C)	22-29 Btu/hr-ft-°F (38-50 J/m-s-°C)
Specific Heat	68-750° (20-400°C)	0.19-0.12 Btu/lbm-°F (0.8-0.5 J/gm-°C)
Density		0.12 lbm/in^3 (3.3 gm/cm^3)

2.4.1.6 *Silicon carbide*

Various super-refractory fibers were first produced in the early 1950's based upon work by the Arthur D. Little Co. by various production methods. The primary of these based upon:

1. Evaporation for polycrystalline fiber process.

2. HITCO continuous process for polycrystalline fibers.

3. Vapor deposition of aluminum oxide single crystals (Reference 2.4.1.6(a)).

The most recent advances in the CVD type process in use by AVCO consist of substrate wires drawn through glass reaction tubes at high temperature.

Silicon carbide fibers are produced with a nominal 0.0055 in. (140 μm) filament diameter and are characteristically found to have high strength, modulus and density. Fiber forms are oriented toward the strengthening of aluminum or titanium alloys for both longitudinal and transverse properties. Additional forms are also produced as polycrystalline fiber whiskers of varying length and diameters (Reference 2.4.1.6(b)).

Several systems for describing the material morphology exist, the alpha and beta forms designated by Thibault and Lindquist being the most common (Reference 2.4.1.6(c)).

Practically all silicon carbide monofilament fibers are currently produced for metal composite reinforcement. Alloys employing aluminum, titanium, and molybdenum have been produced (Reference 2.4.1.6(b)).

General processing for epoxy, bisimide, and polyimide resin can be either via a solvated or solventless film impregnation process, with cure cycles equivalent to those provided for carbon or glass reinforced products. Organic matrix silicon carbide impregnated products may be press, autoclave, or vacuum bag oven cured. Lay-up on tooling proceeds as with carbon or glass composite products with all bleeding, damming, and venting as required for part fabrication. General temperature and pressure ranges for the cure of the selected matrix resins used in silicon carbide products will not adversely affect the fiber morphology.

Silicon carbide ceramic composites engineered to provide high service temperatures (in excess of 2640°F or 1450°C) are unique in several thermal properties. The overall thermal resistance is determined by the through conductivity, thermal expansion, thermal shock and creep resistance. Thermal conductivities of silicon carbide ceramics have a range in Btu-in/s-ft^2-°F of 0.12 at room temperature to 0.09 at 1470°F (W/m·K of 60 at room temperature to 48 at 800°C). Expansion values range, in percentage of original dimension, from 0.05 at 390°F (200°C) to 1470°F (0.30% at 800°C). The creep resistance of the silicon carbide ceramic will vary as the percentage of intra-granular silicon phase increases. In general, the creep rate is very low when compared to aluminum oxide or zirconium oxide materials.

Mechanical properties of silicon carbide materials are shown in Table 2.4.1.6(a). Fracture toughness as measured by double torsion analysis has reported literature values for K_{Ic} ranging from 0.55 ksi Jm (0.6 MPa $\sqrt{m}$) for monocrystalline SiC/Si to 5.5 ksi Jm (6.0 MPa Jm) for hot pressed SiC ceramics (Reference 2.4.1.6(g)). Corrosion resistance, of consideration in advanced structural material design, has been evaluated with a variety of mineral acids on the basis of corrosive weight loss as shown in Table 2.4.1.6(b).

General cost ranges for the CVD processed fibers are currently in the $100.00 per lb., with the control in crystalline form requiring additional expense (Reference 2.4.1.6(e)).

TABLE 2.4.1.6(a) *Material properties of silicon carbide materials.*

Property	Reported Values (ksi)	(MPa)	Reference Information
FLEXURAL STRENGTH	100-1000	700-7000	single crystal, 99+% purity (1)
	10-60	70-400	polycrystalline materials, 78-99% purity, with < 12+% free silicon, sintered (1)
	5-8	30-60	sintered SiC - graphite composites - epoxy, imide, polyimide matrix. (2)
COMPRESSIVE STRENGTH	500-1000	3000-7000	single crystal, 99+% purity (1)
	10-25	70-170	polycrystalline materials, 78-99% purity, with < 12+% free silicon, sintered.(2)
	14-60	97-400	Sintered SiC - graphite composites - epoxy, imide, polyimide matrix. (2)
TENSILE STRENGTH	~20	~140	single crystal, 99+% purity (1)
	5-20	30-140	polycrystalline materials, 78-99% purity, with < 12+% free silicon, sintered.(2)
	2.5-25	17-170	sintered SiC - graphite composites - epoxy, imide, polyimide matrix. (2)
MODULUS OF ELASTICITY	~9.5	~66	single crystal, 99+% purity (1)
	~7.0	~48	Polycrystalline materials, 78-99% purity, with < 12+% free silicon, sintered.(2)

(1) Reference 2.4.1.6(b)
(2) Reference 2.4.1.6(d)

TABLE 2.4.1.6(b) *Corrosive weight loss at 212 °F (100 °C) (Reference 2.4.1.6(e)).*

TEST REAGENT	Si/SiC COMPOSITES 12% Si	SiC - NO FREE Si
	mg/cm²·yr	*mg/cm²·yr*
98% Sulfuric Acid	55	1.8
50% Sodium Hydroxide	complete within days	2.5
53% Hydrofluoric Acid	7.9	< 0.2
70% Nitric Acid	0.5	< 0.2
25% Hydrochloric Acid	0.9	< 0.2

2.4.1.7 Quartz

Quartz fiber is very pure (99.95%) fused silica glass fiber. Typical fiber properties are shown in Table 2.4.1.7(a). Quartz is produced as continuous strands consisting of 120 or 240 individual filaments of 9 micron nominal diameter. These single strands are twisted and plied into heavier yarns. Quartz fibers are generally coated with an organic binder containing a silane coupling agent which is compatible with many resin systems. Strands for rovings are combined into multiple ends without applied twist. These strands are coated with a "direct size" which is compatible with many resins. Woven fabrics may be used as woven or may be "scoured" (washed) to remove the nonfunctional components of the binder and some, but not all, of the silane coupling agent. Following scouring, the fabric may be finished with a variety of silane coupling agent finishes having specific resin compatibility.

Quartz fiber nomenclature is the same as that for E or S glass fibers except that the glass composition is designated by the letter Q as shown in Table 2.4.1.7(b). Commonly used quartz fabrics are listed in Table 2.4.1.7(c). Quartz rovings are continuous reinforcements formed by combining a number of 300 2/0 zero twist strands. End counts of 8, 12, and 20 are available having yields from 750 to 1875 yards per pound (660 to 264 g/km). Quartz fibers are also available in the form of chopped fiber in cut lengths from 1/8 inch to 2 inches (3 to 50 mm).

Quartz fibers with a filament tensile strength of 850 ksi (5,900 MPa) have the highest strength-to-weight ratio, virtually exceeding all other high temperature materials. The quartz fibers can be used at temperatures much higher than "E" glass or "S" glass fiber with service temperatures up to 1920°F (1050°C) possible. Quartz fibers do not melt or vaporize until the temperature exceeds 3000°F (1650°C), providing potential in ablative applications. Additionally, these fibers retain virtually all of the characteristics and properties of solid quartz.

The quartz fibers are chemically stable. They are not affected by halogens or common acids in the liquid or gaseous state with the exception of hydrofluoric and hot phosphoric acids. Quartz fibers should not be used in environments where strong concentrations of alkalies are present.

Quartz fibers, when combined with certain matrix systems, offer potential advantages in stealth application due to their high electrical resistivity properties. Quartz does not form paramagnetic centers, nor does it capture neutrons in high energy applications. These fibers offer a low dielectric constant and loss tangent providing excellent properties as electrical insulators. Typical properties for quartz fibers combined with three different polymer matrix systems are shown in Tables 2.4.1.7(d) - (f). Quartz products are relatively expensive compared to "E" or "S_2" glass products, with prices ranging from $45 to $150 per pound. Additional information can be found in Reference 2.4.1.7.

Table 2.4.1.7(a) *Properties of quartz fiber.*

Specific gravity	2.20
Density, lb/in^3	0.0795
g/cm^3	2.20
Tensile strength	
Monofilament, ksi	870
GPa	6.0
Roving, ASTM D2343 Impregnated	
Strand Test -	
Astroquartz II 9779, ksi	530.5
GPa	3.6
Modulus, Msi	10.0
GPa	72.0
Elongation, percent	
$\frac{\text{Monofilament Tensile Strength} \times 100}{\text{Modulus}}$	8.7
Thermal	
Coefficient of expansion	
10^{-6} in/in/°F	0.3
10^{-6} cm/cm/°C	0.54
Thermal conductivity	
Cal/sec/cm/°C	0.0033
Btu/hr/ft/°F	0.80
Btu/hr/sq ft/in/°F	9.5
Specific heat	
Joules/Kg/°C	7500
Btu/lb/°F	1.80
Electrical	
Dielectric constant, 10 GHz, 75°F (24°C)	3.78
Loss tangent, 10 GHz, 75°F (24°C)	0.0001

TABLE 2.4.1.7(b) *Quartz continuous strands.*

Strand Number	Number of filaments	Strand Count		Filament Diameter	
		yds/lb	g/km	10^{-5} in.	µm
QCG 300 1/0	119[a]	30,000	6.5	45	1.1
QCG 300 2/0	240[b]	15,000	33	35	0.89
QCG 300 1/2	240[a]	15,000	33	35	0.89
QCG 300 2/2	480[a]	7,500	66	35	0.89
QCG 300 2/8	1920[a]	1,875	264	35	0.89

[a]Used for fabric yarns.
[b] Used for roving and fabric yarns.

TABLE 2.4.1.7(c) *Construction of woven fabrics for aerospace applications.*

Style	Count	Warp Fill	Fill Yarn	Weave	Weight Oz/Sq.Yd.
503	50x50	300 1/2	300 1/2	plain	3.5
507	27x25	300 1/2	300 1/2	plain	2.0
525	50x 50	300 1/0	300 1/0	plain	2.0
527	42x32	300 2/2	300 2/2	plain	5.6
531	68x65	300 1/2	300 1/2	8HS	5.1
557	57x31	300 2/2	300 1/0	crowfoot	5.0
570	38x24	300 2/8	300 2/8	5HS	19.3
572	17x16	300 2/8	300 2/8	plain	9.9
581	57x54	300 2/2	300 2/2	8HS	8.4
593	49x46	300 2/2	300 2/2	5HS	7.5

TABLE 2.4.1.7(d) *Typical properties for quartz/epoxy.*

PROPERTY	Room Temperature U.S.	Room Temperature SI	1/2 hr at 350°F (180°C) U.S.	1/2 hr at 350°F (180°C) SI
Tensile Strength (ksi, MPa)	74.9 - 104	516 - 717	65.4 - 92.2	451 - 636
Tensile Modulus (Msi, GPa)	3.14 - 4.09	21.7 - 28.2	2.83 - 3.67	19.5 - 25.3
Flexural Strength (ksi, MPa)	95.5 - 98.9	658 - 682	53.9 - 75.9	372 - 523
Flexural Modulus (Msi, GPa)	3.27- 3.46	22.5 - 23.8	2.78 - 3.08	19.2 - 21.2
Compressive Strength (ksi, MPa)	66.4 - 72.4	458 - 499	42.6 - 49.9	294 - 344
Compressive Modulus (Msi, GPa)	3.43 - 3.75	23.6 - 25.9	3.10 - 3.40	21.4 - 23.4
Laminate Resin Content (wt%)	33.5 - 32.0			
Specific Gravity (g/cm^3)	1.73 - 1.77			

TABLE 2.4.1.7(e) *Typical properties for quartz/toughened epoxy.*

PROPERTY	Room Temperature U.S.	Room Temperature SI	180°F (82°C) U.S.	180°F (82°C) SI
Flexural Strength (ksi,MPa)	129.0	889	111.7	770
Flexural Modulus (Msi,GPa)	4.0	27.6	3.9	26.9
Compressive Strength (ksi,MPa)	88.2	608	77.5	534
Compressive Strength, Wet (ksi,MPa)	76.6	528	70.8	488
Compressive Modulus (Msi,GPa)	4.2	29.0	3.8	26.2
Compressive Modulus, Wet (Msi,GPa)	3.7	25.5	4.0	27.6
Short Beam Strength (ksi,MPa)	13.2	91.0	11.8	81.4
Short Beam Strength, Wet (ksi,MPa)	9.2	63.4	9.3	64.1
Resin Content (wt%)	32.0			
Ply Thickness (in,mm)	0.009	0.23		

TABLE 2.4.1.7(f) *Typical properties for quartz/polyimide.*

PROPERTY	Room Temperature U.S.	Room Temperature SI	1/2 Hour at 350°F (177°C) U.S.	1/2 Hour at 350°F (177°C) SI
Tensile Strength (ksi,MPa)	79.1 - 105	545 - 724		
Tensile Modulus (Msi,GPa)	3.9	27		
Flexural Strength (ksi,MPa)	93.7 - 102	646 - 703	62.4 - 68.3	430 - 471
Flexural Modulus (Msi,GPa)	3.2	22	2.6 - 2.8	18 - 19
Compressive Strength (ksi,MPa)	67.0 - 67.4	462 - 465	38.6 - 45.2	266 - 312
Compressive Modulus (Msi,GPa)	3.5 - 3.7	24 - 26	2.8	19
Laminate Resin Content (wt%)	36.2 - 36.2			

2.4.2 RESIN MATERIALS

2.4.2.1 Overview

Resin is a generic term used to designate the polymer, polymer precursor material, and/or mixture or formulation thereof with various additives or chemically reactive components. The resin, its chemical composition and physical properties, fundamentally affect the processing, fabrication and ultimate properties of composite materials. Variations in the composition, physical state, or morphology of a resin and the presence of impurities or contaminants in a resin may affect handleability and processab ility, lamina/laminate properties, and composite material performance and long-term durability. This section describes resin materials used in polymer matrix composites and adhesives, and considers possible sources and consequences of variations in resin chemistry and composition, as well as the effects of impurities and contaminants, on resin processing characteristics and on resin and composite properties.

2.4.2.2 Epoxy

The term epoxy is a general description of a family of polymers which are based on molecules that contain epoxide groups. An epoxide group is an oxirane structure, a three-member ring with one oxygen and two carbon atoms. Epoxies are polymerizable thermosetting resins containing one or more epoxide groups curable by reaction with amines, acids, amides, alcohols, phenols, acid anhydrides, or mercaptans. The polymers are available in a variety of viscosities from liquid to solid.

Epoxies are used widely in resins for prepregs and structural adhesives. The advantages of epoxies are high strength and modulus, low levels of volatiles, excellent adhesion, low shrinkage, good chemical resistance, and ease of processing. Their major disadvantages are brittleness and the reduction of properties in the presence of moisture. The processing or curing of epoxies is slower than polyester resins. The cost of the resin is also higher than the polyesters. Processing techniques include autoclave molding, filament winding, press molding, vacuum bag molding, resin transfer molding, and pultrusion. Curing temperatures vary from room temperature to approximately 350 °F (180°C). The most common cure temperatures range between 250° and 350°F (120° and 180°C). The use temperatures of the cured structure will also vary with the cure temperature. Higher temperature cures generally yield greater temperature resistance. Cure pressures are generally considered as low pressure molding from vacuum to approximately 100 psi (700 kPA).

2.4.2.3 Polyester (thermosetting)

The term thermosetting polyester resin is a general term used for orthophthalic polyester resin or isophthalic polyester resin. Polyester resins are relatively inexpensive and fast processing resins used generally for low-cost applications. In combination with certain fillers, they can exhibit resistance to breakdown under electrical arc and tracking conditions. Isophthalic polyester resins exhibit higher thermal stability, dimensional stability, and creep resistance. In general, for a fiber-reinforced resin system, the advantage of a polyester is its low cost and its ability to be processed quickly.

Fiber-reinforced polyesters can be processed by many methods. Common processing methods include matched metal molding, wet lay-up, press (vacuum bag) molding, injection molding, filament winding, pultrusion, and autoclaving. Polyesters can be formulated to cure more rapidly t han do phenolics during the thermoset molding process. While phenolic processing, for example, is dependent on a time/temperature relationship, polyester processing is primarily dependent on temperature. Depending on the formulation, polyesters can be processed from room temperature to 350 °F (180°C). If the proper temperature is applied, a quick cure will occur. Without sufficient heat, the resin/catalyst system will remain plasticized. Compared to epoxies, polyesters process more easily and are much tougher, whereas phenolics are more difficult to process and brittle, but have higher service temperatures.

2.4.2.4 Phenolic

Phenol-formaldehyde resins and their direct precursors were first produced commercially in the early 1900's for use in the commercial market. Ureaformaldehyde and melamine-formaldehyde appeared in the 1920 - 1930's as a less expensive alternative for lower temperature use. Phenolics, in general, cure by a condensation route with the off-gassing of water. The resulting matrix is characterized by both chemical and thermal resistance as well as hardness, and low smoke and toxic degradation products.

The phenolic polymers, often called either phenolic resole or novolacs resins, are c ondensation polymers based upon either a reaction of excess formal dehyde with a base catalyst and phenol (resole), or a reaction of excess phenol with an acidic catalyst and formaldehyde (novolac). The basic difference between resoles and novolacs consist of no methylol groups in the novolacs and the resulting need for an extension agent of paraformaldehyde, hexamethylenetetraamine, or additional formaldehyde as a curative. These resins have higher molecular weights and viscosities than either parent material. Consequently, they are optimal for processing parts of unusual conformations and complex curvature. The resins allow either press or autoclave cure and allow relatively high temperature free-standing postcures.

2.4.2.4.1 Resoles

The reaction of phenol and excess formaldehyde in the presence of base is characterized by low-molecular-weight prepolymers that are soluble in base and contain a large degree of methylol groups (-CH_2OH). These prepolymers are processed to a workable viscosity (resites) and then cured to an intractable solid of high crosslink density. Water is lost as a volatile (as much as 10-12% of the resin by weight).

2.4.2.4.2 Novolacs

The second type of phenolic consists of excess phenol reacted in the presence of an acid catalyst with formaldehyde. These prepolymer resins are complex mixtures of low molecular weight materials slightly soluble in acids and exhibiting random methylene (-CH_2) at the ortho-, para-, and ortho-para-positions on the aromatic ring. Unless a large excess of phenol is present, the material will form an infusible resin. The excess phenol used to moderate the processing viscosity can be varied as the application requires. Both water and formaldehyde are volatile products.

2.4.2.5 Bismaleimide

Bismaleimides are a class of thermosetting resins only recently available commercially in prepreg tapes, fabrics, rovings, and sheet molding compound (SMC). Bismaleimide resins, as the term implies, are the maleimide formed from the reaction of a diamine and maleic anhydride. Typically the diamine is aromatic, with methylenedianiline (MDA), the most common by far.

Bismaleimides form useful polymers by homopolymerization or by polymerization with diamines, epoxies, or unsaturated compounds, singular or in mixtures. A wide range of materials like allyl-, vinyl-, acrylate-, epoxy-, and polyester-, and phenolic-type reactive dil uents and resins can be used to tailor the properties of the bismaleimide system. However, attention to the specific components is required for useful polymers.

The physical form of the bismaleimide resin depends on the requirement of the final application. The form can vary from a solid to a pourable liquid at room temperature. For aerospace prepregs, sticky resins are required resulting in proprietary specific formulations.

The advantages of BMI resins are best discussed in the relation to epoxy resins. Emerging data s uggests that BMI's are versatile resins with many applications in the electronic and aerospace industries. Their primary advantage over epoxy resins is their high glass transition temperature, in the 500-600°F range (260-320°C). Glass transition temperatures for high temperature epoxies are generally less than 500 °F (260°C). The second advantage of BMI resins is high elongation with the corresponding high service temperature capabilities. While the high temperature epoxies have approximately one percent elongation when cured with diaminodiphenylsulfone (DDS), BMI's can have two-three percent elongation. Thus, bismaleimide resins deliver higher temperature capability and higher toughness providing excellent performance at ambient and elevated temperatures.

The processing of bismaleimide resins are essentially like that of epoxy resins. BMI's are suitable for standard autoclave processing, injection molding, resin transfer molding, and SMC, among others. The processing time of BMI's are similar to epoxies, except that for the additional higher service temperature, a free-standing post-cure is required. The only limitation is that room temperature curing BMI's have not yet been developed.

The cost of current BMI's is generally higher than the high temperature epoxies. The main disadvantage of bismaleimide resins is their recent commercial introduction. This results in few literature sources or authoritative reviews. Additionally, the suppliers are as limited as the types of BMI's. This latter disadvantage is partially offset by the wide variety of suitable co-monomers.

2.4.2.6 Polyimides

The polyimide resin family comprises a diverse number of polymers all of which contain an aromatic heterocyclic ring structure. The bismaleimides discussed in Section 2.4.2.5 are a subset of this family. Other polyimides are synthesized from a variety of cyclic anhydrides or their diacid derivatives thr ough reaction with a diamine. This reaction forms a polyamic acid which then undergoes condensation by the removal of water and/or alcohol.

Polyimide matrix composites excel in high temperature environments where their thermal resistance, oxidative stability, low coefficient of thermal expansion and solvent resistance benefit the design. Their primary uses are circuit boards and hot engine and aerospace structures.

A polyimide may be either a thermoset resin or a thermoplastic. The thermoplastic varieties are discussed in Section 2.4.2.7.2. Thermosetting polyimides characteristically have crosslinkable end-caps and/or a rigid polymer backbone. A few thermoplastic polyimides can become thermoset polymers if a sufficiently high postcure temperature is employed during part processing. Alternately, partially cured thermoset polyimides containing residual plasticizing solvents can exhibit thermoplastic behavior. Thus, it is difficult to state with certainty that a particular polyimide is indeed a thermoset or thermoplastic. Polyimides, therefore, represent a transition between these two polymer classifications.

Polyimide properties, such as toughness and thermal resistance, are influenced by the degree of crosslinking and chain extension. Molecular weight and crosslink density are determined by the specific end cap group and by the stoichiometry of the anhydride:amine mixture which produces the polyamic acid by stepwise chain growth, after which the polyamic acid is recyclized by continued thermal cure to form the final polymer structure. The choice of solvent employed in the resin formulation has a significant impact on crosslinking and chain extension. Solvents such as N-methyl 2-pyrroli done (NMP), promote chain extension by increasing resin flow, chain mobility and molecular weight prior to formation of a substantial crosslink network. From a practical standpoint, these solvents are beneficial to polymerization, but they are detrimental to part manufacture because of their tendency to cause ply delaminations.

Most polyimide resin monomers are powders. Some bismaleimides are an exception. As a result, solvents are also added to the resin to enable impregnation of unidirectional fiber and woven fabrics. Commonly, a 50:50 by weight mixture is used for fabrics, and a 90:10 by weight high solids mixture is used to produce a film for unidirectional fiber and low areal weight fabric prepregs. Solvents are further used to control prepreg handling qualities, such as tack and drape. Most of the solvents are removed in a drying process during impregnation, but total prepreg volatiles contents typically r ange between 2 and 8% by weight. This includes all volatiles, including those produced by the condensation cure reactions.

Polyimides require high cure temperatures, usually in excess of 550 °F (~90°C). Consequently, normal epoxy composite consumable materials are not usable, and steel tooling becomes a necessity. Polyimide bagging and release films, such as Kapton and Upilex, replace the lower cost nylon bagging and polytetrafluoroethylene (PTFE) release films common to epoxy composite processing. Fiberglass fabrics must be used for bleeder and breather materials instead of polyester mat materials.

2.4.2.7 Thermoplastic materials

2.4.2.7.1 Semi-crystalline

Semi-crystalline thermoplastics are so named because a percentage of their volume consists of a crystalline morphology. The remaining volume has a random molecular orientation termed amorphous, the name given to thermoplastics containing no crystalline structure. The total percentage of volume which can become crystalline depends on the polymer. Low density polyethylene, for example, can be as high as 70% crystalline (Reference 2.4.2.7.1(a)). Semi-crystalline thermoplastics are characterized by the ability of their molecules to form three-dimensionally ordered arrays (Reference 2.4.2.7.1(b)). This is in contrast to amorphous polymers that contain molecules which are unable to pack in an ordered crystalline structure. A partial list of semi-crystalline thermoplastics includes polyethylene, polypropylene, polyamides, poly phenylene sulfide, polyetheretherketone, (polyetherketoneketone) and polyarylketone.

Semi-crystalline thermoplastics can be converted into several physical forms, including films, powders and filaments. Combined with reinforcing fibers, they are available in injection molding compounds, compression-moldable random sheets, unidirectional tapes, towpregs, and woven prepregs. Fibers impregnated include carbon, nickel-coated carbon, aramid, glass, quartz, and others.

Semi-crystalline thermoplastics reinforced with short fibers have been used for over two decades in the injection molding industry. The inherent speed of processing, ability to produce complicated, detailed parts, excellent thermal stability, and corrosion resistance have enabled them to become established in the automotive, electronic, and chemical processing industries (Reference 2.4.2.7.1(c)).

The combination of long and continuous fibers with higher performance semi-crystalline thermoplastics is a recent development, but these composites have already shown several advantages over existing materials. The chemical stability of the materials provides for unlimited shelf life. Pot life problems and the need for cold storage are eliminated. The semi-crystalline materials usually possess better corrosion and solvent resistance than amorphous polymers, exceeding that of thermosets in some cases (Reference 2.4.2.7.1(c)). This corrosion resistance is exploited in chemical processing industry equipment. Another benefit of the crystal structure is retention of properties above the glass transition temperature (T_g) of the material. These materials may be used in applications above their T_g depending on loading requirements. One example is down-hole oil field sucker rod guides (Reference 2.4.2.7.1(d)).

Some semi-crystalline thermoplastics possess properties of inherent flame resistance, superior toughness, good mechanical properties at elevated temperatures and after impact, and low moisture absorption which have led to their use in the aerospace industry in secondary and primary structures (References 2.4.2.7.1(e)-(f)). Inherent flame resistance has made these materials good candidates for aircraft interiors and for ship and submarine applications. The superior toughness makes them viable candidates for aircraft leading edges and doors where impact damage resistance is required (Reference 2.4.2.7.1(g)). Low moisture absorption and low outgassing has stimulated interest in space structures where moisture swelling is a problem (Reference 2.4.2.7.1(h)). Also nickel-coated carbon/thermoplastic systems are finding uses in EMI shielding applications.

The primary disadvantages of semi-crystalline thermoplastic composites are lack of a design data base, 0° compression properties that are lower than those of 350 °F (180°C) epoxy systems, and creep resistance (Reference 2.4.2.7.1(c)). The creep resistance of semi-crystalline thermoplastics is superior to that of amorphous thermoplastics. Creep resistance in the fiber direction of a laminate is not expected to be a problem.

Processing speed is the primary advantage of thermoplastic materials. Chemical curing of the material does not take place during processing. Therefore, r educed cycle times compared to thermoset composites are experienced (References 2.4.2.7.1(i) and (j)). However, thermoplastic prepregs are typically boardy and do not exhibit the tack and drape of thermosets. Forms are available that consist of thermoplastic and reinforcing fibers interlaced together, known as commingled which are dr apeable. The present costs of high performance engineering thermoplastic materials are slightly higher than equivalent performance epoxies, and tooling costs may be higher. However, final part cost may be reduced, due to the decreased processing time. The ability to postform or reprocess molded parts also offers cost saving advantages.

A wide variety of methods and techniques are available for processing semi-crystalline thermoplastics, including stamp molding, thermoforming, autoclave molding, diaphragm forming, roll forming, filament winding, and pultrusion. Semi-crysta lline thermoplastics differ from amorphous ones in that the morphology can change based on the time/temperature history of the material during molding. Therefore, the degree of crystallinity can be controlled by controlling the cooling rate. The material must be processed above its melt temperature, which requires temperatures ranging from 500 to 700 °F (260 - 370°C) for the higher performance materials. Thermal expansion differences between the tool and the thermoplastic material should be addressed, due to the high processing temperature. The actual pressure r equired varies with the process, but can be as high as 5000 psi (34 MPa) for stamp molding and as low as 100 psi (0.7 MPa) for thermoforming. Once formed, semi-crystalline thermoplastics can be joined by a variety of methods, including ultrasonic welding, infrared heating, vibration, hot air and gas, resistance heating, and conventional adhesives.

2.4.2.7.2 Amorphous

The majority of thermoplastic polymers are composed of a random molecular orientation and are termed amorphous. The molecules are unable to align themselves in an ordered manner, since they are non-uniform or composed of units which have large side groups. In contrast, semi-crystalline thermoplastics have molecules that form ordered three-dimensional arrays (Reference 2.4.2.7.1(b)). Some amorphous thermoplastics include polysulfone, polyamide-imide, polyphenylsulfone, polyphenylene sulfide sulfone, polyether sulfone, polystyrene, polyetherimide, and polyarylate.

Amorphous thermoplastics are available in several physical forms, including films, filaments, and powders. Combined with reinforcing fibers, they are also available in injection molding compounds, compression moldable random sheets, unidirectional tapes, woven prepregs, etc. The fibers used are primarily carbon, aramid, and glass.

Amorphous thermoplastics are used in many applications; the specific use depends on the polymer of interest. Their applications are well established in the medical, communication, transportation, chemical processing, electronic, and aerospace industries. The majority of applications use the unfilled and short fiber form. Some uses for the unfilled polymers include cookware, power tools, business machines, corrosion resistant piping, medical instruments, and aircraft canopies. Uses for short-fiber-reinforced forms include printed circuit boards, transmission parts, under-the-hood automotive applications, electrical connections, and jet engine components (Reference 2.4.2.7.2(a)).

The use of amorphous thermoplastics as matrix materials for continuous fiber reinforced composites is a recent development. The properties of composites have led to their consideration for primary and secondary aircraft structures, including interior components, flooring, fairings, wing skins, and fuselage sections (References 2.4.2.7.2(b) and (c)).

The specific advantages of amorphous thermoplastics depend upon the polymer. Typically, the resins are noted for their processing ease and speed, high temperature capability, good mechanical properties, excellent toughness and impact strength, and chemical stability. The stab ility results in unlimited shelf life, eliminating the cold storage requirements of thermoset prepregs. Several amorphous thermoplastics also have good electrical properties, low flammability and smoke emission, long term thermal stab ility, and hydrolytic stability (Reference 2.4.2.7.2(a)).

Amorphous thermoplastics generally have higher temperature capabilities than semi-crystalline thermoplastics. Polymers with glass transition temperatures as high as 500 °F (260°C) are available. Also, processing is simplified, because the formation of a crystalline structure is avoided, resulting in less shrinkage due to their lower melt viscosities. Amorphous polymers generally have lower solvent and creep resistances and less property retention above the glass transition temperature than semi-crystalline thermoplastics (Reference 2.4.2.7.1(f)).

The primary advantages of amorphous thermoplastics in continuous fiber reinforced composites are potential low cost process at high production rates, high temperature c apability, good mechanical properties before and after impact, and chemical stability. High temperature capability and retention of mechanical properties after impact have made amorphous thermoplastics attractive to the aerospace i ndustry. A service temperature of 350°F and toughness two to three times that of conventi onal thermoset polymers are typical (Reference 2.4.2.7.1(f)). The most significant advantage of thermoplastics is the speed of processing, resulting in lower costs. Typically, cycle times in production are less than for thermosets since no chemical reaction occurs during the forming process (References 2.4.2.7.1(i) and (j)).

Amorphous thermoplastics share many of the disadvant ages of semi-crystalline thermoplastics, such as a lack of an extensive database and reduced 0° compression properties compared to 350°F (180°C) cure thermosets. Solvent resistance, which is good for semi-crystalline thermoplastics, is a concern for most amorphous ones. They can be attacked to varying degrees, depending on the polymers and solvents of interest. The creep resistance of the polymer is a concern, but should be good for composite forms loaded in the fiber direction. The materials do not have tack and drape as thermosets do; however, some amorphous thermoplastics are available in commingled forms, which are drapable.

The costs of amorphous thermoplastics prepreg used for advanced composites are higher than equivalent performance epoxies. Finished part costs may be lower due to the processing advantages discussed above. Reprocessibility of material results in reduced scrap rates, translating into additional cost savings. For example, the same sheet laminate can be thermoformed several times until the desired configuration is achieved. In addition, certain forms can be recycled.

The processes used with continuous reinforced composites include stamp molding, thermoforming, autoclave molding, diaphragm forming, roll forming, filament winding, and pultrusion. The high melting temperatures require process temperatures ranging from 500 °F to 700°F (260 to 370°C). Thermal expansion differences between the tool and the thermoplastic material s hould be addressed due to the high processing temperatures. Forming pressures range from 100 psi (0.7 MPa) for thermoforming to 5000 psi (35 MPa) for stamp molding. Several amorphous thermoplastics that are hygroscopic must be dried before processing. Hot molds are also recommended to increase material flow. The materials can be joined by several methods, including common adhesives, or fusion bonding such as; ultrasonic welding, infrared heating, hot air and gas, and resistance heating. Surface pr eparation techniques for using adhesives can be different from those for thermosets. Solvent bonding techniques can be used for joining amorphous thermoplastics but not most semi-crystalline thermoplastics.

One important class of amorphous thermoplastic matrices is the condensation cure polyimides. Examples include polyamideimides, such as Torlon, and polyimides having more flexible back bones, such as AvimidR K3B, NR 150B2 and the LaRC polymers developed by NASA. As stated in 2.4.2.1.6, polyimides represent a transition between thermoset and thermoplastic polymers. Thus, these thermoplastics also have many characteristics typical of epoxy and phenolic thermoset polymers (e.g., excellent solvent resistance and high maximum operating temperature limits).

Due to negligible crosslink density, these polymers impart some toughness to composite laminates and permit limited flow during processing, although this flow is more like the high creep rates exhibited by superplastic metals. Unlike other thermoplastics, these polymers do not produce liquid flows, even under high consolidation pressures. Typical processing conditions for the condensation cure thermoplastics are 550 °F (290°C) and greater temperatures with consolidation pressures starting at 200 psig (1.4 MPa).

Many of these thermoplastic polymers have been developed with the intent to rapidly stamp or compression mold structural composites parts at low cost. However, this potential has yet to be realized because of low production volumes, high capital equipment and tooling costs as well as excessive fiber distortion in the formed part. The most successful structural applications of these polymers have utilized autoclave processing to reduce tooling costs and fiber distortion. Other polymers in this class have been developed for use in circuit boards because of their low dielectric constant, low moisture absorption and low coefficient of thermal expansion. In these applications, compression molding had been found to be advantageous and cost effective.

Compared to other thermoplastic polymers, the condensation cure thermoplastics have not found a wide variety of applications. Their processability is very similar to the thermosetting polyimides, and this has been a limiting factor. Volatiles are produced by the condensation reaction, and they cause laminate porosity unless consolidation pressures are high enough to suppress void nucleation and growth. Costly high temperature tooling and consumable materials (e.g., vacuum bags and release films) are also required for part processing. While the toughness and processability of many of these condensation cured thermoplastic polyimides are slightly better than those of competing thermosetting polyimides, their maximum operation temperature limit is somewhat lower. For the present, these thermoplastic polymers are limited to special niche markets which take advantage of their unique performance capabilities.

2.4.2.8 Specialty and emerging resin systems

2.4.2.8.1 Silicone

The silicones are a synthetic resin, composed primarily of organos ilicon. The term silicone resin is a general term used for high temperature poly methyl siloxane. Silicone resins are available from a low viscosity liquid to a solid friable resin.

The silicone resin is used where high temperature stab ility, weatherability, good electrical properties and moisture resistance are required. These excellent properties have allowed the silicone resin to be used in laminates, varnishes, mineral filled molding compounds, and long glass fiber molding compounds. The silicone resin has been used as an impregnant for mica paper, flexible glass tape, glass cloth, and mica products. The molding compounds may be processed by conventional methods: injection, compression, and transfer molding. The cure temperature varies from 250 °F to 450°F (120°C to 230°C). The cure time varies from 30 minutes to 24 hours, depending upon cure temperature, wall thickness of molded part, and the desired cured properties. In some applications, additional post cure will be required.

2.5 PROCESSING OF PRODUCT FORMS

2.5.1 Fabrics and preforms

2.5.1.1 Woven fabrics

Woven or knitted fabric product forms, unlike tapes and rovings, are in most circumstances produced prior to the resin impregnation step. Therefore, these product forms, in most part, offer product continuity or retention of fiber placement prior to, during, and after the impregnation step. Most fabric constructions offer more flexibility for lay-up of complex s hapes than straight unidirectional tapes offer. Fabrics offer the option for resin impregnation either by solution or the hot melt process. Generally, fabrics used for structural applications use like fibers or strands of the same weight or yield in both the warp (longitudinal) and fill (transverse) directions. However, this is not a set rule as the number of combinations of reinforcement fibers and weave styles are essentially unlimited for custom applications. Also some fabrics are produced which incorporate thermoplastic strands that then become the resin matrix when the fabric is processed to its final state.

Woven fabric selections for structural applications have several parameters which may be considered. These variables are strand weight, tow or strand count, weave pattern, and fabric finish. The variables for glass fabrics are considerably greater than carbon fabrics due to the availability of a greater range of yarn weights. The availability of carbon tow weights or filament count tows are few in comparison. Generally, the lighter or thinner the fabric, the greater the fabric cost. Also factored into the cost is the complexity of the weave pattern or machine output for heavy fabrics. For aerospace structures, tightly woven fabrics are usually the choice for areal weight considerations, minimizing resin void size, and maintaining fiber orientation during the fabrication process.

2.5.1.1.1 Conventional woven fabrics

Woven structural fabrics are usually constructed with reinforcement tows, strands, or yarns interlocking upon themselves with over/under placement during the weaving process. The more common fabrics are plain or satin weaves. The plain weave construction results from each fiber alternating over and then under each intersecting strand (tow, bundle, or yarn). With the common satin weaves, such as 5 harness or 8 harness, the fiber bundles traverse both in warp and fill directions changing over/under position less frequently. (See Figures 2.5.1.1.1(a) and (b).)

These satin weaves have less crimp and are easier to distort than a plain weave. With plain weave fabrics and most 5 or 8 harness woven fabrics the fiber strand count is equal in both warp and fill directions. Example: 3K plain weave often has an additional designation such as 12 x 12, meaning there are twelve tows per inch in each direction. This count designation can be varied to increase or decrease fabric areal weight or to accommodate different fibers of varying weight.

2.5.1.1.2 Stitched or knitted fabrics

These fabrics can offer many of the mechanical advantages of unidirectional tapes. Fiber placement can be straight or unidirectional without the over/under turns of woven fabrics. The fibers are held in place by stitching with fine yarns or threads, after preselected orientations or one or more layers of dry plies. This product form, much like preplied unidirectional tapes, offers a wide range of multi-ply orientations. Although there may be some added weight penalties or loss of some ultimate reinforcement fiber properties, some gain of interlaminar shear and toughness properties may be realized. Some common stitching yarns are polyester, aramid, or thermoplastics.

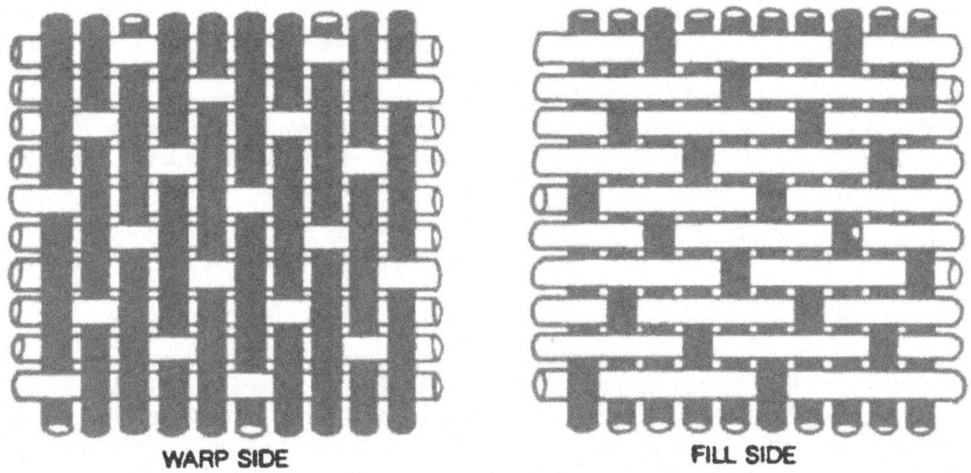

FIGURE 2.5.1.1.1(a) *5 Harness Satin weave construction. In this weave construction each yarn goes over 4 and under 1 yarn in both directions.*

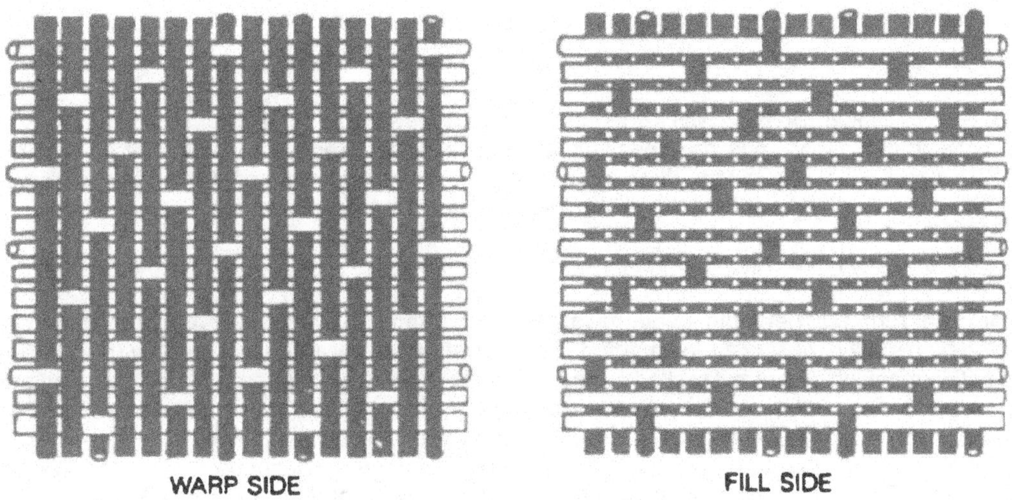

FIGURE 2.5.1.1.1(b) *8 Harness Satin weave construction. In this weave construction each yarn goes over 7 and under 1 yarn in both directions.*

2.5.1.1.3 Specialty fabrics

To list all the possible woven or knitted fabric forms would require space beyond the scope of this document. As an example, there are in excess of one hundred glass fabrics listed in a standard weaver's handbook. These fabrics vary in weight from 0.55 oz./square yard (18.65 gm/m^2) to 53 oz./square yard (1796 gm/m^2) and vary in thickness from 0.0012 in (0.0305 mm) to 0.0450 in (1.143 mm). Such an industrial listing is limited to but a few basic patterns such as plain, basket, Leno, harness, and twill weaves. There are many other fabrics such as triaxial, orthogonal, knitted bidirectional, stitched multilayer, and angle interlock, to name a few. From these also arise combinations and three-dimensional weaves.

2.5.2 Preimpregnated forms

2.5.2.1 Prepreg roving

This impregnated product form generally applies to a single grouping of filament or fiber ends, such as 20 end or 60 end glass rovings. Carbon rovings are usually identified as 3K, 6K, or 12K rovings. Other counts are available. It is possible, preferably during the resin impregnation step, to combine two or more counts or filaments or ends to increase the rovings weight, width, etc. per linear length. For mechanical testing purposes individual rovings are usually wound, side by side, to form single ply tapes and processed as such. The roving product form, with its packaging on individual spools, offers the means for automated fiber placement during the manufacture of parts. The rovings can be placed in a unidirectional pattern, like tapes, or to generate a crossover interlocking effect. Most applications for roving products utilize mandrels for filament winding and then resin cure to final configuration. In addition, this product form is used for efficient build-up of oriented filaments to create preforms. The preforms are combined with other lay-ups or processed individually in closed tools rather than the conventional mandrel cure process. Most rovings are supplied untwisted, in nearly flat continuous bands. Band widths can be controlled to a degree during the impregnation step. Compared to tapes or fabrics, roving areal weights for individual plies or wraps are more dependent on the winding process than the impregnation step. However, resin control of the preimpregnated rovings shares a like degree of accuracy.

2.5.2.2 Prepreg tape

All product forms generally begin with spooled unidirectional raw fibers packaged as continuous strands. Normally, untwisted tows or ends are specified for unidirectional product forms to obtain ultimate fiber properties. This particular product form depends on the proper fiber wet-out and the tenacity of the uncured resin to maintain proper fiber placement until the tape reaches the curing procedure.

2.5.2.2.1 Conventional unidirectional tapes

This particular form has been the standard within the user industry for many years and is common with thermosetting resins. The most common method of manufacture is to draw collimated raw (dry) strands into the impregnation machine where hot melted resins are combined with the strands using heat and pressure. The combination of fibers and resin usually travels through the machine between coated carrier papers or films for ease of release. The tapes are usually trimmed to specified widths in line. One side of the carrier is usually removed prior to the roll-up position to facilitate continuous visual inspection. The remaining carrier is usually left in place with the tape to serve as a separator on the roll and as a processing aid for fabrication purposes. The tape manufacturing process is continuous within the linear limits of the raw strands creeled to the machine or specified lot size or availability of resin. Most impregnation machines are designed to permit in-line change over to new rolls (take-ups) without interruption. Raw strand collimation is adjusted to control specified areal weight (dry weight/area). Resin filming for tape machine operations is often done as a separate controlled operation. Some machines accommodate in-line filming that permit resin content adjustments during the impregnation process. Tapes as wide as 60 inches (1.5 m) are commercially available.

2.5.2.2.2 Two-step unidirectional tapes

Although not a general practice within the prepreg industry, there are unidirectional tapes manufactured from preimpregnated rovings. The collimation of these rovings to make tapes allow the use of solution impregnated resins, rather than hot melt systems. Although the pr oduct form may be similar to conventional tapes, thin uniform flat tapes may be difficult to produce.

2.5.2.2.3 Supported unidirectional tapes

To enhance specific mechanical properties or part manufacturing handling operations, it is sometimes advantageous to add product form during the manufacture of unidirectional tapes. Generally, these added fibrous forms are lightweight to be accommodated during the normal tape manufacture operation. The added form may be combined in the machine dry or preimpregnated prior to the tape production. More common added forms are lightweight mats or scrim fabrics of the same or unlike fiber type. The added product form will affect material properties compared to tapes without the supporting material.

2.5.2.2.4 Coated unidirectional tapes

Some tape suppliers offer the option of added tape surface coating. These resi nous coatings of films are usually of different rheology or viscosities from the fiber impregnation resin to remain as distinct boundaries between plies of the cured tapes. As with supported unidirecti onal tapes, the added layer may be combined during the tape manufacturing operation.

2.5.2.2.5 Preplied unidirectional tapes

These tapes originate as any of the above-described tape forms in single-ply form. Then through a process of stacking, two or more layers of individual tapes are oriented at predetermined angles in relation to the centerline of the new progressively generated tape or broadgoods form. The original individual tapes are located side to side in each angled layer to form a continuous linear form. The original single-ply tapes are usually precut in segments at angles to correspond to the new product form's edges. The progressive stacking sequence usually takes place on a continuous carrier (paper or film) atop a flat surface much like the fabrication process. The carrier, with the preplied form in place, is utilized to take up the preplied tapes onto a shipping/handling core. The predetermined length of the individual precut segments will generally regulate the width of the preplied tapes. However, a final trim of both edges to control specified widths can be incorporated during the take-up step. For economic purposes the preplying operation usually is done in widths of approximately 24 inches (0.6 m) or greater. Should narrow widths be required, they can be accommodated with a secondary slitting operation. To some extent the retention of this product form's continuity is, like single ply tapes, dependent on the tack or tenacity of the uncured resin.

2.5.2.3 Prepreg fabric

2.5.2.4 Preconsolidated thermoplastic sheet

2.6 SHIPPING AND STORAGE PROCESSES

Composite precursor materials and adhesives can be very sensitive to how they are stored and shipped. Contamination must be avoided, as it will invariably reduce properties. Materials that have been pre-impregnated (prepreg), film adhesives, and other resins are temperature variation sensitive. They can also be very sensitive to moisture and humidity before they are cured. As a result, these materials need special handling and storage in order to provide desired results.

2.6.1 Packaging

Prepreg and film adhesive should be supported on cardboard rolls, or in some other manner. They s hould be sealed in moisture-proof bags, with desiccant packages if possible. Once packaged, they should be stored in conditions as recommended by the manufacturer, usually at or below 0 °F (-18°C) for a shelf life of six months or longer. Since the cure of thermoset materials continues to progress at room temperature, and even these lower storage temperatures, a record must be kept of the time exposed at room and storage temperatures. This record will be used to establish the useful life of the material and to determine when retesting is required. The time that material can be at room temperature and still usable, known as the out-time, can range from minutes to thirty days or longer. For some materials the processing characteristics can change dramatically depending on how much storage and out-time they have experienced.

2.6.2 Shipping

Since these materials require a carefully controlled environment, maintaining that environment while shipping the product can be challenging. Usually the material, still in its moisture-proof sealed bag, is placed in a shipping container approved for use with dry ice. Enough dry ice is placed in the container to allow some to be remaining upon the scheduled arrival, plus about 24 hours. Chemically based temperature sensitive materials, or electronic temperature recording devices can be placed in the container to assure material integrity upon delivery.

2.6.3 Unpackaging and storage

Upon receipt the material should be placed in a freezer to maintain the recommended storage temperature. Any time during shipping where the material temperature has exceeded this storage temperature is deducted from the out-time for the material. When the material is needed for use it needs to be allowed to reach room temperature before the moisture-proof bag is opened. If this is not done moisture will condense on the cool material, and may result in prebond moisture problems with the material.

2.7 CONSTRUCTION PROCESSES

Construction processes are those used to bring various forms of fiber and fabric reinforcement together to produce the reinforcement pattern desired for a given composite part or end item. The resin may or may not be in its final chemical or physical form during placement of the reinforcement. Construction processes include both manual and automated methods of fiber placement, as well as adhesive bonding and sandwich construction.

2.7.1 Hand lay-up

2.7.2 Automated tape lay-up

2.7.3 Fiber placement

Fiber placement is the automated application of multiple preimpregnated or coated fibers directly to a tool surface at zero tension. The application roller remains in intimate contact with the tool surface to provide ply compaction and allows lay-up on complex contours. The use of multiple fibers and the ability to drop off and restart individual fiber forms allows for significant changes in part contour and size. In addition, fiber placement may refer to three dimensional preform fabrication which may utilize dry fibers.

2.7.4 Braiding

The braiding process fabricates a preform or final shape at the same time that it generates the woven form. This product form is a unique fiber reinforcement which can use preimpregnated yarn as well as dry fibers. The main advantage of the braiding process is its ability to conform to odd shapes and maintain fiber continuity while developing high damage tolerance compared to unidirectional and laminated products. This allows formation of square, oval, and other constant cross-section shapes. The three-dimensional form of braiding has evolved to the point of allowing the non-uniform cross-sections to be fabricated while maintaining weaving in all three planes.

The uses of braiding have varied during its development. The best known example of braided structure is the fiberglass and carbon fishing rods that became popular in the 1980's. Braiding has also found uses in pressurized piping and complex ducting. A demonstration of its versatility is the open-wheel race car body which was fabricated by braiding. The process has also been used in rocket applications for motorcases and launchers.

In biaxial and triaxial braiding, a mandrel is usually used to form the braid. The mandrel also acts as the mold for the final product. The braiding machine controls the rate of feed of the mandrel and the rotational speed of the carriers. The combination of these parameters and the size of the mandrel controls the braid angle. The braid angle, along with the effective yarn, tape, or tow width (width of the specific size yarn, tape, or tow on the mandrel as placed by the braiding process), ultimately controls the coverage of the braid on the surface of the fabricated form. As the braid angle increases, the maximum size of the mandrel which can be covered with a specific yarn, tape, or tow size decreases. For complicated forms, expendable mandrels may be used. These include mandrels made from low melting temperature metal alloys and water-dissolvable casting materials, and collapsible mandrels.

In three-dimensional (3D) braiding, the weaving process itself is used to control the shape of the fabricated product. The typical 3D braiding process involves a bed of cops, or weaving loops, which are moved in a systematic manner. This systematic movement creates an interwoven product in the x-y plane. As the yarns, tapes, or tows are pulled into the weaving process, the z-direction is also intertwined. The resulting product is essentially self-supporting due to interweaving in three directions. For precision exterior dimension, matched metal molds can be used during the resin matrix curing process. The following are the general steps involved in the braiding process:

1. Set the feed speed, cop speed, and weave pattern (3D braiding).
2. Run the braiding machine until the product is finished.
3. If prepreg material is not being used, use an appropriate resin impregnation process - RTM, wet resin impregnation, etc.
4. Cure according to the appropriate process determined by the impregnation method - autoclave cure, vacuum bag, RTM, etc.
5. Remove the part from the mold or mandrel.

2.7.5 Filament winding

Filament winding is an automated process in which a continuous fiber bundle (or tape), either preimpregnated or wet impregnated with resin, is wound on a removable mandrel in a pattern. The filament winding process consists of winding onto a male mandrel that is rotating while the winding head moves along the mandrel. The speed of the winding head as it moves along the mandrel in relation to the rotation of the mandrel controls the angular orientation of the fiber reinforcement. Filament winding can be done using wet resin winding, preimpregnated yarns and tapes. The following general steps are used for filament winding:

The construction of the mandrel is critical to the process and the materials of choice are dependent upon the use and geometry of the finished part. The mandrel must be capable of withstanding the applied winding tension, retaining sufficient strength during intermediate vacuum compaction procedures. In addition, if the outer surface of the part is dimensionally critical, the part is generally transferred from the male winding mandrel to a female tool for cure. If the internal surface of the part is dimensionally critical, the part is usually cured using the male winding mandrel as the cure tool. Metal is used in segmented collapsible m andrels or in cases where the domes are removed to leave a cylindrical part. Other mandrel material choices are low melt alloys, soluble or frangible plaster, eutectic salts, sand and inflatables.

The following general steps are used for filament winding:

1.) The winder is programmed to provide correct winding pattern.
2.) The required number of dry fiber or prepreg roving/slit tape spools for the specified band width are installed on the winding machine
3.) When wet winding, the fiber bundle is pulled through the resin bath.
4.) The fiber bundle is pulled through the eye, attached to the mandrel, the winding tension is set and the winding program is initiated..
5.) When winding is complete, the mandrel is disassembled as required and removed from the part if the part is to be cured on a female tool., otherwise the part is trimmed and prepared for cure on the male mandrel.
6.) Elevated temperature cure of thermosets resin parts is usually performed in an oven or autoclave, room temperature cure resin parts are usually placed under vacuum to provide compaction during cure. During cure the male mandrel or female tool is often rotated to maintain resin distribution.
7.) After cure the mandrel is removed from the part (for male tooled parts)

Cured product characteristics can be affected by both the winding process and design features such as:

1. Uniformity of the fiber to resin ratio (primarily wet winding)
2. Wind angle
3. Layer sequence
4. Effective fiber bandwidth (tight fiber weave or loose/open fiber weave pattern)
5. End closure.

The cure cycle and compaction procedure affects such cured product characteristics as described in the applicable cure and consolidation process section - 2.8.1 (for vacuum bag molding for room temperature cure resins), 2.8.2 (for oven cure), or 2.8.3 (for autoclave cure).

2.7.6 Pultrusion

The pultrusion process consists of passing a continuous resin-impregnated fiber bundle through a heated die for part shape and cure. This process is limited to constant cross-sections such as rods, t ubes, I-beams, and channels. The pultrusion process works well with quick-curing resins and is a very low-cost method for high-production parts with constant cross-sections. For a discussion of cure and consolidation during pultrusion, see Section 2.8.6 below.

2.7.7 Sandwich construction

Sandwich construction, as applied to polymer matrix composites, is a structural panel concept consisting in its simplest form of two relatively thin, parallel sheets of structural laminated materials bonded to and separated by a relatively thick, lightweight core. The following information is limited to non-metallic sandwich construction used for structural applications. Sandwich construction provides a method to obtain high bending stiffness at minimal weight in comparison to monolithic laminate construction. This advantage must be weighed against the risk of increased processing difficulty that can increase pr oduction costs over monolithic construction. Damage tolerance and ease of repair should also be considered when selecting sandwich panel or monolithic laminate construction. Good structural practice requires selection of skin, core and adhesive materials to be strategically based on overall part quality considerations including:

1. Surface quality (pinholes, mark-off, etc.)
2. Skin quality (porosity, consolidation, waviness, resin loss)
3. Adhesive bond and fillet quality (strength, fillet size)
4. Core strength, cell size, bonding preparation
5. Resistance to moisture ingress

Polymer matrix composite sandwich construction is most often fabricated using autoclave cure, press cure or vacuum bag cure. Skin laminates may be pre-cured and subsequently bonded to core, co-cured to core in one operation, or a combination of the two methods. Pre-cured skin sandwich construction insures a high quality surface, but adequate fit-up to core must be addressed. Co-curing often results in poor panel surface quality which is prevented by using a secondary surfacing material co-cured in the standard cure cycle or a subsequent "fill-and-fair" operation. Co-cured skins may also have poorer mechanical properties, and this may require the use of reduced design values.

Cure cycles can be developed to reliably produce good quality sandwich panels. For co-cured sandwich construction, this is essential. Some primary cure cycle considerations are transport of volatiles, core evacuation and/or pressurization, adhesive and prepreg resin viscosity profiles, and compatibility to monolithic structure co-cured with the sandwich structure.

Skin materials for co-cure processing have a "low flow" resin material system that prevents resin running down the cell walls into the core. A compatible adhesive must be selected that develops an adequate fillet bond to the selected core whether co-cured or secondarily bonded. For co-cured construction, prepreg resin to adhesive compatibility must be demonstrated.

Core should be selected according to the required characteristics of the application often including surface quality, shear stiffness and strength, compression strength, weight, water absorption, and damage tolerance. Currently available core materials include metallic and non-metallic honeycomb core and a variety of non-metallic foams. Honeycomb core selection can be made from a range of common carbon, glass or aramid fiber reinforced matrix materials including phenolics, epoxies, polyimides, or thermoplastics.

Additional information may be found in References 2.7.7(a)-(d).

2.7.8 Adhesive bonding

Three types of adhesive bonding are commonly employed with composite structures. These are cocuring, secondary bonding and cobonding. A typical cocure application is the simultaneous cure of a stiffener and a skin. Adhesive film is frequently placed into the interface between the stiffener and the skin to increase fatigue and peel resistance. Principal advantages derived from the cocure process are excellent fit between bonded components and guaranteed surface cleanliness

Secondary bonding utilizes precured composite detail parts. Honeycomb sandwich assemblies commonly use a secondary bonding process to ensure optimal structural performance. Laminates cocured over honeycomb core may have distorted plies which have dipped into the core cells. As a result, compression stiffness and strength can be reduced as much as 10 and 20 percent, respectively. While secondary bonding avoids this performance loss, care must be exercised prior to bonding in order to ensure proper fit and surface cleanliness. In some applications, aluminum foil layers or an adhesive sandwiched between two layers of polyester release film is placed into the bonded joint. The assembly is then bagged and run through a simulated bonding cycle using the same temperatures and pressures as those in the actual cycle. The foil or film is removed, and its thickness is measured. Based upon these measurements, additional adhesive can be added to the bondline to ensure proper fit; or detail parts can be reworked to eliminate interference fits.

Precured laminates undergoing secondary bonding usually have a thin nylon or fiberglass peel ply cured onto the bonding surfaces. While the peel ply sometimes hampers nondestructive inspection of the precured laminate, it has been found to be the most effective m eans of ensuring surface cleanliness prior to bonding. When the peel ply is stripped away, a pristine surface becomes available. Light scuff s anding removes high resin peak impressions, produced by the peel ply weave which, if they fracture, create cracks in the bondline.

Peel plies are generally not useful for thermoplastic composite laminates. Instead, plasma technologies such as flame spray are employed to remove minor amounts of contaminants and to increase surface reactivity. Thermosetting adhesives are sometimes used with pre-consolidated thermoplastic composites, but more commonly melt fusible thermoplastic films are utilized. Amorphous thermoplastics (e.g., polyetherimide) are superior choices for an adhesive film because of their wide processing latitude. In some instances, nichrome wire or ferromagnetic particles are placed into the film to resistively heat the film and effect flow within the bondline. Reference 2.7.8(a) provides an excellent overview of this technology.

Cobonding is a combination of secondary bonding and cocuring in which one detail part, usually a skin or spar web, is precured. Adhesive is placed into the bondline and additional composite plies for another detail part (e.g., a blade or hat stiffener) are laid up over the adhesive. The adhesive and composite plies are then concurrently cured together. The cobonding process has the advantage of avoiding expensive matched metal tooling that may be required for a cocured integrally stiffened composite part having the same geometry .

Whether cobonded joints develop the same structural performance levels as cocured joints is a matter of conjecture. The high cost of matched metal tooling has m ade conclusive testing prohibitive. Presently, there is no proof that cobonding is inferior to cocuring.

Historically, secondary bonding has been very susceptible to bondline failure as a result of improper cleaning and contamination (e.g., silicones). Cocured joints have demonstrated significantly less susceptibility to shop contaminants; therefore, it is anticipated that cobonding will be somewhat less susceptible to improper surface preparation than secondary bonding.

In many applications, composites are secondarily bonded or cocured with metals. Common examples are stepped lap splices and closure ribs and spars. Special attention must be given to minimizing thermal mismatch in composite to metal bonding. Carbon/epoxy and aluminum have been successfully b onded using adhesives which cure at 250°F (121°C) or less. With 350°F (177°C) curing adhesives, titanium is recommended because its coefficient of thermal expansion more closely matches that of carbon fiber composites.

Surface cleanliness is more critical for metals than composites in a bonded assembly. Aluminum, stainless steel and titanium detail parts require solvent vapor degreasing, alkaline cleaning and acid etch to produce an oxide layer with a controlled thickness and reactivity. The Forest Products Laboratory (FPL) etch, phosphoric acid anodize and chromic acid anodize processes are commonly employed as aluminum surface pre-treatments. Titanium pre-treatments include chromic acid anodize or chromated hydrofluoric acid etch processes. Phosphate solutions have proven successful in pretreating stainless steel surfaces.

In all instances, metal surfaces must be sprayed with a thin coat of adhesive bonding primer within few hours of pre-treatment. For best environmental resistance and bondline durability, a chromated epoxy primer is recommended. However, environmental regulations will restrict both the usage of chromium containing compounds and the application of primers with high volatile solvent contents. The challenge then for the coming decade is to develop environmentally friendly pre-treatment processes and primers while retaining or improving bondline durability under adverse environmental conditions and cyclic loading.

Additional information on joint design, adhesive materials selection processing, testing and quality assurance may be found in MIL-HDBK-691, Adhesive Bonding (Reference 2.7.8(b)).

2.8 CURE AND CONSOLIDATION PROCESSES

Resin consolidation and cure processes are required to ensure that the individual sections or layers of a composite part are properly bonded, and that the matrix is intact and capable of maintaining the placement of the fibrous reinforcement which will carry the loads applied to the part. These processes are among the most sensitive in the materials processing pipeline. As a thermosetting composite part is formed during cure, the material is undergoing extensive chemical and morphological change. As a result, there are many actions occurring simultaneously. Some of these actions can be controlled directly, others only indirectly, and some of them interact. Such actions as evolution of voids or shifting of reinforcing fibers during matrix flow may result in large changes in properties of the cured composite.

In the case of a thermoplastic matrix composite, the matrix is not intended to undergo chemical change, during consolidation, but changes such as chain scissions resulting in production of volatiles may occur inadvertently. In addition, resin flow is required for consolidation, and semicrystalline thermoplastics may undergo morphological changes such as changes in the degree of crystallinity upon melting, flow and recrystallization, particularly in the fiber/matrix interphase. These c hanges can cause significant changes in mechanical and physical properties of the consolidated composite. In amorphous thermoplastics, segregation of varying molecular weight materials in the interphase may also result in changes in composite properties.

2.8.1 Vacuum bag molding

Vacuum bag molding is a process in which the lay-up is cured under pressure generated by drawing a vacuum in the space between the lay-up and a flexible sheet placed over it and sealed at the edges. In the vacuum bag molding process, the reinforcement is generally placed in the mold by hand lay-up using prepreg or wet resin. High flow resins are preferred for vacuum bag molding. The following steps are used in vacuum bag molding:

1. Place composite material for part into mold.
2. Install bleeder and breather material.
3. Place vacuum bag over part.
4. Seal bag and check for leaks.
5. Place tool and part in oven and cure as required at elevated temperature.
6. Remove part from mold.

Parts fabricated using vacuum bag oven cure have lower fiber volumes and higher void contents. Vacuum bag molding is a low-cost method of fabrication and uses low-cost tooling for short production runs.

2.8.2 Oven cure

Composite material can be cured in ovens using various pressure application methods. Vacuum bagging, as described in the above section, can be used to remove volatiles and trapped air, and utilize atmospheric pressure for consolidation. Another method of pressure application for oven cures is the use of shrink wrapping or shrink tape. This method is commonly used with parts that have been filament wound, because some of the same rules for application apply. The tape is wrapped around the completed lay-up, usually with only a layer of release material between the tape and the lay-up. Heat is applied to the tape, usually using a heat gun, to make the tape shrink, and can apply a tremendous amount of pressure to the lay-up. After shrinking the part is placed in the oven for cure. High quality parts can be made inexpensively using shrink tape, with a couple of caveats. First, the part must be of a configuration where the t ape can apply pressure at all points. Second, flow of the resin during cure must be limited, because the tape will not continue to shrink in the oven. If the resin flows excessively, the pressure applied by the shrink tape will be reduced substantially.

2.8.3 Autoclave cure

This process uses a pressurized vessel to apply both pressure and heat to parts that have been sealed in a vacuum bag. Autoclaves in general operate at 10 - 300 psi (70-2000 kPa) and up to 800 °F (420°C). Heat transfer and pressure application to the part is achieved by circulation (convection) of pressurized gas, usually air, nitrogen, or carbon dioxide, with the autoclave. The following steps are typically used in autoclave curing:

1. Composite prepreg is applied to the released mold by one of the applicable construction processes described in Section 2.7.
2. Cure monitoring devices such as thermocouples are installed.
3. Release ply, bleeder, and breather materials are placed over the part and edge dams are installed onto the mold if required.
4. The vacuum bag with vacuum fittings installed is placed over the part and sealed. The bag is then checked for any leaks.

Composite materials that are typically processed in autoclaves include adhesives, reinforced thermoset matrix (epoxy, bismaleimide etc.) laminates and reinforced thermoplastic matrix laminates. In the case of thermoset resin systems, the cure cycle is developed to induce specific chemical reactions within the polymer matrix by exposing the material to elevated temperatures while simultaneously applying vacuum and pressure to consolidate individual plies and compress voids. The cure cycle and vacuum bagging procedure affect such cured product characteristics as:

1.) Degree of cure
2.) Glass transition temperature
3.) Void content%
4.) Cured resin content/fiber volume
5.) Residual stress
6.) Dimensional tolerances and
7.) Mechanical properties.

The cure cycle and vacuum bagging procedures have a significant effect.

2.8.4 Press molding

Press curing uses heated platens to apply both pressure and heat to the part. Presses, in general, operate at 20 - 1000 psi (140 - 7000 kPa) and up to 600 °F (320°C). Press curing is very economical for flat parts and high production rates. Tooling requires matched die molds for contoured parts. The following steps are used in press molding:

1. Composite material is placed in the mold cavity.
2. Cure monitoring devices are installed.
3. Parts are placed into press and cured. Pressure, temperature, and time are monitored during the cure cycle to ensure curing parameters are met.

Press curing produces high quality parts with low void content.

2.8.5 Integrally heated tooling

With integrally heated tooling the heat required for cure is provided through the tool itself, rather than through the use of external heating in an oven or autoclave. This can be used to make high quality parts without using an autoclave if matched mold tools are used. The heat is usually provided by imbedding electrical resistance elements or hot oil circulation channels within the tool. This can result in hot and cold spots within the tool. Heat surveys are necessary to ensure that all parts of the tool perform with a heat profile that allows the part to be cured completely and with high quality.

2.8.6 Pultrusion die cure and consolidation

Pultrusion is an automated process for the continuous manufacture of composites with a constant cross-sectional area. A continuous reinforcing fiber is integral to the process and the finished product. Pultrusion can be dry, employing prepreg thermosets or thermoplastics, or wet, where the continuous fiber bundle is resin-impregnated in a resin bath. The wet resin process was devel oped around the rapid addition reaction chemistry exhibited by thermoset polyester resins, although advances in resin and catalyst systems has made the use of epoxy systems commonplace.

In pultrusion the material is cured in a continuous process that can provide large quantities of high quality cured shapes. The material is drawn through a heated die that is specially designed for the shape being made. The tool is designed such that the volume of the cavity for cure causes the resin pressure to build, allowing consolidation of the material to occur. This cure cavity pressure is built up against the cured material that is downstream of it, and induced by the new material upstream which is continuously being drawn into the cavity. As a result this process can be very sensitive to variation in the tow and rate used for pultrusion.

The resins used for pultrusion are also very specialized. There is little time for volatile removal, consolidation, and other activities that can take considerably longer using other cure processes. The resin must be able to cure very rapidly, sometimes in less than a second, when exposed to the proper temperature. The resin must also be very consistent. Disruptions to this process can be very time consuming and expensive. Like most continuous processes, much of the operating expenses are associated with starting up and stopping the line.

The key elements in the process consist of a reinforcement delivery platform, resin bath (for wet pultrusion), preform dies, a heated curing die, a pulling system and a cut-off station. A wide range of solid and hollow profiles can be produced by the process and stitched fabrics, random mats and bidirectional reinforcements can be used in the process. The die employs a bell section opening to help reduce hydraulic resin pressures which build up in the die. The die is also plated to help eliminate die wall adhesion as well as hardened to counteract the abrasive action of the fibers.

In general the following process is used:

1. The reinforcements are threaded though the reinforcement delivery station.
2. The fiber bundle is pulled through the resin bath (if using a wet process) and die preforms.
3. A strap is used to initiate the process by pulling the resin impregnated bundle through the preheated die.
4. As the impregnated fiber bundle is pulled through the heated die, the die temperature and pulling rate are controlled such that the cure of the product (for thermosets) is completed prior to exiting the heated die.
5. The composite parts are cut off by the saw at the desired length as the continuous pultr uded product exits the heated die.

The most critical process variable in pultrusion is the temperature control of the product which is a function of the temperature profile of the heated die and the line speed. Temperature control is critical because the product must achieve full cure just prior to exiting the pultrusion die. Other variables which affect cured properties are fiber tension which directly influences the fiber alignment of the final product, and resin bath viscosity which contributes to the completeness of fiber wet-out and the uniformity of the fiber to resin ratio of the final product.

2.8.7 Resin transfer molding

The RTM process resin is pumped into a matc hed die mold which contains the fiber reinforcement. The process requires low viscosity resins and tools designed with bleed holes to allow air to escape during resin transfer. The holes are closed when the cavity is filled to allow pump to apply pressure to the resin. The following general steps are used for resin transfer molding:

1. Place fiber reinforcement into the mold.
2. Close mold and check that all air vent holes are open.
3. Pump resin into mold until full, then close air vents and pressurize resin.
4. Cure at room temperature or elevated temperature.
5. After cure, remove part from mold.

RTM can yield good quality parts similar to compression molded types.

2.8.8 Thermoforming

Thermoforming fiber-reinforced thermoplastics. The thermoforming process, as applied to thermoplastic composite materials, is generally divided into two categories: melt-phase forming (MPF) and solid-phase forming (SPF). Thermoforming capitalizes on the rapid processing characteristics of thermoplastics. The composite thermoforming process can be broken down to four basic steps:

1. The material is heated to its processing temperature external to the forming tool. This can be accomplished with radiant heat.
2. The oven-heated material is rapidly and accurately transferred to the forming tool.
3. The heated material is pressure-formed with matched die set tooling into desired shape.
4. The formed laminate is cooled and its shape is set by sinking the heat into the tooling.

MPF is performed at the melting point of the thermoplastic matrix and requires sufficient pressure and/or vacuum application during the forming process to provide complete consolidation. The MPF process is preferred when sharp contour changes requiring some level of resin flow are a characteristic of the part geometry.

SPF is generally performed at temperatures between the onset of crystallization and below the peak melting point. This temperature range provides sufficient formability while the material remains in a solid form. SPF allows forming of preconsolidated sheet to be performed without a consolidation phase, but it is limited to part geometries exhibiting gentle curvatures.

The processing time for thermoforming is governed by the rates at which heat can be added to the material and then removed. This is primarily a function of the material thermal properties, material thickness, forming temperature, and tooling temperature. The pressures required to shape the material are dependent on various factors including part geometry, material thickness, and formability. The general deformability behavior of thermoplastics also depends on the strain-rate used during forming and the thermal history of the thermoplastic matrix. The forming process can affect such final properties as:

1.) Degree of crystallinity,
2.) Glass transition temperature,
3.) Fiber orientation/alignment,
4.) Uniformity of the fiber to resin ratio,
5.) Residual stress,
6.) Dimensional tolerances, and
7.) Mechanical Properties.

The forming process has s significant effect on the quality of the finished part. High quality parts with predictable engineering properties require that a well controlled thermoforming process developed for specific applications be utilized.

2.9 ASSEMBLY PROCESSES

Assembly processes are not conventionally covered within composite material characterization, but can have a profound influence on the properties obtained in service. As seen with test coupons, edge and hole quality can dramatically affect the results obtained. While these effects are not usually covered as material properties, it should be noted that there is an engineering trade-off between part performance and the time and effort expended toward edge and hole quality. These effects need to be considered along with the base material properties.

2.10 PROCESS CONTROL

Composite structures have the potential to provide higher performance in many applications. In order for this potential to be fulfilled, it must be possible to cost effectively manufacture parts of high, uniform quality. During cure of composite parts, the material is being made at the same time as the part. As a result, there are many actions happening at the same time. Some of these actions can be controlled directly, others only indirectly, and some of them interact. Process control is one of the methods used to manage the variability associated with composites.

2.10.1 Common process control schemes

Process control is used to attempt to direct these many changes during cure to reach many objectives. The manufacture of high quality parts is one objective. Others include exotherm avoidance, minimization of cure times, and addressing part specific m anufacturing problems. Several different approaches to process control can be pursued: empirical, active, and passive. The most common is empirical, or trial and error. Many different sets of cure conditions are attempted, with the cure conditions providing the best results being picked for manufacturing. The second is active, or real-time process control. Here data is acquired during the cure from the part in question. Data that can be acquired includes temperature, pressure, resin viscosity, resin chemical characteristics (degree of cure), and average ply thickness. An expert system is used to analyze the cure information, and direct the autoclave how to proc eed with the cure. The third is passive, or off-line process control. Here mathematical models are used to predict the res ponse of the part during cure. Many different cure approaches can be simulated, and the one that best meets the needs at hand are applied.

Each of these process control approaches benefit from an understanding of the effects and interrelationships that are occurring during the cure of the resin. This understanding is referred to as a process control model. The model remains the same r egardless of which particular type of process control is attempted for a particular application.

2.10.1.1 Empirical methods

2.10.1.2 Active sensor-based control

2.10.1.3 Passive model-based control

2.10.2 Example - autoclave cure of a thermoset composite

A generic process control model can be used to evaluate and develop composite cures that pr oduce high quality parts. When the resin is heated and has begun to flow, the system can be divided into gas (volatiles or trapped air), liquid (resin), and solid (reinforcement) phases. All void producing gas phase material should be either eliminated or absorbed by the liquid phase. The liquid phase should be uniformly distributed throughout the part, maintaining or pr oducing the desired resin content. The solid phase should maintain its selected orientation. There are several initial factors that must be determined in order to cure parts, and which are used as input for the process model. These initial factors have been broken down into the following categories: resin, time, heat, applied pressure, process materials, design, and reinforcement. It is well known that different resins, even within the same general material family, do not always provide equivalent results when processed in the same manner. The cure times and temperatures, including dwell(s) and heat up rates, usually control heat flow. For thick structures heat from resin exotherm can be domi nant. The pressure to be used during cure must be determined, and may be c hanged substantially during the cure. Vacuum bagging or other process materials may be used to perform actions such as resin bleed, but can also have other effects, especially when they fail. Design choices such as the use of sandwich construction and radii affect the results obtained with the cure. Finally, although the reinforcement is usually intended to just maintain its orientation, it does influence gas and liquid flow, and picks up some of the applied pressure.

The number of initial factors alone makes composite processing difficult. What makes it even more complicated is that these initial factors affect the desired results and interact with each other in complex, non-linear relationships. Because of this, adjusting one factor in a seemingly logical fashion often does not obtain the desired results. A diagram of such a process model is shown in Figure 2.10.2. This particular model was designed for autoclave cure of thermoset composites. However, this m odel would also be largely applicable to most other composite and adhesive cure processes with slight modification. The initial factors are shown at the top of the figure, and the desired output at the bottom. The center area between the initial factors and the desired outputs represents the process interactions. These process interactions are: degree of cure, viscosity, resin pressure, void prevention, and flow. By using this model, cure process changes and optimization can be performed in a logical progression rather than a hit-or-miss fashion. Each of these process interactions is discussed in turn.

2.10.2.1 Degree of cure

The resin degree of cure acts primarily as an i nput to the viscosity interaction. Determining the rate of change of degree of cure for a resin r equires a knowledge of the particular response for the individual resin and the temperature history for the resin. The resin heat of reaction is used as an index of degree of cure. The rate of change of degree of cure is then calculated as a function of the current degree of cure and the temperature. The rate of change of degree of cure is often not linear, which is why it is difficult to estimate the response of a resin to a new temperature profile without a model. In addition, in thick structures the heat of reaction may contribute significantly to the temperature of the resin, in turn affecting the degree of cure, and the viscosity. After the resin has gelled , the glass transition temperature is often used as an index of degree of cure.

2.10.2.2 Viscosity

The resin viscosity is a function of the resin degree of cure and temperature. The resin viscosity res ponse function does vary from resin to resin. Thermoplastic resins do not chemically react during the fabrication process ("cure"), but do flow upon melting of the resin. Because the chemical makeup of the resin is not changing, the viscosity of a thermoplastic resin is strictly a function of the temperature. In other words, the viscosity effects are entirely physical, and no chemical interactions come into play. However, two different thermoplastic resins may have different viscosities at the same temperature due to chain length or other chemical differences.

A thermoset resin does react, so its chemical makeup is constantly changing during cure. Because of increases in chain length and crosslink density, the viscosity of the resin at a given temperature will increase over time. This is because there is increased interaction between the chains, and they become increasingly entangled with each other. Once chain extension and cross-linking have extended sufficiently, a thermoset resin will gel. The reason that viscosity effects for thermoset resins are much more difficult

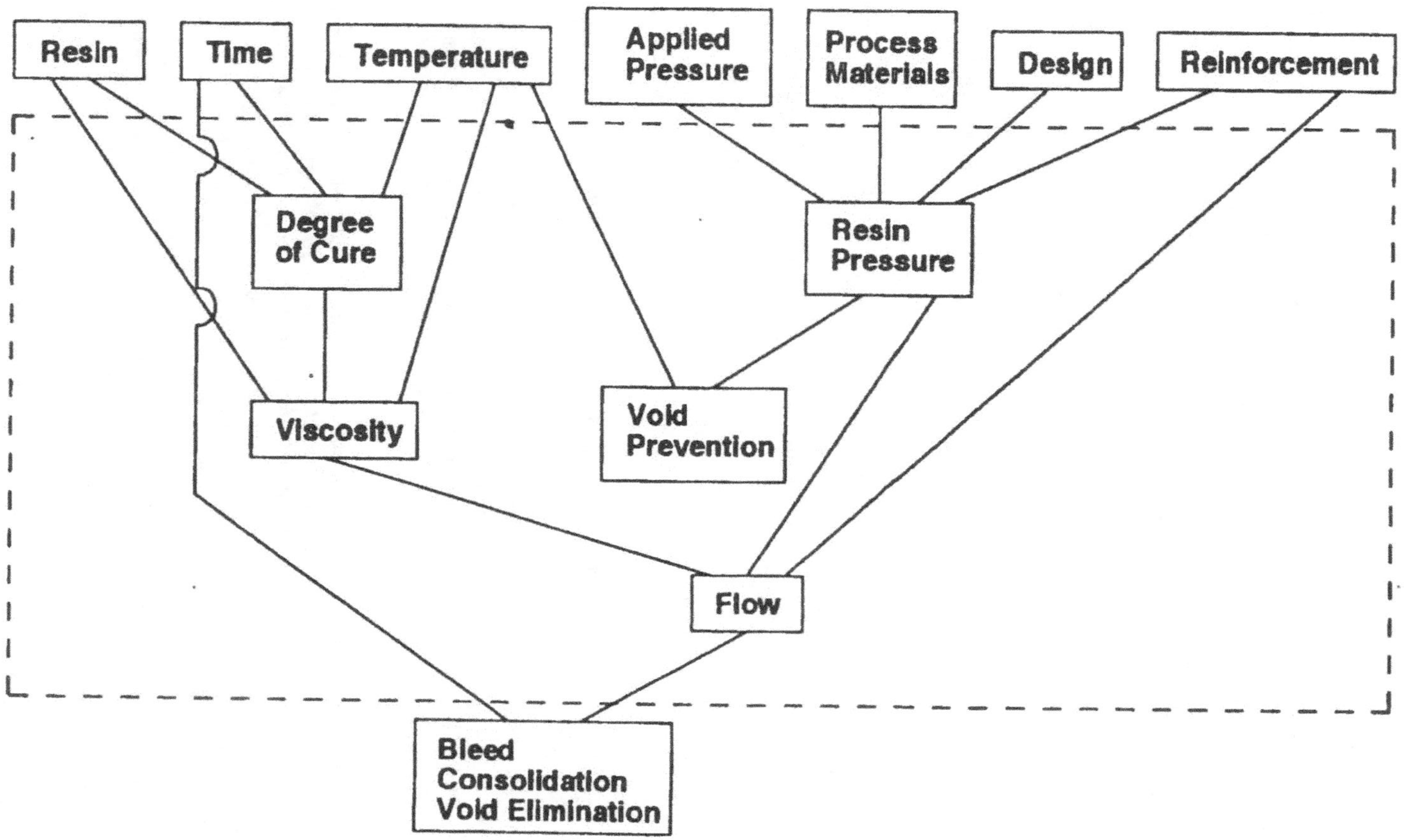

FIGURE 2.10.2 *Composite cure process model.*

to predict than for thermoplastics is this continuous, sometimes rapid, c hange in the chemical makeup of the system.

2.10.2.3 Resin pressure

The pressure applied to a laminate is usually not the same as that which is experienced by the resin, referred to as resin pressure. The concept of resin pressure is frequently conceptualized with a spring and dashpot type model, with resin as the fluid, and the fiber pack as the spring. If the spring is completely surrounded by the fluid, it cannot pick up any of the applied load. If there is not enough resin to surround the spring, due perhaps to resin bleed, the spring (fiber pack) will pick up an increasing percentage of the load. The resin loses the corresponding amount of pressure. A diagram of this model can be seen in Figure 2.10.2.3.

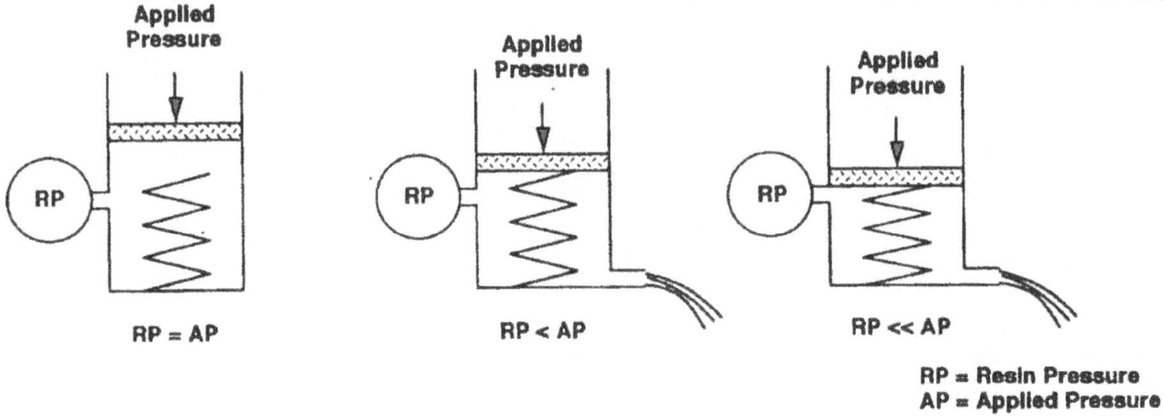

FIGURE 2.10.2.3 *Spring and dashpot model for resin pressure.*

Resin pressure is important because it is the driving force for moving resin and gas phase material from one place to another, and because it helps prevent formation of voids. Resin pressure is a function of the applied pressure, how and what process materials are used for cure, the design, and the reinforcement. If there is not sufficient resin to completely surround the reinforcement, then the reinforcement will pick up some or all of the load.

Just as the reinforcement can pick up applied pressure, so can the other process materials, especially the breather and bleeder. These items act as additional springs in the dashpot/spring model, and can absorb a significant amount of the applied pressure, especially for lower pressure cures. One of the design factors affecting the resin pressure is the use of materials such as honeycomb and some types of foam core. With co-cured skins, if a force is applied to the tool or bag side of the skin, resin pressure will be created, but all the resin has to do is flow slightly into the cell to relieve this pressure. This results in quality problems with honeycomb parts, especially if the skins are fairly thin, such as less than five plies. If the skins are fairly thick, then through the thickness resin pressure variations could be present. This would allow the surface of the parts at the tools surface to be under appreciable resin pressure, while at the honeycomb side the resin pressure would be near zero. Given an infinite amount of time, these pressures would equalize, but not in the time frame for many cures. When the skins are thin, the resin pressure is near zero. Thus the skins on thin skin honeycomb are cured with near zero resin pressure, essentially a contact lay-up, and the quality of the skins is often reduced. Because of the resistance that the reinforcement provides, some interesting resin pressure effects can be noted, along with their consequences on part quality. Just as through the thickness variations in resin pressure can be established, they can also be present in the plane of the reinforcement.

This helps explain why widely different laminate quality can be present on the same part cured at the same time. Consider the bridging of the fiber reinforcement in a tight corner. Unless the plies can slide past each other to contact the tool in this corner, the reinforcement is, by definition, picking up all of the pressure applied by the autoclave, and the resin pressure is zero. At the location of bridges it can often be s een that increased porosity has occurred, voids at the tool interface are present, and excess resin has built up. These are all due to the fact that the resin pressure in this location is near zero.

Areas surrounding this bridging may have adequate resin pressure. A series of experiments on honeycomb panels revealed that while the resin pressure in the skins (co-cured) was near zero, the resin pressure in the edge band (laminate) was significantly higher. The quality of the laminate in the edge band area was significantly higher even though the two points were only inches apart. This demonstrates the concept of differing resin pressures in close proximity.

2.10.2.4 Void prevention

Some resin systems, especially the higher temperature systems such as polyimides and phenolics, produce volatiles as a part of the cure reaction chemistry. While these bypr oducts are being evolved, the applied pressure should be minimal, and vacuum applied. As soon as all the volatiles have been created, then resin pressure can be used to drive out any volatile products remaining prior to gel. Once the resin has gelled, flow of the resin has been completed, and issues such as resin content, bleed, volatile content have been settled. The continuing cure advances the cure of the resin, but the physical configuration of the resin is locked in.

Some volatiles may be present in the prepreg, the most common being absorbed moisture. If resin pressure is maintained above the volatile vapor pressure until gel, these compounds cannot volatilize, increasing their volume many-fold, and forming additional voids and/or porosity. This functions in the same manner as a car radiator, as diagramed in Figure 2.10.2.4.

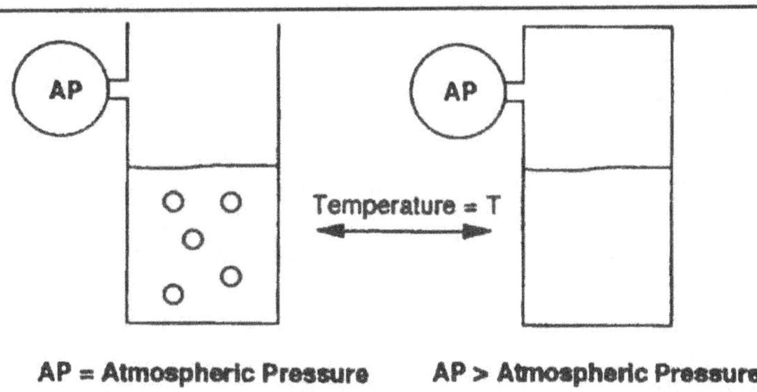

FIGURE 2.10.2.4 *Void prevention resin pressure model.*

2.10.2.5 Flow

The viscosity, resin pressure and reinforcement factors feed into the flow factor. The viscosity and reinforcement can be thought of as resistances to flow, while the resin pressure can be thought of as the driving force for flow. The amount of flow that occurs due to these factors is then a function of time. This is consistent with experience. If the resin is more viscous, less flow would be expected with the same resin pressure and reinforcement. If the reinforcement is changed, perhaps to a tighter weave, t hen the resistance to resin and gas phase flow is increased. Once these flow characteristics have been established, then they and the time available for flow determine how bl eed of the laminate takes place, how the laminate is consolidated, and the elimination of voids present in the lay-up or formed during the cure.

2.11 PREPARING MATERIAL AND PROCESSING SPECIFICATIONS

Requirements for materials and processes are frequently so specific and extensive, a special type of engineering drawing format was developed. Material and process specifications are one of the ways used to control composite material variability. Specifications are usually E-sized engineering drawings (see MIL-STD-490 (Reference 2.11)). They are part of the engineering package that defines a particular product, whether an airplane or a golf club.

2.11.1 Types of specifications

Material and process specifications are similar but do have some differing requirements.

2.11.1.1 Material specifications

The primary purpose of material specifications is to control the purchase of critical materials. The properties and values contained in the specification will relate to, but not necessarily be identical to, the properties used for activities such as design and structure testing. The properties and values contai ned in the specification are used to assure that the material does not change substantially with time. This is especially critical for materials used in primary applications, and which have undergone expensive qualifications. Material specifications are included in relevant contracts, and are part of the purchase order requirements to purchase material.

2.11.1.2 Process specs - controls end product

Process specifications establish the procedures that are required to control the end product. The more process dependent the materials and/or end product are, the more detailed and complex the process requirements. On the other hand, if there is a wide window of acceptable product produced by the process, the requirements may be minimal. Composite and adhesive bonding processing specifications are usually detailed because the materials are very sensitive to process variations, and the aerospace end item requirements are usually very stringent.

2.11.2 Format for specifications

Most specifications follow a similar format, based on guidelines contained in documents such as MIL-STD-490 (Reference 2.11) and DOD-STD-100 (Reference 2.11.2). The sections of a material or process specification are generally as follows: scope, applicable documents, technical requirements or process controls, receiving inspection and quality control, delivery, notes, and approved sources and other. Each is covered in more detail in the following sections.

2.11.2.1 Scope

The first section is the scope, which generally describes the materials or processes covered by the specification in a few sentences. Also covered in this first section are any types, classes, or forms, of the materials that are governed by the specification. Another method for handling different material configurations is the use of slash sheets. These slash sheets are part of the base document, but provide the additional information that is specific to that particular material. For example, one material specification may cover several different thicknesses of the same film adhesive, each thickness being a different class. The scope section establishes the shorthand terminology, or callout, which is used to identify the material on other engineering and procurement documents. A process specification may cover multiple processes, such as anodizing, with minor process variations based on the type of alloy being processed.

2.11.2.2 Applicable documents

The second section identifies all the other documents that are referenced within the specification. Testing procedures, and other material and process specifications may be called out. A trade-off is made between a specification being self contained, and redundancy between multiple specifications for similar materials or processes. For example, if a change to a testing procedure is required, only a change to the referenced testing specification is required. If the specifications are all self contained, the test procedure within each specification must be revised. The time and expense associated with changes to common materials and procedures can be substantial. However, when only a limited am ount of information is required, the modular approach can bring along a great deal of unused information. These configuration management issues are discussed in more detail in a following section.

2.11.2.3 Technical requirements / process controls

The third section covers the technical requirements for the material or controls for the process. For a materials specification, these requirements can include physical, chemical, mechanical, shelf and work life, toxicity, environmental stability, and many other characteristics. The requirements can be minimum values, maximum values, and/or ranges. Sometimes it is only required that the data obtained from the test be submitted. Only the test result requirements are contai ned in this section. The test procedure used to obtain this result is covered in the next section. For a process specification, the controls required to ensure the product produced is consistent are specified.

2.11.2.4 Receiving inspection and qualification testing

The fourth section covers testing. Receiving inspection testing is that which is performed each time a quantity of the material is purchased, or a lot of pr oduct is processed. Although it is required that all the requirements of the specification be met at all times, only a fraction of the tests are performed routinely. Qualification testing usually involves testing to all of the requirements of the specification to insure that the supplier or processor is capable of meeting the requirements, and is performed only once unless there is cause.

Responsibility for the testing required is also delineated. The manufacturer may do all their receiving inspection testing, or the user may perform additional testing upon receipt of the material. Required reports are defined, as well as requirements for resampling and retesting if a requirement is initially failed.

Sampling and the specific test procedures to be used to determine conformance to the technical requirements are contained in this section. Testing procedures can be critical. In most cases, the value obtained cannot be used unless the specific test used to generate the value is documented. Test results can change when test procedures change, even though the material being tested has not changed itself. Also important is the preparation of the test specimens. Test results can vary widely depending on the configuration and condition of the test specimens. The conditions under which the test is performed can dramatically change the results. Preconditioning of the specimen prior to test is also important, such as exposure to elevated temperature and humidity prior to test.

2.11.2.5 Delivery

Delivery requirements are covered in the fifth section. Issues such as packaging and identification, storage, shipping and documentation must be established. Packaging is especially critical for temperature sensitive materials such as prepreg and film adhesive.

2.11.2.6 Notes

The sixth section is usually notes, although sixth and later section formats can vary substantially. Notes are additional information for reference, and are not requirements unless specifically stated in the requirements section.

2.11.2.7 Approved sources and other

Seventh and additional sections can include information such as what materials are qualified to the specification. This section may reference a separate document that lists the qualified materials. Because of the substantial expense that can be experienced as a result of qualification, normally only materials that are currently qualified are used for production applications.

2.11.3 Specification examples

Specifications in common use are generally released by industry associations or the military. Industry associations common to composite and adhesive bonded structure are SAE, ASTM, and SACMA. In addition, companies may develop their own internal specifications for materials or processes that are not adequately covered by industry/military specifications, or to protect proprietary information. Com pany specifications may be similar in style and content to industry and military specifications, but can vary widely in approach and level of control. Each has advantages and disadvantages.

2.11.3.1 Industry

Examples of industry specifications are as follows:

AMS 3897	Cloth, Carbon Fiber, Resin Impregnated
AMS 3894	Carbon Fiber Tape and Sheet, Epoxy Resin Impregnated

AMS specifications are available from SAE, 400 Commonwealth Drive, Warrendale, PA 15096-001.

2.11.3.2 Military

Examples of military specifications are as follows.

MIL-A-83377	Adhesive Bonding (Structural) for Aerospace and Other Systems, Requirements for
MIL-P-9400	Plastic Laminate and Sandwich Construction Parts and Assembly, Aircraft Structural, Process Specification Requirements
MIL-T-29586	Thermosetting Polymer Matrix, Unidirecti onal Carbon Fiber Reinforced Prepreg Tape (Widths up to 60 Inches), General Specification for

Military specifications are available from DODSSP, Standardization Document Order Desk, 700 Robbins Ave., Bldg. 4D, Philadelphia, PA 19111-5094.

2.11.4 Configuration management

Most major aerospace companies use many materials and process specifications to control and define their products, and those made by their subcontractors. Many companies prefer to have a company controlled specification for some materials and processes, even when equivalent industry or military specifications are available. Industry and military specifications are, by definition, consensus documents. Reaching this consensus can take a good deal of time, and may conflict with a specific company's objectives. With company control, tailoring of the specification to company requirements can be relatively easily effected. Company specifications do allow tailoring, but at the cost of standardization. Company specific tests and procedures incur additional expense. There may be many specifications that govern essentially the same material. Sometimes this is because different specifications offer different levels of control (testing). The amount and complexity of testing required for procurement of material can soon account for a large percentage of the total cost of the material. If only minor changes to the specification are made, amendments or supplements can be released. Some specification changes may only be in effect for limited periods of time, or restricted to certain facilities. Control of the current and prior versions of specifications is an important issue. Specification changes can have a great influence on manufacturing operations, and if additional expenses are associated with the changes in the revision, prices and timing for implementation may have to be negotiated. Not all operations and subcontractors may start work per a new revision of a specification at the same time. In addition, confusion frequently arises from different parties unintentionally using different revisions, or versions, of the same specification.

REFERENCES

2.4.1.2(a) *Data Manual for Kevlar™49 Aramid*, E.I. Du Pont de Nemours & Co., Wilmington, DE 1989.

2.4.1.2(b) Krueger, W.H., et al., "High Performance Composites of J2 Thermoplastic Matrix Reinforced with Kevlar™ Aramid Fiber," *Proceedings of the 33rd International SAMPE Symposium*, March 1988, pp. 181-193.

2.4.1.2(c) Gruber, M.B., "Thermoplastic Tape Laydown and Consolidation," SME Technical Paper EM86-590.

2.4.1.2(d) Okine, R.K., et al., "Properties and Formability of a Novel Advanced Thermoplastic Composite Sheet Product," *Proceedings of the 32nd International SAMPE Symposium*, April 1987, pp. 1413-1425.

2.4.1.2(e) Egerton, M.W. and Gruber, M.B., "Thermoplastic Filament Winding Demonstrating Economics and Properties via In-Situ Consolidation," *Proceedings of the 33rd International SAMPE Symposium*, March 1988, pp. 35-46.

2.4.1.2(f) Khan, S., "Environmental Effects on Woven Thermoplastic Composites," *Proceedings, Materials Week™, 87*, ASM International, Cincinnati, OH, October 10-15, 1987.

2.4.1.3.1(a) "PPG Fiber Glass Yarn Products/Handbook," PPG Industries, Inc., 1984

2.4.1.3.1(b) *Engineered Materials Handbook*, ASM, International, Metals Park, Ohio, 1987, pp. 107-100.

2.4.1.3.1(c) "Textile Fibers for Industry," Owens-Corning Fiberglas Corporation, Pub. No. 5-TOD-8285-C, September 1985.

2.4.1.3.2(a) ASTM Specification D 579, "Greige Woven Glass Fabrics Fiber Yarns," Annual Book of ASTM Standards, Vol 7.01, American Society for Testing and Materials, West Conshohocken, PA.

2.4.1.3.2(b) ASTM Specification D 578, "Glass Fiber Yarns," Annual Book of ASTM Standards, Vol 7.01, American Society for Testing and Materials, West Conshohocken, PA.

2.4.1.3.4(a) "Discover S-2 Glass Fiber," Owens-Corning Fiberglas Corp., Pub. No. 5-ASP-13101-A, 1986.

2.4.1.3.4(b) "Industrial Fabrics," Clark - Schwebel Fiber Glass Corp.

2.4.1.3.4(c) "Product Bulletin," Stratifils H_R/P109.

2.4.1.4(a) Shoenberg, T., "Boron and Silicon Carbide Fibers," *Engineered Materials Handbook, Volume 1, Composites*, ASM International, Metals Park, Ohio, 1987, pp. 58-59.

2.4.1.4(b) Bascom, W.D., "Other Continuous Fiber," *Engineered Materials Handbook, Volume 1, Composites*, ASM International, Metals Park, Ohio, 1987, pp. 117-118.

2.4.1.4(c) Krukonis, V., "Boron Filaments," *Handbook of Fillers and Reinforcements for Plastics*, H. Katz and J. Milewski, eds., Van Nostrand Reinhold, Co., 1978.

2.4.1.4(d) DeBolt, H.E., "Boron and Other High Strength, High Modulus Low-Density Filamentary Reinforcing Agents," *Handbook of Composites*, G. Lubin, ed., Van Nostrand Reinhold Co., 1982, pp. 171-195.

2.4.1.4(e) MIL-HDBK-727, *Design Guide for Producibility*, Chapter 6.

2.4.1.4(f) "Boron Monofilament, Continuous, Vapor Deposited," MIL-B-83353.

2.4.1.4(g) "Filaments, Boron - Tungsten Substrate, Continuous," AMS 3865B.

2.4.1.6(a) *Advanced Materials*, Carrol-Porczynski, Chemical Publishing, 1962, pp. 117,118.

2.4.1.6(b) *Engineered Materials Handbook*, ASM International, Vol I, Metals Park, Ohio, 1987, p.58-59.

2.4.1.6(c) Lynch, et al., "Engineered Properties of Selected Ceramic Materials," *Journal of American Ceramic Society*, 1966, pp. 5.2.3-5.2.6.

2.4.1.6(d) *Advanced Materials and Processes*, "Guide to Selected Engineered Materials," Vol 2, No 1, June 1987, pp. 42-114.

2.4.1.6(e) BP Chemical, Advanced Materials Division, Carborundum.

2.4.1.7 "Quartz Yarn of High Purity," AMS 3846A.

2.4.2.7.1(a) Hertzberg, R.W., *Deformation and Fracture Mechanics of Engineering Materials*, John Wiley & Sons, New York, 1976 p. 190.

2.4.2.7.1(b) Kaufman, H.S., "Introduction to Polymer Science and Technology," Lecture Notes.

2.4.2.7.1(c) "The Place for Thermoplastics in Structural Components". Contract MDA 903-86-K 0220, pp. 1-2, 20, 29.

2.4.2.7.1(d) Murtha, T.P., Pirtle, J., Beaulieu, W.B., Waldrop, J.R., Wood, H.D., "New High Performance Field Installed Sucker Rod Guides," *62 Annual Technical Conference and Exhibition of SPE*, Dallas, Texas, September 1987, pp. 423-429.

2.4.2.7.1(e) Brown, J.E., Loftus, J.J., Dipert, R.A., "Fire Characteristics of Composite Materials - a Review of the Literature," NAVSEA 05R25, Washington, DC, August 1986.

2.4.2.7.1(f) "The Place for Thermoplastics in Structural Compounds," contract MOA 903-86-K-0Z20, Table 3-5, p. 1.

2.4.2.7.1(g) Horton, R.E., McCarty, J.E., "Damage Tolerance of Composites," *Engineered Materials Handbook*, Volume 1, Composites, ASTM International, 1987, pp. 256-267.

2.4.2.7.1(h) Silverman, E.M., Griese, R.A., Wright, W.F., "Graphite and Kevlar Thermoplastic Composites for Space Craft Applications," *Technical Proceedings, 34th International SAMPE Symposium*, May 8-11, 1989, pp. 770-779.

2.4.2.7.1(i) Shahwan, E.J., Fletcher, P.N., Sims, D.F., "The Design, Manufacture, and Test of One-Piece Thermoplastic Wing Rib for a Tiltrotor Aircraft," *Seventh Industry/Government Review of Thermoplastic Matrix Composites*, San Diego, California, February 26-March 2, 1991, pp. III-C-1 - III-C-19.

2.4.2.7.1(j) Lower Leading Edge Fairing to AH-64A Apache Helicopter-Phillips 66-McDonnell Douglas Helicopter Prototype Program.

2.4.2.7.2(a) *Modern Plastics Encyclopedia*, Materials, McGraw-Hill, 1986-1987, p. 6-112.

2.4.2.7.2(b) Koch, S., Bernal, R., "Design and Manufacture of Advanced Thermoplastic Structures," *Seventh Industry/Government Review of Thermoplastic Matrix Composites Review*, San Diego, California, February 26-March 1, 1991, pp. II-F1 - II-F-22.

2.4.2.7.2(c) Stone, R.H., Paul, M.L., Gersten, H.E., " Manufacturing Science of Complex Shape Thermoplastics," *Eighth Thermoplastic Matrix Composites Review*," San Diego, California, January 29-31, 1991, pp. I-B-1 - I-B-20.

2.7.7(a) Lubin, G., *Handbook of Composites*, Van Nostrand Reinholt, 1982, Chapter 21, "Sandwich Construction".

2.7.7(b) MIL-P-9400C, *Plastic Laminate and Sandwich Construction, Parts and Assembly, Aircraft Structural, Process Specification Requirements.*

2.7.7(c) MIL-HDBK-23 (Canceled), *Structural Sandwich Construction.*

2.7.7(d) MIL-STD-401, *Sandwich Constructions and Core Materials: General Test Methods.*

2.7.8(a) Kinloch, A.J., Kodokian, G.K.A., and Watts, J.F., "The Adhesion of Thermoplastic Fiber Composites," *Phil. Trans. Royal Soc: London A*, Vol. 338, pp. 83-112, 1992.

2.7.8(b) MIL-HDBK-691, Adhesive Bonding

2.11 MIL-STD-490

2.11.2 MIL-STD-100

CHAPTER 3 QUALITY CONTROL OF PRODUCTION MATERIALS

Page

3.1 INTRODUCTION

Quality conformance tests are needed to assure the continued integrity of a previously characterized material system. The tests performed must be able to characterize each batch/lot of material so a proper assessment of critical properties of a material system can be made. These critical properties provide information on the integrity of a material system with regard to material properties, fabrication capability, and usage. Additionally, the test matrix must be designed to economically and quickly evaluate a material system.

Quality control in a production environment involves inspection and testing of composites in all stages of prepreg manufacture and part fabrication. Tests must be performed by the material s upplier on the fiber and resin as separate materials, as well as on the composite prepreg material. The user of the prepreg must perform receiving inspection and revalidation tests, in-process control tests, and nondestructive inspection tests on finished parts. These tests are described in the following sections and normal industry practice is discussed.

3.2 QUALITY ASSURANCE PROCEDURES

3.2.1 Receiving Inspection

The composite material user typically prepares material specifications which define incoming material inspection procedures and supplier controls that ensure the materials used in composite construction will meet the engineering requirements. These specifications are based on material allowables generated by allowables development programs. The acceptance criteria for mechanical tests must be specified to assure that production parts will be fabricated with materials which have equivalent properties as the materials used to develop the allowables.

The user material specifications typically require the suppliers to provide evidence that each production lot of material in each shipment meets the material specification requirements. This evi dence will include test data, certification, affidavits, etc., depending upon the user quality assurance plan and purchase contract requirements for a particular material. The test r eports contain data to verify the conformance of material properties to user specifications and acceptance standards.

Acceptance test requirements may vary from user to user. However, the tests must be sufficient to assure the material will meet or exceed the engineering requirements. A typical example of acceptance tests required for carbon/epoxy unidirectional tape is shown in Table 3.2.1. Note that Table 3.2.1 is divided into two parts. The first part concerns uncured prepreg properties. The purpose of these tests is to assure that the resin and fibers materials are within acceptable limits. The second part involves tests on cured laminates or laminae. The mechanical property tests s hould be selected to reflect important design properties. They can be direct tests of a property or a basic test that correlates with critical design properties. The 90 °/0° tension test evaluates the fiber strength and modulus. The 90 °/0° compression test evaluates the reinforced fiber/resin combination. The compression testing also includes hot dry tests since one resin-dependent mechanical property should include elevated temperature tests to ensure the material's temperature c apability . A shear test should be run as a resin evaluation. The short beam shear test or the ±45 ° tension test should be used depending on the end product's emphasis on interlaminar or in-plane properties.

TABLE 3.2.1 *Typical acceptance and revalidation tests required for suppliers and users.*

PROPERTY	TESTING REQUIRED			SPECIMENS REQUIRED PER SAMPLE
	PRODUCTION ACCEPTANCE (SUPPLIER)(3)	PRODUCTION ACCEPTANCE (USER)(3)	REVALIDATION (USER)(3)	
Prepreg Properties				
Visual & Dimensional	X	X		-
Volatile Content	X	X		3
Moisture Content	X	X	X	3
Gel Time	X	X	X	3
Resin Flow	X	X	X	2
Tack	X	X	X	1
Resin Content	X	X		3
Fiber Areal Weight	X	X		3
Infrared Analysis	X			1
Liquid Chromatograph	X	X	X	2
Differential Scanning Calorimetry	X	X	X	2
Lamina Properties				
Density	X			3
Fiber Volume	X			3
Resin Volume	X			3
Void Content	X			3
Per Ply Thickness	X	X	X	1
Glass Transition Temp	X	X	X	3
SBS or ±45° Tension	X(2)	X(2)	X(2)	6
90°/0° Compression Strength	X(1)	X(1)	X(2)	6
90°/0° Tension Strength & Modulus	X(2)	X(2)	X(2)	6

(1) Tests shall be conducted at RT/Dry and Maximum Temperature/Dry (See Volume I, Section 2.2.2).
(2) Tests shall be conducted RT/Dry.
(3) Supplier is defined as the prepreg supplier. User is defined as the composite part fabricator. Production acceptance tests are defined as tests to be performed by the supplier or user for initial acceptance. Revalidation tests are tests performed by the user at the end of guaranteed storage life or room temperature out time to provide for additional use of the material after expiration of the normal storage or out time life.

Receiving inspection test requirements should address test frequency and, in the event of initial failure to satisfy these requirements, retest criteria. Test frequency is a function of the quantity of material (weight and rolls) in a batch. Typical testing may include specimens from first, last and random rolls. A retest criteria should be included for the cured lamina tests so that the material is not rejected because of testing anomalies. If a material fails a test, a new panel from the same suspect roll of material should be fabricated and used to rerun that specific test. If a batch has multiple rolls, that test should run on material from the roll before and after the suspect roll in order to isolate the potential problem. If the material fails the retest, the entire batch should be reviewed by material engineering. As use and confidence increase, the receiving inspection procedure can be modified. For example, the test frequency can be decreased or certain tests can be phased out.

3.2.2 Process verification

The quality assurance department for the user generally has the responsibility for verifying that the fabrication processes are carried out according to engineering process specification requirements. This encompasses a wide range of activities described below to control the fabrication process.

Material Control: The user process specifications must set the material control for the following items as a minimum.

1. Materials are properly identified by name and specification.
2. Materials are stored and packaged to preclude damage and contamination.
3. Perishable materials, prepregs and adhesives, are within the allowable storage life at the time of release from storage and the allowed work life at time of cure.
4. Prepackaged kits are properly identified and inspected.
5. Acceptance and reverification tests are identified.

Materials Storage and Handling: The user material and process specifications set procedures and requirements for storage of prepregs, resin systems and adhesives to maintain acceptable material quality. Storing these materials at low temperatures, usually 0 °F or below, retards the reaction of the resin materials and extends their useful life. Negotiations between the supplier and user result in an agreement on how long the supplier will guarantee the use of these perishable materials when stored under these conditions. This agreed to time is incorporated as one of the requirements in the user material specification.

Materials are generally stored in sealed plastic bags or containers to prevent moisture from condensing on the cold material and migrating into the polymer when it is removed from the freezer and allowed to warm up to ambient temperature. The time interval between material removal from the freezer and when the material bag or container may be opened is generally empirically determined. Physical characteristics such as material roll, stacking height thickness, or material type (e.g., tape vs broadgoods) are considered when determining this time interval. Therefore, the user should have procedures that prevent premature removal of materials from storage bags or containers before material temperature stabilization occurs.

Tooling: The tooling (molds) to be used for lay-up are subject to tool proofing/qualification procedures. This demonstrates that the tooling is capable of producing parts that conform to drawing and specification requirements, when used with the specified materials, lay-up and bagging methods, and cure profile. Also, cured material specimens made from the tool should be tested to ensure they meet specified mechanical and physical properties. Tool surfaces must be inspected before each use to ensure the tool surface is clean and free of conditions which could contaminate or damage a part.

Facilities and Equipment: The user will establish requirements to control the composite work area environment. These requirements are a part of the user's process specifications. The requirements should be commensurate with the susceptibility of materials to contamination by the shop environment. Inspection and calibration requirements for autoclaves and ovens must be defined.

Contamination restrictions in environmentally-controlled areas typically prohibit the use of uncontrolled sprays (e.g., silicon contamination), exposure to dust, handling contamination, fumes, oily vapors, and the presence of other particulate or chemical matter which may affect the manufacturing process. Conditions under which operators may handle materials should also be defined. Lay-up and clean room air filtrations and pressurization systems should be capable of providing a slight positive overpressure.

In-Process Control: During lay-up of composite parts, certain critical steps or operations must be closely controlled. Requirements and limits for these critical items are stated in the user process specifications. Some of the steps and operations to be controlled are listed below:

1. Verification that the release agent has been applied and cured on a clean tool surface.

2. Verification that perishable materials incorporated into the part comply with the applicable material specifications.

3. Inspection of prepreg lay-ups to assure engineering drawing requirements for number of plies and orientation are met.

4. Inspection of honeycomb core installation, if applicable, and verification that positioning meets the engineering drawing requirements.

5. The user paperwork should contain the following information.

    a. Material supplier, date of manufacturer, batch number, roll number, and total accumulated hours of working life.

    b. Autoclave or oven pressure, part temperatures, and times.

    c. Autoclave or oven load number.

    d. Part and serial number.

Part Cure: Requirements must be defined in user process specifications for the operating parameters for autoclaves and ovens used for curing parts. These include heat rise rates, times at temperature, cool-down rates, temperature and pressure tolerance, and temperature uniformity surveys in the autoclave or ovens.

Process Control Specimens: Many manufacturers require special test panels to be laid up and cured along with production parts. After cure, these panels are tested for physical and mechanical properties to verify the parts they represent meet the engineering properties.

The requirements for physical and mechanical testing are frequently defined by drawing notes which designates a type or class for each part. Non-critical or sec ondary structure may require no test specimens and no testing. Critical or safety-of-flight parts may require complete physical and mechanical testing.

During early composite material production, most users required tests for 0 ° flexure strength and modulus and short beam shear strength. However, in recent years these tests have been changed by many manufacturers to require glass transition temperature, per ply thickness, fiber volume, void content, and ply count on samples taken from designated areas on the production part.

3.2.3 Final Inspection

Having assured in-process control, the detail composite parts must also be inspected for conformance to dimensional and workmanship requirements and nondestructively inspected for processing-induced defects and damage.

Assembly Inspection: Laminates are prone to particular types of defects unless they are machined and drilled properly. Workmanship standards, required by manufacturer's process specifications, are needed to control the quality of trimmed edges and drilled holes. These standards establish visual acceptance/rejection limits for the following typical defects: splintering, delamination, loose surface fibers, overheating, surface finish, off-axis holes, and surface cratering. Typical defects in the drilling operations are delaminations and broken fibers which start at the hole boundary. Since these defects are internal in nature, an evaluation of the seriousness of the flaws is not possible by visual inspection alone. It should be backed up by nondestructive inspection techniques. Internal defect acceptance and rejection limits must be establis hed for nondestructive inspection.

3.2.4 Nondestructive Inspection

The extent of nondestructive (NDI) inspection on composite parts is dependent on whether the parts are primary structure, safety-of-flight or secondary structure, non-safety-of-flight. The type or class of part is usually defined on the engineering drawing. The engineering drawing also references a process specification which defines the NDI tests and the accept/reject criteria. The NDI tests are used to find flaws and damage such as voids, delaminations, inclusions, and micro-cracks in the matrix.

NDI techniques commonly used in production include visual, ultrasonic and X-ray inspection. Other methods, such as infrared, holographic, and acoustic inspection are being developed and may be used in production applications in the future.

Visual inspection is an NDI technique involving checks to assure the parts meet drawing requirements and to evaluate the surface and appearance of the part. The inspection includes examination for blisters, depressions, foreign material inclusions, ply distortions and folds, surface roughness, surface porosity, and wrinkles. Accept/reject criteria for such defects are given in the manufacturer's process specifications.

The most widely used nondestructive inspection technique for composites production is ultrasonic thru-transmission C-scan inspection, followed by ultrasonic pulse echo A-scan inspection. Since the subject is so broad, the engineering requirements and criteria are usually contained in a document that is referenced in the user's process specification. The principal defects evaluated by ultrasonics are internal voids, delaminations, and porosity. These inspections require fabrication of standards with built-in known defects. The output is in the form of charts which shows the sound attenuation variations over the entire part. The charts are compared to the part to show the locations of the sound attenuation variations. If defects are found outside the limits allowed by the specification, the parts are rejected and dispositioned by Engi neering. Parts may be dispositioned 1) acceptable as is, 2) subjected to further rework or r epair to make the part acceptable or 3), scrapped.

X-ray inspection is frequently used in NDI testing to evaluate bonding of inserts in laminate panels and honeycomb core to facesheet bonds in sandwich panels. The extent of testing r equired is designated on the engineering drawing by type or class of inspection. The type or class is usually defi ned in a separate document that is referenced in the manufacturer's process specification. As with ultrasonic inspection, standards with built-in defects are usually required to evaluate the radiographic film properly.

3.2.5 Specifications and documentation

The specification for materials, fabrication processes, and material testing techniques must ensure compliance with the engineering requirements.

Chapters 3, 4, and 6 in Volume 1 of this handbook describe acceptance test methods for characterizing fiber, matrix, and resin-impregnated fiber materials by their chemical, physical, and mechanical properties. Sections 3.3 and 3.4 of this volume provide information on variable statistical sampling plans that are based on MIL-STD-414 (Reference 3.2.5(a)). These plans control the frequency and extent of material property verification testing to achieve targeted quality levels.

The specifications for destructive and nondestructive test equipment and test methods should contain test and evaluation procedures. These procedures need to describe the means by which the equipment will be calibrated to maintain the required accuracy and repeatability; they should also establish the calibration frequency. Information on the standards to be used in the calibration of chemical analysis equipment will be found in preceding sections of this handbook which deal with the particular test technique.

The standards for quality control documentation requirements are found in military and federal specifications such as MIL-Q-9858A used by the Department of Defense in their hardware procurement contracts and Federal Aviation Regulation Part 21 "Certification Proc edures for Products and Parts" used by the Federal Aviation Administration production approval holders (References 3.2.5(b) and (c)).

3.2.6 Destructive tests

3.2.6.1 Background

Destructive tests are often used to ensure the structural integrity of a component whenever assurance cannot be gained by nondestructive techniques alone. These tests include periodic dissection of the part to examine the interior of complex structures and mechanical testing of coupons cut from excess parts of the component.

3.2.6.2 Usage

The goal of destructive tests is to supplement nondestructive testing to assure the structural integrity of composite parts. As more complex composite parts are used, destructive tests provide a means to examine areas not adequately inspected by other methods. Destructive tests of first articles can be used to verify structural concepts, tooling and fabrication processes. Destructive testing is necessary w hen nondestructive inspection is not sufficient to assure part quality and there is potential for undetectable manufacturing defects. For more complex composite configurations, nondestructive inspection cannot adequately inspect all features of the part. Use and selection of destructive testing is illustrated in Figure 3.2.6.2.

3.2.6.3 Destructive test approaches

There are two primary categories of destructive tests: dissection of the full part or examination of trim sections of the part. Full dissection, generally done for the first part from a new tool, gives a complete examination of the part, but is expensive to perform. Examination of excess trim sections is the preferable approach whenever possible. The part is not destroyed, structural details can still be examined and mechanical test coupons can be obtained.

Full Part Dissection: Full part dissection is the approach often envisioned when the term "destructive testing" is mentioned. Since it prevents future use of the part,full part dissection should be reserved for parts that meet the following criteria:

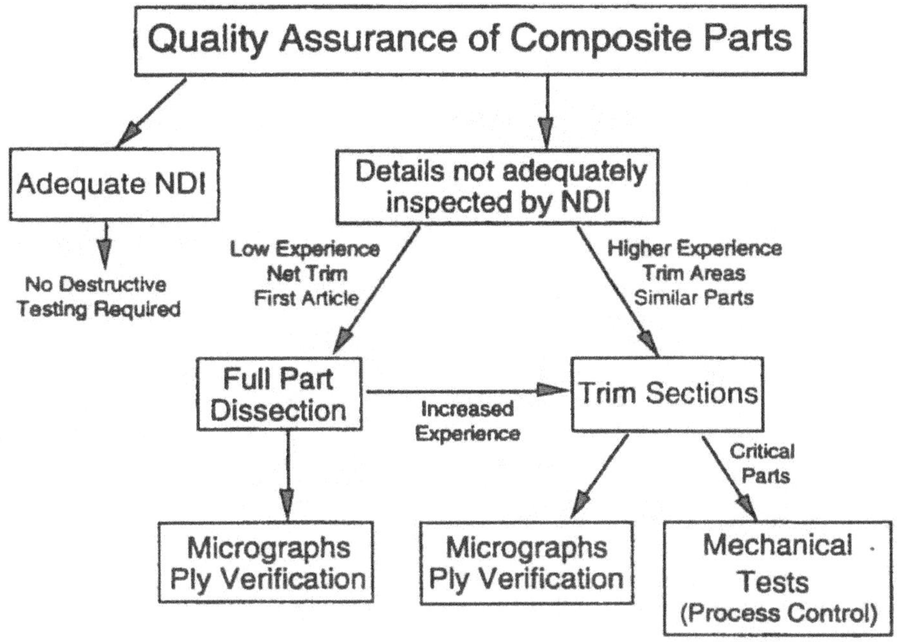

FIGURE 3.2.6.2 *Use of destructive tests.*

- Areas cannot be adequately inspected by NDI
- Part is complex *and* there is a low experience level for working with the structural configuration or fabrication process
- Part is net trim; detail areas of interest cannot be examined using excess trim areas or part extensions.

Trim Sections: Examination and testing of trim sections offers a balance of quality assurance and cost. Trim sections can be part extensions that are intentionally designed to go beyond the trim line or can be taken from cutout areas inside the part. Section cuts from detail areas can be examined for discrepancies. Test c oupons can be machined from the sections and mechanically tested to ensure the structural capability of the part and verify the quality of the fabrication process. Using coupons in this way can satisfy destructive testing requirements and process control requirements (Ref. Section 3.2.2).

3.2.6.4 Implementation guidelines

The frequency of destructive tests are dependent on part type and experience. If the producer has significant fabrication experience, complex parts may not require periodic destructive testing, but only a first article dissection. For low experience with complex parts, periodic inspection with increasing intervals may be preferable. Critical (safety of flight) parts warrant consideration for destructive testing.

Examination and testing of trim sections can be carried out on a more frequent basis and at less cost than full part dissection. Quality assurance can be enhanced by using more frequent and less elaborate trim section examinations.

Destructive tests should be conducted before the part leaves the factory. Periodic destructive tests monitor the manufacturing processes to assure the quality of parts. If a problem does occur, the periodic inspections bracket the number of suspect parts. Not every part series needs to be examined. If many parts reflect the same type of configurations and complexity, they can be pooled together for sampling purposes. Parts made on tools fabricated from one master splash can also be grouped together.

Sampling: A typical sampling plan might include first article full part dissection followed by periodic inspections employing dissection of trim sections. The periodic inspection intervals can vary depending on success rate. After a few successful destructive tests, the interval can be increased. If nonconforming areas are found in destructive tests, the inspection interval can be tightened up. If problems are found in service, additional components from the same production series can be dissected to assure that the problem was isolated.

For the trim section approach, periodic destructive tests can be conducted at smaller intervals since the cost is much less. Small intervals may be especially desirable in the case of critical parts.

For first article inspection, one of the first few articles may be chosen to represent first article. Some of the reasons for not stipulating the very first structure built are: (1) it may not be as representative of the production run because of lessons learned and special handling; and (2) another part with processing problems or discrepancies may reveal far more information.

Potential areas: Potential areas and items to examine include:

Primary load paths within the part,
Areas that showed indications from non-destructive inspection,
Tool markoff near cocured details,
Ply drop offs at a taper,
Ply wrinkles,
Resin starved and resin rich areas,
Corner radii and cocured details,
Core to face sheet fillets,
Tapered core areas.

3.2.6.5 Test types

Both full part dissection and trim sections involve examination of detail areas. After machining the detail areas, photomicrographs can be obtained to examine the microstructure. Another type of destructive testing is ply verification. Only a small section is need to perform a deply or grind down to verify that the plies are laid up in the correct stacking sequence and orientation. For machine layup, this procedure should not be necessary after initial validation. To investigate items such as ply layup, potential ply wrinkles and porosity, initial core plugs can be taken at fastener hole locations and photomicrographs can be developed.

When mechanically testing coupons that were machined from trim sections, the coupons should be tested for the critical failure mode for that part or that area of the part. Tests addressing typical failure modes are unnotched compression, open hole compression and interlaminar tension and shear.

3.3 MATERIAL PROPERTY VERIFICATION

This section is reserved for future use.

3.4 STATISTICAL PROCESS CONTROL

This section is reserved for future use.

REFERENCES

3.2.5(a) MIL-STD-414, *Sampling Procedures and Tables for Inspection by Variables for Percent Defective*, Change Notice 1, 8 May 1968.

3.2.5(b) MIL-Q-9858A, *Quality Program Requirements*

3.2.5(c) Federal Aviation Regulation Part 21 "Certification Procedures for Products and Parts"

This page intentionally left blank

CHAPTER 4 DESIGN AND ANALYSIS

Page

4.1 INTRODUCTION

The concept of designing a material to yield a desired set of properties has received impetus from the growing acceptance of composite materials. Inclusion of material design in the structural design process has had a significant effect on that process, particularly upon the preliminary design phase. In this preliminary design, a number of materials will be considered, including materials for which experimental materials property data are not available. Thus, preliminary material selection may be based on analytically-predicted properties. The analytical methods are the result of studies of micromechanics, the study of the relationship between effective properties of composites and the properties of the composite constituents. The inhomogeneous composite is represented by a homogeneous anisotropic material with the effective properties of the composite.

The purpose of this chapter is to provide an overview of techniques for analysis in the design of composite materials. Starting with the micromechanics of fiber and matrix in a lamina, analyses through simple geometric constructions in laminates are considered.

A summary is provided at the end of each section for the purpose of highlighting the most important concepts relative to the preceding subject matter. Their purpose is to reinforce the concepts, which can only fully be understood by reading the section.

The analysis in this chapter deals primarily with symmetric laminates. It begins with a description of the micromechanics of basic lamina properties and leads into classical laminate analysis theory in an arbitrary coordinate system. It defines and compares various failure theories and discusses the response of laminate structures to more complex loads. It highlights considerations of translating individual lamina results into predicted laminate behavior. Furthermore, it covers loading situations and structural responses such as buckling, creep, relaxation, fatigue, durability, and vibration.

4.2 BASIC LAMINA PROPERTIES AND MICROMECHANICS

The strength of any given laminate under a prescribed set of loads is probably best determined by conducting a test. However, when many candidate laminates and different loading conditions are being considered, as in a preliminary design study, analysis methods for estimation of laminate strength become desirable. Because the stress distribution throughout the fiber and matrix regions of all the plies of a laminate is quite complex, precise analysis methods are not available. However, reasonable methods do exist which can be used to guide the preliminary design process.

Strength analysis methods may be grouped into different classes, depending upon the degree of detail of the stresses utilized. The following classes are of practical interest:

1. Laminate level. Average values of the stress components in a laminate coordinate system are utilized.

2. Ply, or lamina, level. Average values of the stress components within each ply are utilized.

3. Constituent level. Average values of the stress components within each phase (fiber or matrix) of each ply are utilized.

4. Micro-level. Local stresses of each point within each phase are utilized.

Micro-level stresses could be used in appropriate failure criteria for each constituent to determine the external loads at which local failure would initiate. However, the uncertainties, due to departures from the assumed regular local geometry and the statistical variability of local strength make such a process impractical.

At the other extreme, laminate level stresses can be useful for translating measured strengths under single stress component tests into anticipated strength estimates for combined stress cases. However this procedure does not help in the evaluation of alternate laminates for which test data do not exist.

Ply level stresses are the commonly used approach to laminate strength. The average stresses in a given ply are used to calculate first ply failure and then subsequent ply failure leading to laminate failure. The analysis of laminates by the use of a ply-by-ply model is presented in Section 4.3 and 4.4.

Constituent level, or phase average stresses, eliminate some of the complexity of the micro-level stresses. They represent a useful approach to the strength of a unidirectional composite or ply. Micromechanics provides a method of analysis, presented in Section 4.2, for constituent level stresses. Micromechanics is the study of the relations between the properties of the constituents of a composite and the effective properties of the composite. Starting with the basic constituent properties, Sections 4.2 through 4.4 develop the micromechanical analysis of a lamina and the associated ply-by-ply analysis of a laminate.

4.2.1 Assumptions

Several assumptions have been made for characterizing lamina properties.

4.2.1.1 Material homogeneity

Composites, by definition, are heterogeneous materials. Mechanical analysis proceeds on the assumption that the material is homogeneous. This apparent conflict is resolved by considering homogeneity on microscopic and macroscopic scales. Microscopically, composite materials are certainly heterogeneous. However, on the macroscopic scale, they appear homogeneous and respond homogeneously when tested.

The analysis of composite materials uses effective properties which are based on the average stress and average strain.

4.2.1.2 Material orthotropy

Orthotropy is the condition expressed by variation of mechanical properties as a function of orientation. Lamina exhibit orthotropy as the large difference in properties between the 0° and 90° directions. If a material is orthotropic, it contains planes of symmetry and can be characterized by four independent elastic constants.

4.2.1.3 Material linearity

Some composite material properties are non-linear. The amount of non-linearity depends on the property, type of specimen, and test environment. The stress-strain curves for composite materials are frequently assumed to be linear to simplify the analysis.

4.2.1.4 Residual stresses

One consequence of the microscopic heterogeneity of a composite material is the thermal expansion mismatch between the fiber and the matrix. This mismatch causes residual strains in the lamina after curing. The corresponding residual stresses are often assumed not to affect the material's stiffness or its ability to strain uniformly.

4.2.2 Fiber composites: physical properties

A unidirectional fiber composite (UDC) consists of aligned continuous fibers which are embedded in a matrix. The UDC physical properties are functions of fiber and matrix physical properties, of their volume fractions, and perhaps also of statistical parameters associated with fiber distribution. The fibers have, in general, circular cross-sections with little variability in diameter. A UDC is clearly anisotropic since properties in the fiber direction are very different from properties transverse to the fibers.

Properties of interest for evaluating stresses and strains are:

Elastic properties
Viscoelastic properties - static and dynamic
Thermal expansion coefficients
Moisture swelling coefficients
Thermal conductivity
Moisture diffusivity

A variety of analytical procedures may be used to determine the various properties of a UDC from volume fractions and fiber and matrix properties. The derivations of these procedures may be found in References 4.2.2(a) and (b).

4.2.2.1 Elastic properties

The elastic properties of a material are a measure of its stiffness. This information is necessary to determine the deformations which are produced by loads. In a UDC, the stiffness is provided by the fibers; the role of the matrix is to prevent lateral deflections of the fibers. For engineering purposes, it is necessary to determine such properties as Young's modulus in the fiber direction, Young's modulus transverse to the fibers, shear modulus along the fibers and shear modulus in the plane transverse to the fibers, as well as various Poisson's ratios. These properties can be determined in terms of simple analytical expressions.

The effective elastic stress-strain relations of a typical transverse section of a UDC, based on average stress and average strain, have the form:

$$
\begin{aligned}
\bar{\sigma}_{11} &= n^* \bar{\epsilon}_{11} + \ell^* \bar{\epsilon}_{22} + \ell^* \bar{\epsilon}_{33} \\
\bar{\sigma}_{22} &= \ell^* \bar{\epsilon}_{11} + (k^* + G_2^*) \bar{\epsilon}_{22} + (k^* - G_2^*) \bar{\epsilon}_{33} \\
\bar{\sigma}_{33} &= \ell^* \bar{\epsilon}_{11} + (k^* - G_2^*) \bar{\epsilon}_{22} + (k^* + G_2^*) \bar{\epsilon}_{33}
\end{aligned}
\qquad 4.2.2.1(a)
$$

$$
\begin{aligned}
\bar{\sigma}_{12} &= 2G_1^* \bar{\epsilon}_{12} \\
\bar{\sigma}_{23} &= 2G_2^* \bar{\epsilon}_{23} \\
\bar{\sigma}_{13} &= 2G_1^* \bar{\epsilon}_{13}
\end{aligned}
\qquad 4.2.2.1(b)
$$

with inverse

$$
\begin{aligned}
\bar{\epsilon}_{11} &= \frac{1}{E_1^*}\bar{\sigma}_{11} - \frac{\nu_{12}^*}{E_1^*}\bar{\sigma}_{22} - \frac{\nu_{12}^*}{E_1^*}\bar{\sigma}_{33} \\
\bar{\epsilon}_{22} &= -\frac{\nu_{12}^*}{E_1^*}\bar{\sigma}_{11} + \frac{1}{E_2^*}\bar{\sigma}_{22} - \frac{\nu_{23}^*}{E_2^*}\bar{\sigma}_{33} \\
\bar{\epsilon}_{33} &= -\frac{\nu_{12}^*}{E_1^*}\bar{\sigma}_{11} - \frac{\nu_{23}^*}{E_2^*}\bar{\sigma}_{22} + \frac{1}{E_2^*}\bar{\sigma}_{33}
\end{aligned}
\qquad 4.2.2.1(c)
$$

where an asterisk (*) denotes effective values. Figure 4.2.2.1 illustrates the loadings which are associated with these properties.

The effective modulus k^* is obtained by subjecting a specimen to the average state of stress $\bar{\epsilon}_{22} = \bar{\epsilon}_{33}$ with all other strains vanishing in which case it follows from Equations 4.2.2.1(a) that

$$(\bar{\sigma}_{22} + \bar{\sigma}_{33}) = 2k^*(\bar{\epsilon}_{22} + \bar{\epsilon}_{33}) \qquad 4.2.2.1(d)$$

Unlike the other properties listed above, k^* is of little engineering significance but is of considerable analytical importance.

Only five of the properties in Equations 4.2.2.1(a-c) are independent. The most important interrelations of properties are:

$$n^* = E_1^* + 4k^* \nu_{12}^{*2} \qquad 4.2.2.1(e)$$

$$\ell^* = 2k^* \nu_{12}^* \qquad 4.2.2.1(f)$$

$$\frac{4}{E_2^*} = \frac{1}{G_2^*} + \frac{4\nu_{12}^{*2}}{E_1^*} \qquad 4.2.2.1(g)$$

$$\frac{2}{1 - \nu_{23}^*} = 1 + \frac{k^*}{\left(1 + 4k^* \dfrac{\nu_{23}^{*2}}{E_1^*}\right) G_2^*} \qquad 4.2.2.1(h)$$

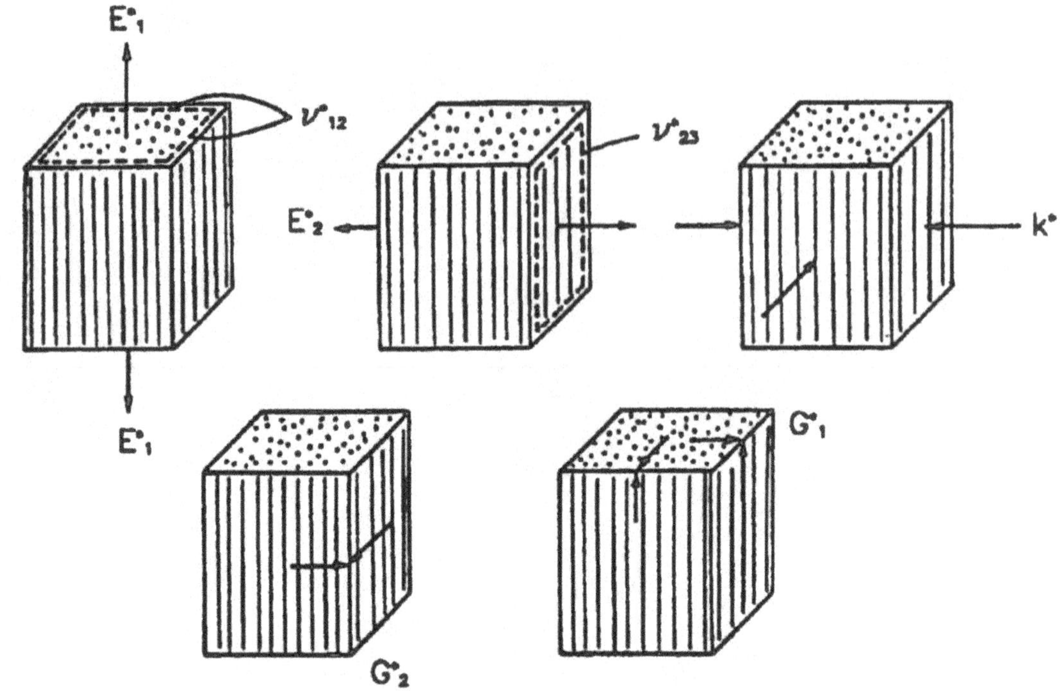

FIGURE 4.2.2.1 *Basic loading to define effective elastic properties.*

$$G_2^* = \frac{E_2^*}{2(1 + \nu_{23}^*)} \qquad 4.2.2.1(i)$$

Computation of effective elastic moduli is a very difficult problem in elasticity theory and only a few simple models permit exact analysis. One type of model consists of periodic arrays of identical circular fibers, e.g., square periodic arrays or hexagonal periodic arrays (References 4.2.2.1(a) - (c)). These models are analyzed by numerical finite difference or finite element procedures. Note that the square array is not a suitable model for the majority of UDCs since it is not transversely isotropic.

The composite cylinder assemblage (CCA) permits exact analytical determination of effective elastic moduli (Reference 4.2.2.1(d)). Consider a collection of composite cylinders, each with a circular fiber core and a concentric matrix shell. The size of the cylinders may vary but the ratio of core radius to shell radius is held constant. Therefore, the matrix and fiber volume fractions are the same in each composite cylinder. One strength of this model is the randomness of the fiber placement, while an undesirable feature is the large variation of fiber sizes. It can be shown that the latter is not a serious concern.

The analysis of the CCA gives closed form results for the effective properties, k^*, E_1^*, ν_{12}^*, n^*, ℓ^*, and G_1^* and closed bounds for the properties G_2^*, E_2^*, and ν_{23}^*. Such results will now be listed for isotropic fibers with the necessary modifications for transversely isotropic fibers (References 4.2.2(a) and 4.2.2.1(e)).

$$k^* = \frac{k_m(k_f + G_m)v_m + k_f(k_m + G_m)v_f}{(k_f + G_m)v_m + (k_m + G_m)v_f}$$

$$= k_m + \frac{v_f}{\dfrac{1}{(k_f - k_m)} + \dfrac{v_m}{(k_m + G_m)}} \qquad 4.2.2.1(j)$$

$$E_1^* = E_m v_m + E_f v_f + \frac{4(\nu_f - \nu_m)^2 v_m v_f}{\dfrac{v_m}{k_f} + \dfrac{v_f}{k_m} + \dfrac{1}{G}} \qquad 4.2.2.1(k)$$

$$\approx E_m v_m + E_f v_f$$

The last is an excellent approximation for all UDC.

$$\nu_{12}^* = \nu_m v_m + \nu_f v_f + \frac{(\nu_f - \nu_m)\left(\dfrac{1}{k_m} - \dfrac{1}{k_f}\right) v_m v_f}{\dfrac{v_m}{k_f} + \dfrac{v_f}{k_m} + \dfrac{1}{G_m}} \qquad 4.2.2.1(l)$$

$$G_1^* = G_m \frac{G_m v_m + G_f(1 + v_f)}{G_m(1 + v_f) + G_f v_m}$$

$$= G_m + \frac{v_f}{\dfrac{1}{(G_f - G_m)} + \dfrac{v_m}{2G_m}} \qquad 4.2.2.1(m)$$

As indicated earlier in the CCA analysis for G_2^* does not yield a result but only a pair of bounds which are in general quite close (References 4.2.2(a), 4.2.2.1(d,e)). A preferred alternative is to use a method of approximation which has been called the Generalized Self Consistent Scheme (GSCS). According to this method, the stress and strain in any fiber is approximated by embedding a composite cylinder in the effective fiber composite material. The volume fractions of fiber and matrix in the composite cylinder are those of the entire composite. Such an analysis has been given in Reference 4.2.2(b) and results in a quadratic equation for G_2^*. Thus,

$$A\left(\frac{G_2^*}{G_m}\right)^2 + 2B\left(\frac{G_2^*}{G_m}\right) + C = 0 \qquad 4.2.2.1(n)$$

where

$$A = 3v_f v_m^2(\gamma - 1)(\gamma + \eta_f)$$

$$+ [\gamma\eta_m + \eta_f\eta_m - (\gamma\eta_m - \eta_f)v_f^3][v_f\eta_m(\gamma - 1) - (\gamma\eta_m + 1)] \qquad 4.2.2.1(o)$$

$$B = -3v_f v_m^2(\gamma-1)(\gamma+\eta_f) + \frac{1}{2}[\gamma\eta_m + (\gamma-1)v_f + 1][(\eta_m - 1)(\gamma+\eta_f) - 2(\gamma\eta_m - \eta_m)v_f^3]$$
$$+ \frac{v_f}{2}(\eta_m + 1)(\gamma-1)[\gamma + \eta_f + (\gamma\eta_m - \eta_f)v_f^3] \qquad 4.2.2.1(p)$$

$$C = 3v_f v_m^2(\gamma-1)(\gamma+\eta_f) + [\gamma\eta_m + (\gamma-1)v_f + 1][\gamma + \eta_f + (\gamma\eta_m - \eta_f)v_f^3] \qquad 4.2.2.1(q)$$

$$\gamma = G_f/G_m \qquad 4.2.2.1(r)$$

$$\eta_m = 3 - 4v_m \qquad 4.2.2.1(s)$$

$$\eta_f = 3 - 4v_f \qquad 4.2.2.1(t)$$

To compute the resulting E_2^* and v_{23}^*, use Equations 4.2.2.1(g-h). It is of interest to note that when the GSCS approximation is applied to those properties for which CCA results are available (see above Equations 4.2.2.1(j-m)), the CCA results are retrieved.

For transversely isotropic fibers, the following modifications are necessary (References 4.2.2(a) and 4.2.2.1(e)):

For k^*	k_f is the fiber transverse bulk modulus
For E_1^*, v_{12}^*	$E_f = E_{1f}$
	$v_f = v_{1f}$
	k_f as above
For G_1^*	$G_f = G_{1f}$
For G_2^*	$G_f = G_{2f}$
	$\eta_f = 1 + 2G_{2f}/k_f$

Numerical analysis of the effective elastic properties of the hexagonal array model reveals that the values are extremely close to those predicted by the CCA/GSCS models as given by the above equations. The results are generally in good to excellent agreement with experimental data.

The simple analytical results given here predict effective elastic properties with sufficient engineering accuracy. They are of considerable practical importance for two reasons. First, they permit easy determination of effective properties for a variety of matrix properties, fiber properties, volume fractions, and environmental conditions. Secondly, they provide the only approach known today for experimental determination of carbon fiber properties.

For purposes of laminate analysis, it is important to consider the plane stress version of the effective stress-strain relations. Let x_3 be the normal to the plane of a thin unidirectionally-reinforced lamina. The plane stress condition is defined by

$$\bar{\sigma}_{33} = \bar{\sigma}_{13} = \bar{\sigma}_{23} = 0 \qquad 4.2.2.1(u)$$

Then from Equations 4.2.2.1(b-c)

$$\bar{\epsilon}_{11} = \frac{1}{E_1^*}\bar{\sigma}_{11} - \frac{\nu_{12}^*}{E_2^*}\bar{\sigma}_{22}$$

$$\bar{\epsilon}_{22} = \frac{\nu_{12}^*}{E_1^*}\bar{\sigma}_{11} + \frac{1}{E_2^*}\bar{\sigma}_{22} \qquad 4.2.2.1(v)$$

$$2\bar{\epsilon}_{12} = \frac{\bar{\sigma}_{12}}{G_1^*}$$

The inversion of Equation 4.2.2.1(v) gives

$$\bar{\sigma}_{11} = C_{11}^*\bar{\epsilon}_{11} + C_{12}^*\bar{\epsilon}_{22}$$

$$\bar{\sigma}_{22} = C_{12}^*\bar{\epsilon}_{11} + C_{22}^*\bar{\epsilon}_{22} \qquad 4.2.2.1(w)$$

$$\bar{\sigma}_{12} = 2G_2^*\bar{\epsilon}_{12}$$

where

$$C_{11}^* = \frac{E_1^*}{1 - \nu_{12}^{*2} E_2^*/E_1}$$

$$C_{12}^* = \frac{\nu_{12}^* E_2^*}{1 - \nu_{12}^{*2} E_2^*/E_1} \qquad 4.2.2.1(x)$$

$$C_{22}^* = \frac{E_2^*}{1 - \nu_{12}^{*2} E_2^*/E_1}$$

For polymer matrix composites, at the usual 60% fiber volume fraction, the square of ν_{12}^* is close enough to zero to be neglected and the ratio of E_2^*/E_1^* is approximately 0.1 - 0.2. Consequently, the following approximations are often useful.

$$C_{11}^* \simeq E_1^* \qquad C_{12}^* \simeq \nu_{12}^* E_2^* \qquad C_{22}^* \simeq E_2^* \qquad 4.2.2.1(y)$$

4.2.2.2 Viscoelastic properties

The simplest description of time-dependence is linear viscoelasticity. Viscoelastic behavior of polymers manifests itself primarily in shear and is negligible for isotropic stress and strain. This implies that the elastic stress-strain relation

$$\sigma_{11} + \sigma_{22} + \sigma_{33} = 3K(\epsilon_{11} + \epsilon_{22} + \epsilon_{33}) \qquad 4.2.2.2(a)$$

where K is the three-dimensional bulk modulus, remains valid for polymers. When a polymeric specimen is subjected to shear strain ϵ_{12}^* which does not vary with time, the stress needed to maintain this shear strain is given by

$$\sigma_{12}(t) = 2G(t)\,\epsilon_{12}^* \qquad 4.2.2.2(b)$$

and G(t) is defined as the shear relaxation modulus. When a specimen is subjected to shear stress, $\overset{\circ}{\sigma}_{12}$, constant in time, the resulting shear strain is given by

$$\epsilon_{12}(t) = \frac{1}{2} g(t)\, \overset{\circ}{\sigma}_{12} \qquad \text{4.2.2.2(c)}$$

and g(t) is defined as the shear creep compliance.

Typical variations of relaxation modulus G(t) and creep compliance g(t) with time are shown in Figure 4.2.2.2. These material properties change significantly with temperature. The relaxation modulus decreases with increasing temperature and the creep compliance increases with increasing temperature, which implies that the stiffness decreases as the temperature increases. The initial value of these properties at "time-zero" are denoted G_o and g_o and are the elastic properties of the matrix. If the applied shear strain is an arbitrary function of time, commencing at time-zero, Equation 4.2.2.2(b) is replaced by

$$\sigma_{12}(t) = 2G(t)\, \epsilon_{12}(0) + 2 \int_o^t G(t-t') \frac{d\epsilon_{12}}{dt'}\, dt' \qquad \text{4.2.2.2(d)}$$

Similarly, for an applied shear stress which is a function of time, Equation 4.2.2.2(c) is replaced by

$$\epsilon_{12}(t) = \frac{1}{2} g(t)\, \sigma_{12}(0) + \frac{1}{2} \int_o^t g(t-t') \frac{d\sigma_{12}}{dt'}\, dt' \qquad \text{4.2.2.2(e)}$$

The viscoelastic counterpart of Young's modulus is obtained by subjecting a cylindrical specimen to axial strain $\overset{\circ}{\epsilon}_{11}$ constant in space and time. Then

$$\sigma_{11}(t) = E(t)\, \overset{\circ}{\epsilon}_{11} \qquad \text{4.2.2.2(f)}$$

and E(t) is the Young's relaxation modulus. If the specimen is subjected to axial stress, $\overset{\circ}{\sigma}_{11}$, constant is space and time, then

$$\epsilon_{11}(t) = e(t)\, \overset{\circ}{\sigma}_{11} \qquad \text{4.2.2.2(g)}$$

and e(t) is Young's creep compliance. Obviously E(t) is related to K and G(t), and e(t) is related to k and g(t). (See Reference 4.2.2.2(a).)

The basic problem is the evaluation of the effective viscoelastic properties of a UDC in terms of matrix viscoelastic properties and the elastic properties of the fibers. (It is assumed that the fibers themselves do not exhibit any time-dependent properties.) This problem has been resolved in general fashion in References 4.2.2.2(b) and (c). Detailed analysis shows that the viscoelastic effect in a UDC is significant only for axial shear, transverse shear, and transverse uniaxial stress.

For any of average strains $\bar{\epsilon}_{22}$, $\bar{\epsilon}_{23}$, and $\bar{\epsilon}_{12}$ constant in time, the time-dependent stress response will be

$$\begin{aligned} \bar{\sigma}_{22}(t) &= E_2^*(t)\, \bar{\epsilon}_{22} \\ \bar{\sigma}_{23}(t) &= 2G_2^*(t)\, \bar{\epsilon}_{23} \\ \bar{\sigma}_{12}(t) &= 2G_1^*(t)\, \bar{\epsilon}_{12} \end{aligned} \qquad \text{4.2.2.2(h)}$$

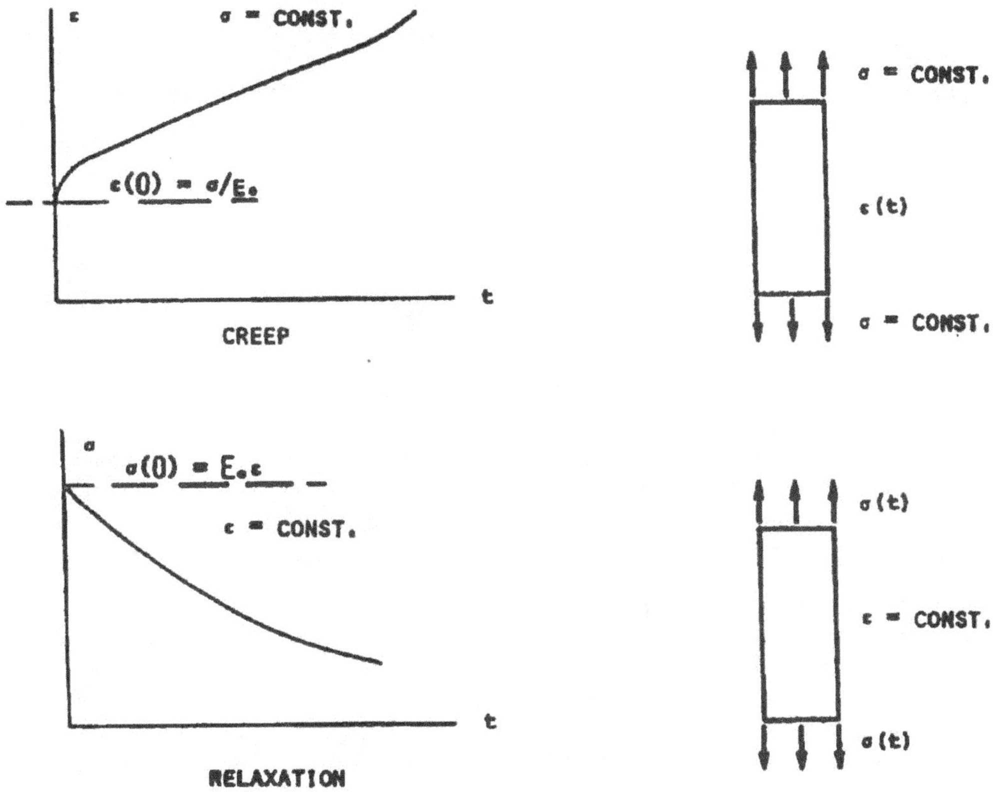

FIGURE 4.2.2.2 *Typical viscoelastic behavior.*

For any of stresses $\bar{\sigma}_{22}$, $\bar{\sigma}_{23}$, and $\bar{\sigma}_{12}$ constant in time, the time-dependent strain response will be

$$\begin{aligned} \bar{\varepsilon}_{22}(t) &= e_2^*(t)\,\bar{\sigma}_{22} \\ \bar{\varepsilon}_{23}(t) &= \frac{1}{2}\,g_2^*(t)\,\bar{\sigma}_{23} \\ \bar{\varepsilon}_{12}(t) &= \frac{1}{2}\,g_1^*(t)\,\bar{\sigma}_{12} \end{aligned} \qquad 4.2.2.2(i)$$

where material properties in Equations 4.2.2.2(h) are effective relaxation moduli and the properties in Equations 4.2.2.2(i) are effective creep functions. All other effective properties may be considered elastic. This implies in particular that if a fiber composite is subjected to stress $\bar{\sigma}_{11}(t)$ in the fiber direction, then

$$\begin{aligned} \bar{\sigma}_{11}(t) &\sim E_1^*\,\bar{\varepsilon}_{11}(t) \\ \bar{\varepsilon}_{22}(t) = \bar{\varepsilon}_{33}(t) &\sim \nu_{12}^*\,\bar{\varepsilon}_{11}(t) \end{aligned} \qquad 4.2.2.2(j)$$

where E_1^* and ν_{12}^* are the elastic results of Equations 4.2.2.2(k) with matrix properties taken as initial (elastic) matrix properties. Similar considerations apply to the relaxation modulus k^*.

The simplest case of the viscoelastic properties entering into Equations 4.2.2.2(h-i) is the relaxation modulus $G_1^*(t)$ and its associated creep compliance $g_1^*(t)$. A very simple result has been obtained for fibers which are infinitely more rigid than the matrix (Reference 4.2.2(a)). For a viscoelastic matrix, the results reduce to

$$G_1^*(t) = G_m(t)\,\frac{1+v_f}{1-v_f}$$

$$g_1^*(t) = g_m(t)\,\frac{1-v_f}{1+v_f} \qquad 4.2.2.2(k)$$

This results in an acceptable approximation for glass fibers in a polymeric matrix and an excellent approximation for boron fibers in a polymeric matrix. However, the result is not applicable to the case of carbon or graphite fibers in a polymeric matrix since the axial shear modulus of these fibers is not large enough relative to the matrix shear modulus. In this case, it is necessary to use the correspondence principle mentioned above (References 4.2.2(a) and 4.2.2.2(b)). The situation for transverse shear is more complicated and involves complex Laplace transform inversion. (Reference 4.2.2.2(c)).

All polymeric matrix viscoelastic properties such as creep and relaxation functions are significantly temperature dependent. If the temperature is known, all of the results from this section can be obtained for a constant temperature by using the matrix properties at that temperature. At elevated temperatures, the viscoelastic behavior of the matrix may become nonlinear. In this event, the UDC will also be nonlinearly viscoelastic and all of the results given here are not valid. The problem of analytical determination of nonlinear properties is, of course, much more difficult than the linear problem (See Reference 4.2.2.2(d)).

4.2.2.3 Thermal expansion and moisture swelling

The elastic behavior of composite materials discussed in Section 4.2.2.1 is concerned with externally applied loads and deformations. Deformations are also produced by temperature changes and by absorption of moisture in two similar phenomena. A change of temperature in a free body produces thermal strains while moisture absorption produces swelling strains. The relevant physical parameters to quantify these phenomena are thermal expansion coefficients and swelling coefficients.

Fibers have significantly smaller thermal expansion coefficients than do polymeric matrices. The expansion coefficient of glass fibers is 2.8×10^{-6} in/in/F° (5.0×10^{-6} m/m/C°) while a typical epoxy value is 30×10^{-6} in/in/F° (54×10^{-6} m/m/C°). Carbon and graphite fibers are anisotropic in thermal expansion. The expansion coefficients in the fiber direction are extremely small, either positive or negative of the order of 0.5×10^{-6} in/in/F° (0.9×10^{-6} m/m/C°). To compute these stresses, it is necessary to know the thermal expansion coefficients of the layers. Procedures to determine these coefficients in terms of the elastic properties and expansion coefficients of component fibers and matrix are discussed in this section.

When a laminate absorbs moisture, there occurs the same phenomenon as in the case of heating. Again, the swelling coefficient of the fibers is much smaller than that of the matrix. Free swelling of the layers cannot take place and consequently internal stresses develop. These stresses can be calculated if the UDC swelling coefficients are known.

Consider a free cylindrical specimen of UDC under uniform temperature change ΔT. Neglecting transient thermal effects, the stress-strain relations (Equation 4.2.2.1(c)) assume the form

$$\bar{\epsilon}_{11} = \frac{1}{E_1^*}\sigma_{11} - \frac{\nu_{12}^*}{E_1^*}\bar{\sigma}_{22} - \frac{\nu_{12}^*}{E_1^*}\bar{\sigma}_{33} + \alpha_1^* \Delta T$$

$$\bar{\epsilon}_{22} = -\frac{\nu_{12}^*}{E_1^*}\bar{\sigma}_{11} + \frac{1}{E_2^*}\bar{\sigma}_{22} - \frac{\nu_{23}^*}{E_2^*}\sigma_{33} + \alpha_2^* \Delta T \qquad \text{4.2.2.3(a)}$$

$$\bar{\epsilon}_{33} = -\frac{\nu_{12}^*}{E_1^*}\bar{\sigma}_{11} - \frac{\nu_{23}^*}{E_2^*}\bar{\sigma}_{22} + \frac{1}{E_2^*}\bar{\sigma}_{33} + \alpha_2^* \Delta T$$

where

α_1^* - effective axial expansion coefficient

α_2^* - effective transverse expansion coefficient

It has been shown by Levin (Reference 4.2.2.3(a)) that there is a unique mathematical relationship between the effective thermal expansion coefficients and the effective elastic properties of a two-phase composite. When the matrix and fibers are isotropic

$$\alpha_1^* = \alpha_m + \frac{\alpha_f - \alpha_m}{\dfrac{1}{K_f} - \dfrac{1}{K_m}}\left[\frac{3(1-2\nu_{12}^*)}{E_1^*} - \frac{1}{K_m}\right]$$

$$\alpha_2^* = \alpha_m + \frac{\alpha_f - \alpha_m}{\dfrac{1}{K_f} - \dfrac{1}{K_m}}\left[\frac{3}{2k^*} - \frac{3(1-2\nu_{12}^*)}{E_1^*} - \frac{1}{K_m}\right] \qquad \text{4.2.2.3(b)}$$

where

α_m, α_f - matrix, fiber isotropic expansion coefficients

K_m, K_f - matrix, fiber three-dimensional bulk modulus

E_1^*, ν_{12}^*, k^* - effective axial Young's modulus, axial Poisson's ratio, and transverse bulk modulus

These equations are suitable for glass/epoxy and boron/epoxy. They have also been derived in References 4.2.2.3(b) and (c). For carbon and graphite fibers, it is necessary to consider the case of transversely isotropic fibers. This complicates the results considerably as shown in Reference 4.2.2.1(c) and (e).

Frequently thermal expansion coefficients of the fibers and matrix are functions of temperature. It is not difficult to show that Equations 4.2.2.3(b) remain valid for temperature-dependent properties if the elastic properties are taken at the final temperature and the expansion coefficients are taken as secant at that temperature.

To evaluate the thermal expansion coefficients from Equation 4.2.2.3(b) or (c), the effective elastic properties, k^*, E_1^*, and ν_{12}^* must be known. These may be taken as the values predicted by Equations 4.2.2.1(j-l) with the appropriate modification when the fibers are transversely isotropic. Figures 4.2.2.3(a) and (b) shows typical plots of the effective thermal expansion coefficients of graphite/epoxy.

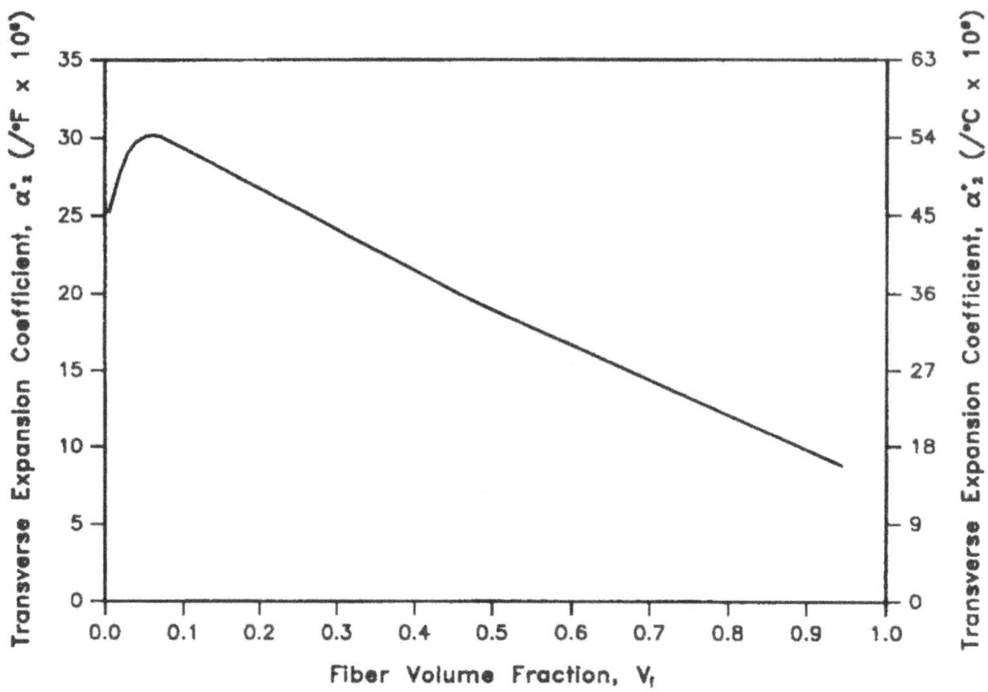

FIGURE 4.2.2.3(a) *Effect of fiber volume on thermal expansion for representative carbon/epoxy composite.* E_{1f} *= 50 Msi (340 GPa).*

When a composite with polymeric matrix is placed in a wet environment, the matrix will begin to absorb moisture. The moisture absorption of most fibers used in practice is negligible; however, aramid fibers alone absorb significant amounts of moisture when exposed to high humidity. The total moisture absorbed by an aramid/epoxy composite, however, may not be substantially greater than other epoxy composites.

When a composite has been exposed to moisture and sufficient time has elapsed, the moisture concentration throughout the matrix will be uniform and the same as the boundary concentration. It is customary to define the specific moisture concentration c by

$$c = C/\rho \qquad 4.2.2.3(c)$$

where ρ is the density. The swelling strains due to moisture are functions of Δc and the swelling coefficients, β_{ij}

$$\epsilon_{ij} = \beta_{ij}\,\Delta c \qquad 4.2.2.3(d)$$

If there are also mechanical stresses and strains, then the swelling strains are superposed on the latter. This is exactly analogous to the thermoelastic stress-strain relations of an isotropic material. The effective swelling coefficients β^*_{ij} are defined by the average strains produced in a free sample subjected to a uniform unit change of specific moisture concentration in the matrix. For discussions of other aspects of moisture absorption, both transient and steady state, see References 4.2.2.3(d) and (e).

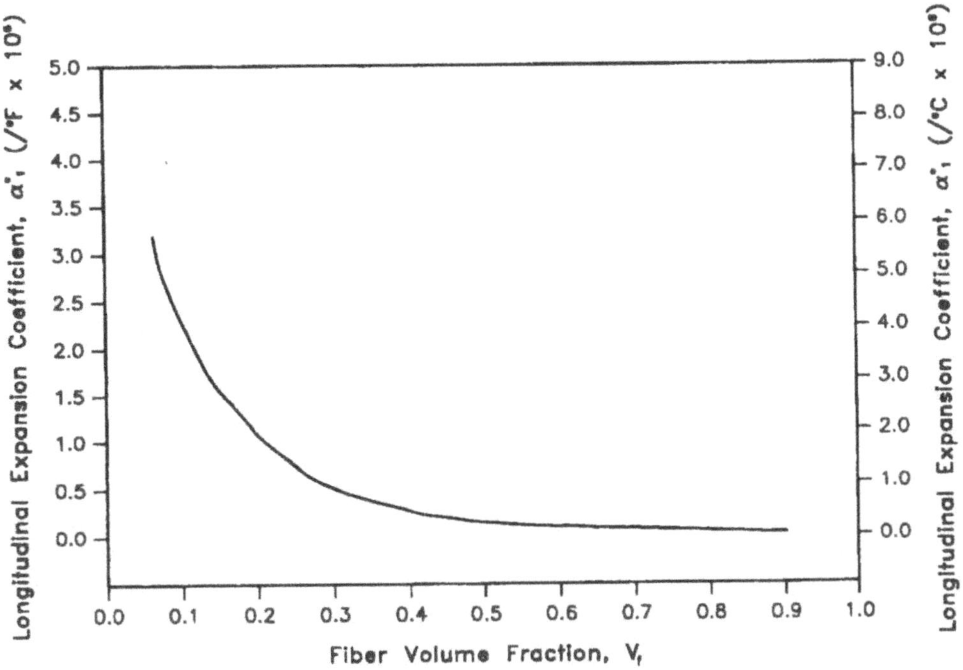

Figure 4.2.2.3(b) *Effect of fiber volume on thermal expansion for representative carbon/epoxy composite.* E_{1f} *= 50 Msi (340 GPa).*

Finally, simultaneous moisture swelling and thermal expansion, or hygrothermal behavior can be considered. The simplest approach is to assume that the thermal expansion strains and the moisture swelling strains can be superposed. For a free specimen,

$$\bar{\epsilon}_{11} = \alpha_1^* \Delta T + \beta_1^* \Delta c$$
$$\bar{\epsilon}_{22} = \bar{\epsilon}_{33} = \alpha_2^* \Delta T + \beta_2^* \Delta c \qquad 4.2.2.3(e)$$

In this event, the matrix elastic properties in Equations 4.2.2.3(a) and (b) may be functions of the final temperature and moisture concentration. This dependence must be known to evaluate α_1^*, α_2^*, β_1^*, and β_2^* in Equation 4.2.2.3(e).

4.2.2.4 Thermal conduction and moisture diffusion

The thermal conduction analysis has many similarities with the analyses for moisture diffusion, as well as electrical conduction, and dielectric and magnetic properties. Since these conductivity problems are governed by similar equations, the results can be applied to each of these areas.

Let T(x) be a steady state temperature field in a homogeneous body. The temperature gradient is given by

$$H_i = \frac{\partial T}{\partial x_i} \qquad 4.2.2.4(a)$$

and the heat flux vector by

$$D_i = \mu_{ij} H_j \qquad 4.2.2.4(b)$$

where μ_{ij} is the conductivity tensor. It may be shown (Reference 4.2.2(a)) that for isotropic matrix and fibers, the axial conductivity μ_1^* is given by

$$\mu_1^* = \mu_m v_m + \mu_f v_f \qquad 4.2.2.4(c)$$

and for transversely isotropic fibers

$$\mu_1^* = \mu_m v_m + \mu_{Lf} v_f \qquad 4.2.2.4(d)$$

where μ_{1f} is the longitudinal conductivity of the fibers. The results of Equations 4.2.2.4(c) and (d) are valid for any fiber distribution and any fiber cross-section.

The problem of transverse conductivity is mathematically analogous to the problem of longitudinal shearing (Reference 4.2.2(a)). All results for the effective longitudinal shear modulus G_1^* can be interpreted as results for transverse effective conductivity μ_2^*. In particular, for the composite cylinder assemblage model

$$\mu_2^* = \mu_m \left[\frac{\mu_m v_m + \mu_f (1 + v_f)}{\mu_m (1 + v_f) + \mu_f v_m} \right] = \mu_m + \frac{v_f}{\dfrac{1}{(\mu_f - \mu_m)} + \dfrac{v_m}{2\mu_m}} \qquad 4.2.2.4(e)$$

These results are for isotropic fibers. For carbon and graphite fibers μ_f should be replaced by the transverse conductivity μ_{2f} of the fibers (Reference 4.2.2.1(e)). As in the elastic case, there is reason to believe that Equation 4.2.2.4(e) accurately represents all cases of circular fibers which are randomly distributed and not in contact. Again the hexagonal array numerical analysis results coincide with the number predicted by Equation 4.2.2.4(e).

To interpret the results for the case of moisture diffusivity, the quantity μ_m is interpreted as the diffusivity of the matrix. Since moisture absorption of fibers is negligible, μ_f is set equal to zero. The results are then

$$\mu_1^* = \mu_m v_m$$

$$\mu_2^* = \mu_m \frac{v_m}{1 + v_f} \qquad 4.2.2.4(f)$$

These equations describe the moisture diffusivity of a composite material.

4.2.3 Fiber composites: strength and failure

The mathematical treatment of the relationships between the strength of a composite and the properties of its constituents is considerably less developed than the analysis for the other physical property relationships discussed in Section 4.2.2. Failure is likely to initiate in a local region due to the influence of the local values of constituent properties and the geometry in that region. This dependence upon local characteristics of high variability makes the analysis of the composite failure mechanisms much more complex than the analyses of the physical properties previously discussed.

Because of the complexity of the failure process, it may be desirable to regard the strength of a unidirectional fiber composite subjected to a single principal stress component as a quantity to be measured experimentally, rather than deduced from constituent properties. Such an approach may well be the practical one for fatigue failure of these composites. Indeed, the issue of determining the degree to which heterogeneity should be considered in the analysis of composite strength and failure is a matter for which there exists a considerable degree of difference of opinion. At the level of unidirectional composites, it is well to examine the effects upon failure of the individual constituents to develop an understanding of the nature of the possible failure mechanisms. This subject is discussed in the following sections. The general issue of the approach to failure analysis is treated further in laminate strength and failure.

The strength of a fiber composite clearly depends upon the orientation of the applied load with respect to the direction in which the fibers are oriented as well as upon whether the applied load is tensile or compressive. The following sections present a discussion of failure mechanisms and composite-constituent property relations for each of the principal loading conditions.

4.2.3.1 Axial tensile strength

One of the most attractive properties of advanced fiber composites is high tensile strength. The simplest model for the tensile failure of a unidirectional fiber composite subjected to a tensile load in the fiber direction is based upon the elasticity solution of uniform axial strain throughout the composite. Generally, the fibers have a lower strain to failure than the matrix, and composite fracture occurs at the failure strain of the fibers alone. This results in a composite tensile strength, F_1^{tu}, given by:

$$F_1^{tu} = v_f F_f^{tu} + v_m \sigma_m^t \qquad 4.2.3.1$$

where F_f^{tu} - the fiber tensile strength

σ_m^t - the stress in the matrix at a strain equal to the fiber failure strain

The problem with this approach is the variability of the fiber strength. Non-uniform strength is characteristic of most current high-strength fibers. There are two important consequences of a wide distribution of individual fiber strengths. First, all fibers will not be stressed to their maximum value simultaneously. Secondly, those fibers which break earliest during the loading process will cause perturbations of the stress field near the break, resulting in localized high fiber-matrix interface shear stresses. These shear stresses transfer the load across the interface and also introduce stress concentrations into adjacent unbroken fibers.

The stress distribution at each local fiber break may cause several possible failure events to occur. The shear stresses may cause a crack to progress along the interface. If the interface is weak, such propagation can be extensive. In this case, the strength of the composite material may differ only slightly from that of a bundle of unbonded fibers. This undesirable mode of failure can be prevented by a strong fiber-matrix interface or by a soft ductile matrix which permits the redistribution of the high shear stresses. When the bond strength is high enough to prevent interface failure, the local stress concentrations may cause the fiber break to propagate through the matrix, to and through adjacent fibers. Alternatively, the stress concentration in adjacent fibers may cause one or more of such fibers to break before failure of the intermediate matrix. If such a crack or such fiber breaks continue to propagate, the strength of the composite may be no greater than that of the weakest fiber. This failure mode is defined as a weakest link failure. If the matrix and interface properties are of sufficient strength and toughness to prevent or arrest these failure mechanisms, then continued load increases will produce new fiber failures at other locations in the material. An accumulation of dispersed internal damage results.

It can be expected that all of these effects will occur before material failure. That is, local fractures will propagate for some distance along the fibers and normal to the fibers. These fractures will initiate and grow at various points within the composite. Increasing the load will produce a statistical accumulation of dispersed damage regions until a sufficient number of such regions interact to provide a weak surface, resulting in composite tensile failure.

4.2.3.1.1 Weakest link failure

The weakest link failure model assumes that a catastrophic mode of failure is produced with the occurrence of one, or a small number of, isolated fiber breaks. The lowest stress at which this type of failure can occur is the stress at which the first fiber will break. The expressions for the expected value of the weakest element in a statistical population (e.g., Reference 4.2.3.1.1(a)) have been applied by Zweben (Reference 4.2.3.1.1(b)) to determine the expected stress at which the first fiber will break. For practical materials in realistic structures, the calculated weakest link failure stress is quite small and, in general, failure cannot be expected in this mode.

4.2.3.1.2 Cumulative weakening failure

If the weakest link failure mode does not occur, it is possible to continue loading the composite. With increasing stress, fibers will continue to break randomly throughout the material. When a fiber breaks, there is a redistribution of stress near the fracture site. The treatment of a fiber as a chain of links is appropriate to the hypothesis that fracture is due to local imperfections. The links may be considered to have a statistical strength distribution which is equivalent to the statistical flaw distribution along the fibers. Additional details for this model are given in References 4.2.3.1.1(a) and 4.2.3.1.2. The cumulative weakening model does not consider the overstress on adjacent fibers or the effect of adjacent laminae.

4.2.3.1.3 Fiber break propagation failure

The effects of stress perturbations on fibers adjacent to broken fibers are significant. The load concentration in the fibers adjacent to a broken fiber increases the probability that a second fiber will break. Such an event will increase the probability of additional fiber breaks, and so on. The fiber break propagation mode of failure was studied by Zweben (Reference 4.2.3.1.1(b)). The occurrence of the first fracture of an overstressed fiber was proposed as a measure of the tendency for fiber breaks to propagate, and, hence, as a failure criterion for this mode. Although the first multiple break criterion may provide good correlations with experimental data for small volumes of material, it gives very low failure stress predictions for large volumes of material. Additional work in this area can be found in References 4.2.3.1.3(a) and (b).

4.2.3.1.4 Cumulative group mode failure

As multiple broken fiber groups grow, the magnitude of the local axial shear stress increases and axial cracking can occur. The cumulative group mode failure model (Reference 4.2.3.1.4) includes the effects of the variability of fiber strength, load concentrations in fibers adjacent to broken fibers, and matrix shear failure or interfacial debonding which will serve to arrest the propagating cracks. As the stress level increases from that at which fiber breaks are initiated to that at which the composite fails, the material will have distributed groups of broken fibers. This situation may be considered as a generalization of the cumulative weakening model. In practical terms, the complexity of this model limits its use.

Each of these models has severe limitations for the quantitative prediction of tensile strength. However, the models show the importance of variability of fiber strength and matrix stress-strain characteristics upon composite tensile strength.

4.2.3.2 Axial compressive strength

Both strength and stability failures must be considered for compressive loads applied parallel to the fibers of a unidirectional composite. Microbuckling is one proposed failure mechanism for axial compression (Reference 4.2.3.2(a)). Small wave-length micro-instability of the fibers occurs in a manner analogous to the buckling of a beam on an elastic foundation. It can be demonstrated that this instability can occur even for a brittle material such as glass. Analyses of this instability were performed independently in References 4.2.3.2(b) and (c). The energy method for evaluation of the buckling stress has been used for these modes. This procedure considers the composite as stressed to the buckling load. The strain energy in this compressed but straight pattern (extension mode) is then compared to an assumed buckling deformation pattern (shear mode) under the same load. The change in strain energy in the fiber and the matrix can be compared to the change in potential energy associated with the shortening of the distance between the applied loads at the ends of the fiber. The condition for instability is given by equating the strain energy change to the work done by the external loads during buckling.

The results for the compressive strength, F_1^{cu}, for the extension mode is given by

$$F_1^{cu} = 2v_f\sqrt{\frac{v_f\,E_m E_f}{3(1-v_f)}} \qquad \text{4.2.3.2(a)}$$

The result for the shear mode is

$$F_1^{cu} = \frac{G_m}{1-v_f} \qquad \text{4.2.3.2(b)}$$

The compressive strength of the composite is plotted as a function of the fiber volume fraction, v_f, in Figure 4.2.3.2 for E-glass fibers embedded in an epoxy matrix. The compressive strength of glass-reinforced plastic, with a fiber volume fraction of 0.6 to 0.7, is on the order of 460 to 600 ksi (3100 to 4100 MPa). Values of this magnitude do not appear to have been measured for any realistic specimens. However, the achievement of a strength of half a million psi in a composite of this type would require an average shortening greater than 5%. For the epoxy materials used in this calculation, such a shortening would result in a decrease in the effective shear stiffness of the matrix material since the proportional limit of the matrix would be exceeded. Hence, it is necessary to modify the analysis to consider the inelastic deformation of the matrix. As a first approximation, the matrix modulus in Equations 4.2.3.2(a) and (b) can be replaced by a reduced modulus. A more general result can be obtained by modeling the matrix as an elastic, perfectly plastic material. For this matrix, the secant value at each axial strain value may be assumed to govern the instability. These assumptions (Reference 4.2.3.2(d)) yield the following result for the shear mode:

$$F_1^{cu} = \sqrt{\frac{v_f E_f F^{cpl}}{3(1-v_f)}} \qquad \text{4.2.3.2(c)}$$

where F^{cpl} is the matrix yield stress level.

For the generally dominant shear mode, the elastic results of Equation 4.2.3.2(b) are independent of the fiber modulus, yet the compressive strength of boron/epoxy is much greater than that of glass/epoxy composites. One hypothesis to explain this discrepancy, is that use of the stiffer boron fibers yields lower matrix strains and less of a strength reduction due to inelastic effects. Thus, the results of Equation 4.2.3.2(c) show a ratio of $\sqrt{6}$ or 2.4 for the relative strengths of boron compared to glass fibers in the same matrix.

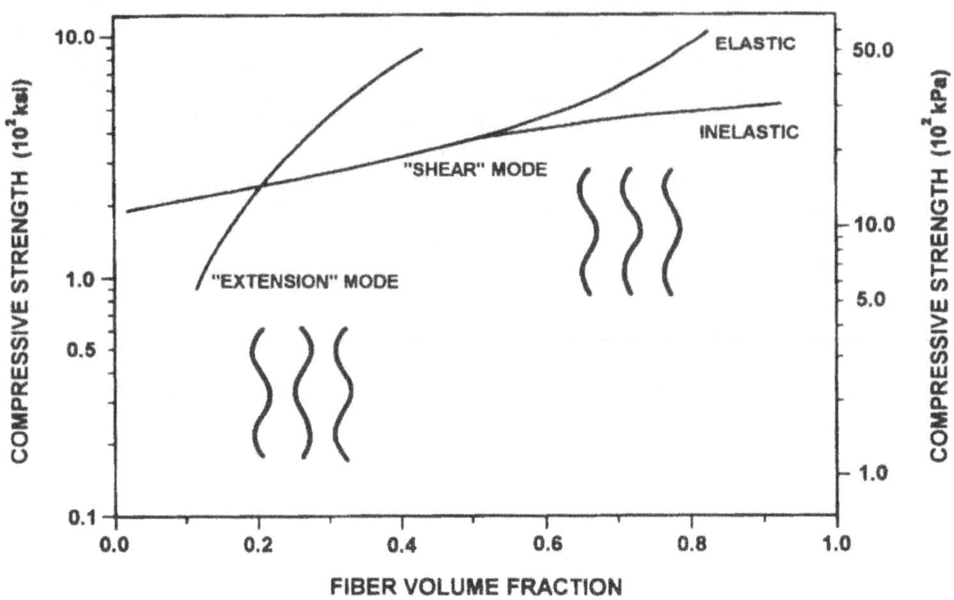

FIGURE 4.2.3.2 *Compressive strength of glass-reinforced epoxy composite.*

All of the analytical results above indicate that compressive strength is independent of fiber diameter. Yet different diameter fibers may yield different compressive strengths for composites because large diameter fibers such as boron (0.005 inch, 0.13 mm D) are better collimated than small diameter fibers, such as glass (0.0004 in, 0.010 mm D). For small diameter fibers, such as aramid and carbon, local out-of-straightness can introduce matrix shear stresses, cause fiber debonding, and produce lower instability stress levels (References 4.2.3.2(e) and 4.2.2.1(d)). Carbon and aramid fibers are anisotropic and have extremely low axial shear moduli. As a result, the elastic buckling stress in the shear mode is reduced to:

$$F^{ccr} = \frac{G_m}{1-v_f(1-G_m/G_{1f})} \qquad 4.2.3.2(d)$$

where G_{1f} is the fiber longitudinal shear modulus (Reference 4.2.3.2(e)). For high fiber shear moduli, this equation reduces to Equation 4.2.3.2(b).

Another failure mechanism for oriented polymeric fibers such as aramid fibers (Reference 4.2.3.2(e)) is a kink-band formation at a specific angle to the direction of compressive stress. The formation of kink-bands is attributed to the fibrillar structure of the highly anisotropic fiber and poor fiber shear strength. Breakup of the fiber into very small diameter fibrils results in degradation of shear stiffness and hence the compressive strength.

The results of the compressive strength analyses indicate that for the elastic case, the matrix Young's modulus is the dominant parameter. For the inelastic case, however, there are strength limitations which depend both upon the fiber modulus and upon the matrix strength. For some materials, performance is limited by a matrix yield strength at a given fiber modulus. For other materials, a gain in compressive strength can be achieved by improving the matrix modulus.

4.2.3.3 Matrix mode strength

The remaining failure modes of interest are transverse tension and compression and axial shear. For each of these loading conditions, material failure can occur without fracture of the fibers, hence the terminology "matrix-dominated" or "matrix modes of failure". Micromechanical analyses of these failure modes are complex because the critical stress states are in the matrix, are highly non-uniform, and are very dependent upon the local geometry. As a result, it appears that the most fruitful approaches will be those that consider average states of stress.

There are two types of shearing stresses which are of interest for these matrix-dominated failures: (1) in a plane which contains the filaments, and (2) in a plane normal to the filaments. In the first case, the filaments provide very little reinforcement to the composite and the shear strength depends on the shear strength of the matrix material. In the second case, some reinforcement may occur; at high volume fractions of filaments, the reinforcement may be substantial. It is important to recognize that filaments provide little resistance to shear in any surfaces parallel to them.

The approach to shear failure analysis is to consider that a uniaxial fibrous composite is comprised of elastic-brittle fibers embedded in an elastic-perfectly plastic matrix. For the composite, the theorems of limit analysis of plasticity (e.g., References 4.2.3.3(a) and (b)) may be used to obtain upper and lower bounds for a composite limit load (Reference 4.2.3.3(c)). The limit load is defined as the load at which the matrix yield stress permits composite deformation to increase with no increase in load. The failure strength of a ductile matrix may be approximated by this limit load.

4.2.4 Strength under combined stress

It is possible to apply the micro-mechanical models for failure described above, to combined stresses in the principal directions. Little work of this type has been done however. Generally the strengths in principal directions have been used in a failure surface for a homogeneous, anisotropic material for estimation of strength under combined loads. The understanding of failure mechanisms resulting from the above micro-mechanical models can be used to define the general form of failure surface to be utilized. This approach is outlined in the following sections.

Knowledge of the different failure mechanisms and quantitative experimental data for a UDC under single stress components can be used to formulate practical failure criteria for combined stresses. Plane stress failure criteria are discussed below with references also given for more complicated stress systems. The stresses considered are averaged over a representative volume element. The fundamental assumption is that there exists a failure criterion of the form:

$$F(\sigma_{11}, \sigma_{22}, \sigma_{12}) = 1 \qquad \text{4.2.4(a)}$$

which characterizes the failure of the UDC. The usual approach to construction of a failure criterion is to assume a quadratic form in terms of stress or strain since the quadratic form is the simplest form which can adequately describe the experimental data. The various failure criteria which have been proposed all use coefficients based on experimental information such as ultimate stresses under single load components (References 4.2.4(a) - (d)). For example, the general quadratic version of Equation 4.2.4(a) for plane stress would be:

$$\begin{aligned} &C_{11}\sigma_{11}^2 + C_{22}\sigma_{22}^2 + C_{66}\sigma_{12}^2 + 2C_{12}\sigma_{11}\sigma_{22} + 2C_{16}\sigma_{11}\sigma_{12} \\ &\quad + 2C_{26}\sigma_{22}\sigma_{12} + C_1\sigma_{11} + C_2\sigma_{22} + C_6\sigma_{12} = 1 \end{aligned} \qquad \text{4.2.4(b)}$$

The material has different strengths in uniaxial, longitudinal, and transverse tension and compression. Evidently the shear strength is not affected by the sign of the shear stress. It follows that all powers of shear stress in the failure criterion must be even. Consequently, the criterion simplifies to

$$C_{11}\sigma_{11}^2 + C_{22}\sigma_{22}^2 + C_{66}\sigma_{12}^2 + 2C_{12}\sigma_{11}\sigma_{22} + C_1\sigma_{11} + C_2\sigma_{22} = 1 \qquad 4.2.4(c)$$

The ultimate stresses under single component stress conditions for each of σ_{11}, σ_{22}, and σ_{12} determine the constants for the failure criterion.

$$C_{11} = \frac{1}{F_1^{tu} F_1^{cu}} \qquad C_{22} = \frac{1}{F_2^{tu} F_2^{cu}}$$

$$C_1 = \frac{1}{F_1^{tu}} - \frac{1}{F_1^{cu}} \qquad C_2 = \frac{1}{F_2^{tu}} - \frac{1}{F_2^{cu}} \qquad 4.2.4(d)$$

$$C_{66} = \frac{1}{(F_1^{su})^2}$$

However, C_{12} cannot be determined from the single component ultimate stresses. Biaxial stress tests must be performed to determine this coefficient. Frequently, the coefficient is established by relating Equation 4.2.4(c) to the Mises-Henky yield criterion for isotropic materials, yielding

$$C_{12} = -\frac{1}{2}(C_{11}C_{22})^{1/2} \qquad 4.2.4(e)$$

The above failure criterion is the two-dimensional version of the Tsai-Wu criterion (Reference 4.2.4(c)). Its implementation raises several problems; the most severe of these is that the failure criterion ignores the diversity of failure modes which are possible.

The identification of the different failure modes of a UDC can provide physically more realistic, and also simpler, failure criteria (Reference 4.2.4(e)). Testing a polymer matrix UDC reveals that tensile stress in the fiber direction produces a jagged, irregular failure surface. Tensile stress transverse to the surface produces a smooth, straight failure surface (See Figures 4.2.4(a) and (b)). Since the carrying capacity deterioration in the tensile fiber mode is due to transverse cracks and the transverse stress σ_{22} has no effect on such cracks, it is assumed that the plane tensile fiber mode is only dependent on the stresses σ_{11} and σ_{12}.

For compressive σ_{11}, failure is due to fiber buckling in the shear mode and the transverse stress σ_{22} has little effect on the compressive failure. In this *compressive fiber mode*, failure again depends primarily on σ_{11}. The dependence on σ_{12} is not known and arguments may be made for and against including it in the failure criterion.

For tension transverse to the fibers, the *tensile matrix mode*, failure occurs by a sudden crack in the fiber direction as shown in Figure 4.2.4(b). Since stress in the fiber direction has no effect on a crack in the fiber direction, this failure mode is dependent only on σ_{22} and σ_{12}.

For compressive stress transverse to the fibers, failure occurs on some plane parallel to the fibers, but not necessarily normal to σ_{22}. This *compressive matrix mode* is produced by normal stress and shear stress on the failure plane. Again, the stress σ_{11} does not effect this failure.

Each of the failure modes described can be modeled separately by a quadratic polynomial (Reference 4.2.4(e)). This approach provides four individual failure criteria. Note the choice of stress components included in each of these criteria, and the particular mathematical form used, are subjects which are not yet fully resolved. The following criteria appear to a reasonable set with which the different modes of failure can be handled separately.

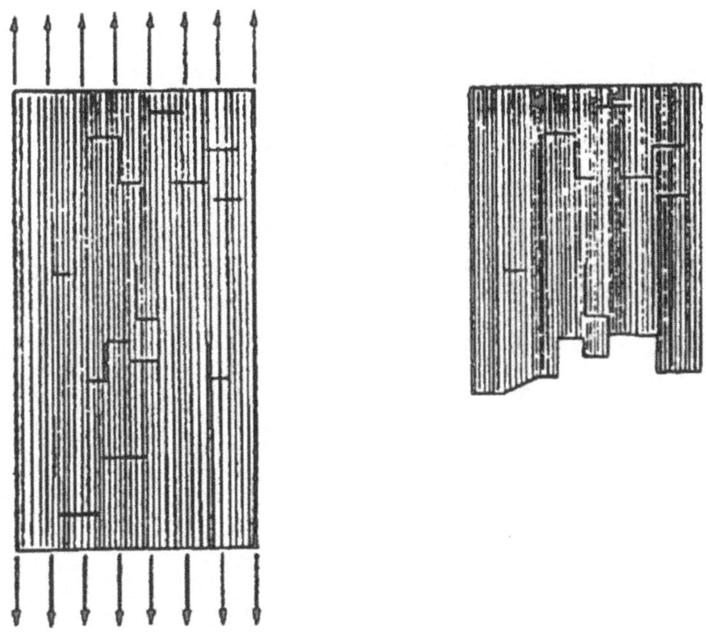

FIGURE 4.2.4(a) *Tensile fiber failure mode.*

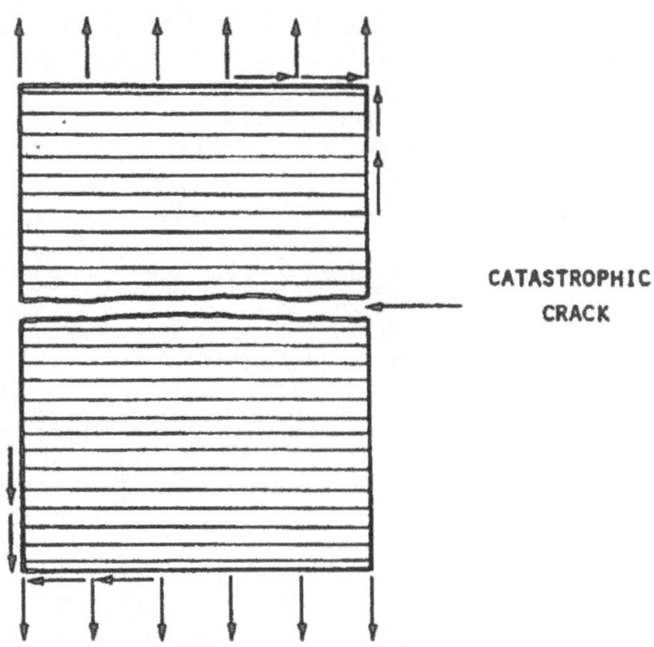

FIGURE 4.2.4(b) *Tensile matrix failure mode.*

Fiber modes

Tensile

$$\left(\frac{\sigma_{11}}{F_1^{tu}}\right)^2 + \left(\frac{\sigma_{12}}{F_{12}^{su}}\right)^2 = 1 \qquad 4.2.4(f)$$

Compressive

$$\left(\frac{\sigma_{11}}{F_1^{cu}}\right)^2 + \left(\frac{\sigma_{12}}{F_{12}^{su}}\right)^2 = 1 \qquad 4.2.4(g)$$

Matrix modes

Tensile

$$\left(\frac{\sigma_{22}}{F_2^{tu}}\right)^2 + \left(\frac{\sigma_{12}}{F_{12}^{su}}\right)^2 = 1 \qquad 4.2.4(h)$$

Compressive

$$\left(\frac{\sigma_{22}}{2F_{23}^{su}}\right)^2 + \left[\left(\frac{F_2^{cu}}{2F_{23}^{su}}\right)^2 - 1\right]\left(\frac{\sigma_{22}}{F_2^{cu}}\right) + \left(\frac{\sigma_{12}}{F_{12}^{su}}\right)^2 = 1 \qquad 4.2.4(i)$$

Note that F_2^{cu} in Equation 4.2.4(i) should be taken as the absolute value. The ultimate transverse shear stress, $\sigma_{23} = F_{23}^{su}$, is very difficult to measure. A reasonable approximation for this quantity is the ultimate shear stress for the matrix. For any given state of stress, one each of Equations 4.2.4(f) and (g) and Equations 4.2.4(h) and (i) are chosen according to the signs of σ_{11} and σ_{22}. The stress components are introduced into the appropriate pair and whichever criterion is satisfied first is the operative criterion.

The advantages are Equations 4.2.4(f) - (i) are:

1. The failure criteria are expressed in terms of single component ultimate stresses. No biaxial test results are needed.

2. The failure mode is identified by the criterion which is satisfied first.

The last feature is of fundamental importance for analysis of fiber composite structural elements, since it permits identification of the nature of initial damage. Moreover, in conjunction with a finite element analysis, it is possible to identify the nature of failure in elements, modify their stiffnesses accordingly, and proceed with the analysis to predict new failures.

4.2.5 Summary

- Composite strength analysis is most commonly performed, by industry, on the macromechanics level given that the analysis of composite materials uses effective lamina properties based on average stress and strain.

- Ply level stresses are the commonly used approach to laminate strength analysis.
- Lamina stress/strain is influenced by many properties of interest, but is dominated by mechanical load and environmental sensitivity.
- Stress-strain elastic behavior, in its simplest form, may be described as a function of a composite materials constitutive properties (i.e., E, G, ν, α).
- Several practical failure criteria exist today that: 1) depend upon cross-plied laminate coupon data to determine lamina stress/strain allowables and 2) identify the failure mode based on the allowable that is first exceeded by its stress/strain counterpart.

4.3 ANALYSIS OF LAMINATES

4.3.1 Lamina stress-strain relations

A laminate is composed of uni-directionally-reinforced laminae oriented in various directions with respect to the axes of the laminate. The stress-strain relations developed in the Section 4.2 must be transformed into the coordinate system of the laminate to perform the laminate stress-strain analysis. A new system of notation for the lamina elastic properties is based on x_1 in the fiber direction, x_2 transverse to the fibers in the plane of the lamina, and x_3 normal to the plane of the lamina.

$$\begin{aligned} E_1 &= E_1^* & \nu_{12} &= \nu_{12}^* \\ E_2 &= E_3 = E_2^* & \nu_{23} &= \nu_{23}^* \\ G_{12} &= G_1^* & G_{23} &= G_2^* \end{aligned} \qquad 4.3.1(a)$$

In addition, the laminae are now treated as effective homogeneous, transversely isotropic materials.

$$\begin{Bmatrix} \epsilon_{11} \\ \epsilon_{22} \\ \epsilon_{33} \\ 2\epsilon_{23} \\ 2\epsilon_{13} \\ 2\epsilon_{12} \end{Bmatrix} = \begin{bmatrix} \frac{1}{E_1} & \frac{-\nu_{12}}{E_1} & \frac{-\nu_{12}}{E_1} & 0 & 0 & 0 \\ \frac{-\nu_{12}}{E_1} & \frac{1}{E_2} & \frac{-\nu_{23}}{E_2} & 0 & 0 & 0 \\ \frac{-\nu_{12}}{E_1} & \frac{-\nu_{23}}{E_2} & \frac{1}{E_2} & 0 & 0 & 0 \\ 0 & 0 & 0 & \frac{1}{G_{23}} & 0 & 0 \\ 0 & 0 & 0 & 0 & \frac{1}{G_{12}} & 0 \\ 0 & 0 & 0 & 0 & 0 & \frac{1}{G_{12}} \end{bmatrix} \begin{Bmatrix} \sigma_{11} \\ \sigma_{22} \\ \sigma_{33} \\ \sigma_{23} \\ \sigma_{13} \\ \sigma_{12} \end{Bmatrix} \qquad 4.3.1(b)$$

It has become common practice in the analysis of laminates to utilize engineering shear strains rather than tensor shear strains. Thus the factor of two has been introduced into the stress-strain relationship of Equation 4.3.1(b).

The most important state of stress in a lamina is plane stress, where

$$\sigma_{13} = \sigma_{23} = \sigma_{33} = 0 \qquad 4.3.1(c)$$

since it occurs from both in-plane loading and bending at sufficient distance from the laminate edges. The plane stress version of Equation 4.3.1(b) is

$$\begin{Bmatrix} \epsilon_{11} \\ \epsilon_{22} \\ 2\epsilon_{12} \end{Bmatrix} = \begin{bmatrix} \frac{1}{E_1} & \frac{-\nu_{12}}{E_1} & 0 \\ \frac{-\nu_{12}}{E_1} & \frac{1}{E_2} & 0 \\ 0 & 0 & \frac{1}{G_{12}} \end{bmatrix} \begin{Bmatrix} \sigma_{11} \\ \sigma_{22} \\ \sigma_{12} \end{Bmatrix} \quad \text{4.3.1(d)}$$

which may be written as

$$\{\epsilon_\ell\} = [S]\{\sigma_\ell\} \quad \text{4.3.1(e)}$$

Here [S], the compliance matrix, relates the stress and strain components in the principal material directions. These are called laminae coordinates and are denoted by the subscript ℓ.

Equation 4.3.1(d) relates the in-plane strain components to the three in-plane stress components. For the plane stress state, the three additional strains can be found to be

$$\epsilon_{23} = \epsilon_{13} = 0$$

$$\epsilon_{33} = -\sigma_{11}\frac{\nu_{13}}{E_1} - \sigma_{22}\frac{\nu_{23}}{E_2} \quad \text{4.3.1(f)}$$

and the complete state of stress and strain is determined.

The relations 4.3.1(d) can be inverted to yield

$$\{\sigma_\ell\} = [S]^{-1}\{\epsilon_\ell\} \quad \text{4.3.1(g)}$$

or

$$\{\sigma_\ell\} = [Q]\{\epsilon_\ell\} \quad \text{4.3.1(h)}$$

The matrix [Q] is defined as the inverse of the compliance matrix and is known as the reduced lamina stiffness matrix. Its terms can be shown to be

$$[Q] = \begin{bmatrix} Q_{11} & Q_{12} & 0 \\ Q_{12} & Q_{22} & 0 \\ 0 & 0 & Q_{66} \end{bmatrix} \quad \text{4.3.1(i)}$$

$$Q_{11} = \frac{E_1}{1-\nu_{12}^2\frac{E_2}{E_1}} \qquad Q_{12} = \frac{\nu_{12}E_2}{1-\nu_{12}^2\frac{E_2}{E_1}}$$

$$Q_{22} = \frac{E_2}{1-\nu_{12}^2\frac{E_2}{E_1}} \qquad Q_{66} = G_{12}$$

In the notation for the [Q] matrix, each pair of subscripts of the stiffness components is replaced by a single subscript according to the following scheme.

$$11 \rightarrow 1 \qquad 22 \rightarrow 2 \qquad 33 \rightarrow 3$$

$$23 \rightarrow 4 \qquad 31 \rightarrow 5 \qquad 12 \rightarrow 6$$

The reduced stiffness and compliance matrices 4.3.1(i) and (d) relate stresses and strains in the principal material directions of the material. To define the material response in directions other than these coordinates, transformation relations for the material stiffnesses are needed.

In Figure 4.3.1(a), two sets of coordinate systems are depicted. The 1-2 coordinate system corresponds to the principal material directions for a lamina, while the x-y coordinates are arbitrary and related to the 1-2 coordinates through a rotation about the axis out of the plane of the figure. The angle θ is defined as the rotation from the arbitrary x-y system to the 1-2 material system.

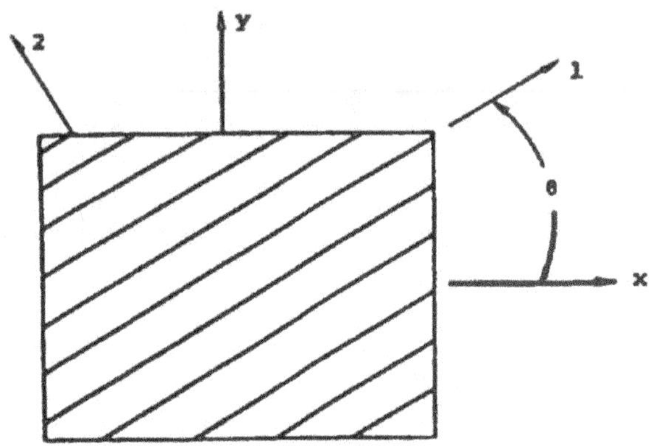

1, 2 Principal Material Coordinates
x, y Laminate, or Arbitrary Coordinates

FIGURE 4.3.1(a) *Coordinate systems.*

The transformation of stresses from the 1-2 system to the x-y system follows the rules for transformation of tensor components.

$$\begin{Bmatrix} \sigma_{xx} \\ \sigma_{yy} \\ \sigma_{xy} \end{Bmatrix} = \begin{bmatrix} m^2 & n^2 & -2mn \\ n^2 & m^2 & 2mn \\ mn & -mn & m^2-n^2 \end{bmatrix} \begin{Bmatrix} \sigma_{11} \\ \sigma_{22} \\ \sigma_{12} \end{Bmatrix} \qquad 4.3.1(k)$$

or

$$\{\sigma_x\} = [\theta]\{\sigma_\ell\} \qquad 4.3.1(l)$$

where $m = \cos\theta$, and $n = \sin\theta$. In these relations, the subscript x is used as shorthand for the laminate coordinate system.

The same transformation matrix [θ] can also be used for the tensor strain components. However, since the engineering shear strains have been utilized, a different transformation matrix is required.

$$\begin{Bmatrix} \epsilon_{xx} \\ \epsilon_{yy} \\ 2\epsilon_{xy} \end{Bmatrix} = \begin{bmatrix} m^2 & n2 & -mn \\ n^2 & m^2 & mn \\ 2mn & -2mn & m^2-n^2 \end{bmatrix} \begin{Bmatrix} \epsilon_{11} \\ \epsilon_{22} \\ 2\epsilon_{12} \end{Bmatrix} \qquad 4.3.1(m)$$

or

$$\{\epsilon_x\} = [\psi]\{\epsilon_\ell\} \tag{4.3.1(n)}$$

Given the transformations for stress and strain to the arbitrary coordinate system, the relations between stress and strain in the laminate system can be determined.

$$\{\sigma_x\} = [\bar{Q}]\{\epsilon_x\} \tag{4.3.1(o)}$$

The reduced stiffness matrix $[\bar{Q}]$ relates the stress and strain components in the laminate coordinate system.

$$[\bar{Q}] = [\theta][Q][\psi]^{-1} \tag{4.3.1(p)}$$

The terms within $[\bar{Q}]$ are defined to be

$$\begin{aligned}
\bar{Q}_{11} &= Q_{11}m^4 + Q_{22}n^4 + 2m^2n^2(Q_{12} + 2Q_{66}) \\
\bar{Q}_{12} &= m^2n^2(Q_{11} + Q_{22} - 4Q_{66}) + (m^4+n^4)Q_{12} \\
\bar{Q}_{16} &= [Q_{11}m^2 - Q_{22}n^2 - (Q_{12} + 2Q_{66})(m^2 - n^2)]\,mn \\
\bar{Q}_{22} &= Q_{11}n^4 + Q_{22}m^4 + 2m^2n^2(Q_{12} + 2Q_{66}) \\
\bar{Q}_{26} &= [Q_{11}n^2 - Q_{22}m^2 + (Q_{12} + 2Q_{66})(m^2 - n^2)]\,mn \\
\bar{Q}_{66} &= (Q_{11} + Q_{22} - 2Q_{12})\,m^2n^2 + Q_{66}(m^2 - n^2)^2
\end{aligned} \tag{4.3.1(q)}$$

$$\bar{Q}_{21} = \bar{Q}_{12} \qquad \bar{Q}_{61} = \bar{Q}_{16} \qquad \bar{Q}_{62} = \bar{Q}_{26}$$

where the subscript 6 has been retained in keeping with the discussion following Equation 4.3.1(j).

$$[\bar{Q}] = \begin{bmatrix} \bar{Q}_{11} & \bar{Q}_{12} & \bar{Q}_{16} \\ \bar{Q}_{21} & \bar{Q}_{22} & \bar{Q}_{26} \\ \bar{Q}_{16} & \bar{Q}_{26} & \bar{Q}_{66} \end{bmatrix} \tag{4.3.1(r)}$$

A feature of $[\bar{Q}]$ matrix which is immediately noticeable is that $[\bar{Q}]$ is fully-populated. The additional terms which have appeared in $[\bar{Q}]$, $\bar{Q}_{16}$ and $\bar{Q}_{26}$, relate shear strains to extensional loading and vice versa. This effect of a shear strain resulting from an extensional strain is depicted in Figure 4.3.1(b). From Equations 4.3.1(q), these terms are zero for θ equal to 0° or 90°. Physically, this means that the fibers are parallel or perpendicular to the loading direction. For this case, extensional-shear coupling does not occur for an orthotropic material since the loadings are in the principal directions. The procedure used to develop the transformed stiffness matrix can also be used to find a transformed compliance matrix.

$$\{\epsilon_\ell\} = [S]\{\sigma_\ell\} \tag{4.3.1(s)}$$

$$\{\epsilon_x\} = [\psi][S][\theta]^{-1}\{\sigma_x\} \tag{4.3.1(t)}$$

$$\{\epsilon_x\} = [\bar{S}]\{\sigma_x\} \tag{4.3.1(u)}$$

Noting that the stress-strain relations are now defined in the laminate coordinate system, lamina stiffnesses can also be defined in this system. Thus, expanding the last of Equations 4.3.1(s) - (u):

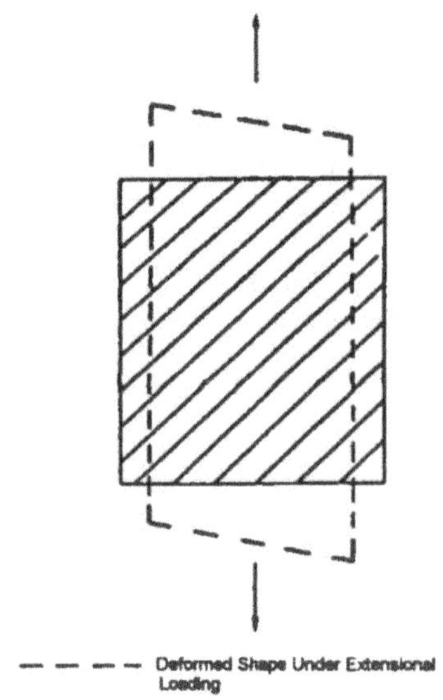

FIGURE 4.3.1(b) *Extensional-shear coupling.*

$$\begin{Bmatrix} \epsilon_{xx} \\ \epsilon_{yy} \\ 2\epsilon_{xy} \end{Bmatrix} = \begin{bmatrix} \bar{S}_{11} & \bar{S}_{12} & \bar{S}_{16} \\ \bar{S}_{21} & \bar{S}_{22} & \bar{S}_{26} \\ \bar{S}_{16} & \bar{S}_{26} & \bar{S}_{66} \end{bmatrix} \begin{Bmatrix} \sigma_{xx} \\ \sigma_{yy} \\ \sigma_{xy} \end{Bmatrix} \qquad 4.3.1(v)$$

The engineering constants for the material can be defined by specifying the conditions for an experiment. For $\sigma_{yy} = \sigma_{xy} = 0$, the ratio $\sigma_{xx}/\epsilon_{xx}$ is Young's modulus in the x direction. For this same stress state, $-\epsilon_{yy}/\epsilon_{xx}$ is Poisson's ratio. In this fashion, the lamina stiffnesses in the coordinate system of Equations 4.3.1(s) - (u) are found to be:

$$E_x = \frac{1}{\bar{S}_{11}} \qquad E_y = \frac{1}{\bar{S}_{22}}$$

$$G_{xy} = \frac{1}{\bar{S}_{66}} \qquad \nu_{xy} = \frac{-\bar{S}_{21}}{\bar{S}_{11}} = \frac{-\bar{S}_{12}}{\bar{S}_{11}} \qquad 4.3.1(w)$$

It is sometimes desirable to obtain elastic constants directly from the reduced stiffnesses, $[\bar{Q}]$, by using Equations 4.3.1(o). In the general case where the $\bar{Q}_{ij}$ matrix is fully populated, this can be accomplished by using Equations 4.3.1(w) and the solution for $\bar{S}_{ij}$ as functions of $\bar{Q}_{ij}$ obtained from the inverse relationship of the two matrices. An alternative approach is to evaluate extensional properties for the case of zero shear strain. For single stress states and zero shear strain, the elastic constants in terms of the transformed stiffness matrix terms are:

$$E_x = \bar{Q}_{11} - \frac{\bar{Q}_{12}^2}{\bar{Q}_{22}}$$

$$E_y = \bar{Q}_{22} - \frac{\bar{Q}_{12}^2}{\bar{Q}_{11}} \qquad 4.3.1(x)$$

$$\nu_{xy} = \frac{\bar{Q}_{12}}{\bar{Q}_{22}} = \frac{\bar{Q}_{21}}{\bar{Q}_{22}}$$

Also,

$$G_{xy} = \bar{Q}_{66}$$

From the terms in the $[\bar{Q}]$ matrix (Equation 4.3.1(q)) and the stiffness relations (Equation 4.3.1(x)), the elastic constants in an arbitrary coordinate system are functions of all the elastic constants in the principal material directions as well as the angle of rotation.

The variation of elastic modulus E_x with angle of rotation is depicted in Figure 4.3.1(c) for a typical graphite/epoxy material. For demonstration purposes, two different shear moduli have been used in generating the figure. The differences between the two curves demonstrate the effect of the principal material shear modulus on the transformed extensional stiffness. The two curves are identical at 0° and 90°, as expected since E_x is simply E_1 or E_2. Between these two endpoints, substantial differences are present. For the smaller shear modulus, the extensional stiffness is less than the E_2 value from approximately 50° to just less than 90°. For these angles, the material stiffness is more strongly governed by the principal material shear modulus than by the transverse extensional modulus. The curves of Figure 4.3.1(c) can also be used to determine the modulus E_y by simply reversing the angle scale.

With the transformed stress-strain relations, it is now possible to develop an analysis for an assemblage of plies, i.e., a laminate.

4.3.2 Lamination theory

The development of procedures to evaluate stresses and deformations of laminates is crucially dependent on the fact that the thickness of laminates is much smaller than the in-plane dimensions. Typical thickness values for individual plies range between 0.005 and 0.010 inch (0.13 and 0.25 mm). Consequently, laminates using from 8 to 50 plies are still generally thin plates and, therefore, can be analyzed on the basis of the usual simplifications of thin plate theory.

In the analysis of isotropic thin plates it has become customary to analyze the cases of in-plane loading and bending separately. The former case is described by plane stress elastic theory and the latter by classical plate bending theory. This separation is possible since the two loadings are uncoupled for symmetric laminates; when both occur, the results are superposed.

The classical assumptions of thin plate theory are:

1) The thickness of the plate is much smaller than the in-plane dimensions;
2) The shapes of the deformed plate surface are small compared to unity;
3) Normals to the undeformed plate surface remain normal to the deformed plate surface;
4) Vertical deflection does not vary through the thickness; and
5) Stress normal to the plate surface is negligible.

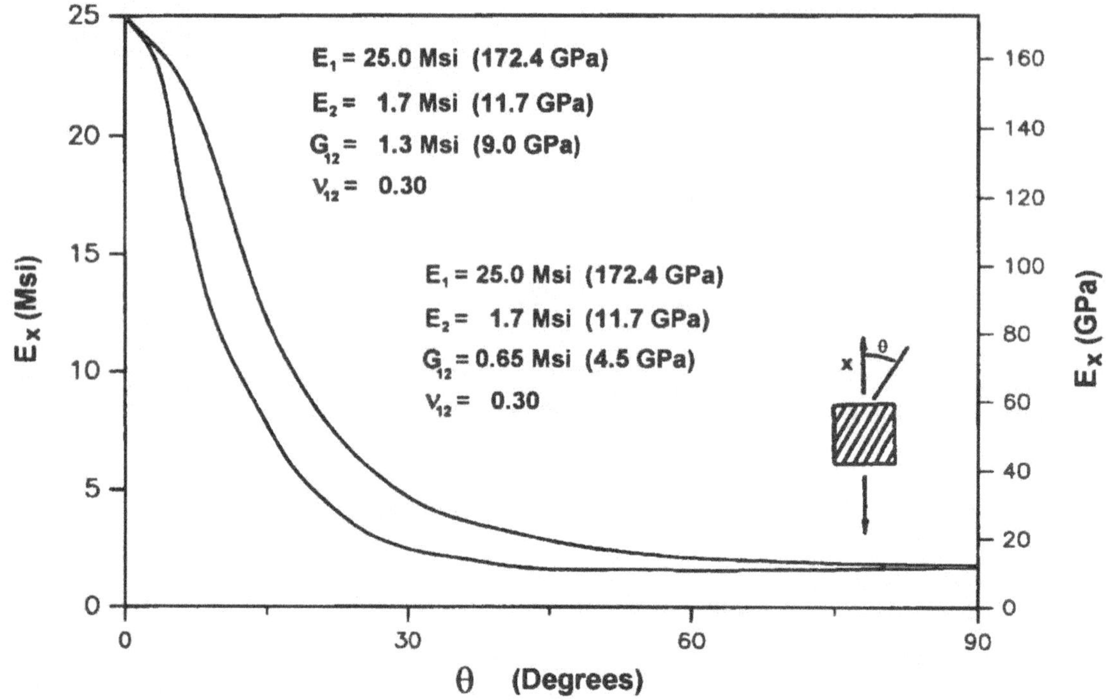

FIGURE 4.3.1(c) *Variation of E_x with angle and G_{12} for typical graphite/epoxy materials.*

On the basis of assumptions (2) - (4), the displacement field can be expressed as:

$$
\begin{aligned}
u_z &= u_z^{\circ}(x,y) \\
u_x &= u_x^{\circ}(x,y) - z\frac{\partial u_z}{\partial x} \\
u_y &= u_y^{\circ}(x,y) - z\frac{\partial u_z}{\partial y}
\end{aligned}
\qquad 4.3.2(a)
$$

with the x-y-z coordinate system defined in Figure 4.3.2(a). These relations (Equation 4.3.2(a)) indicate that the in-plane displacements consist of a mid-plane displacement, designated by the superscript (°), plus a linear variation through the thickness. The two partial derivatives are bending rotations of the mid-surface. The use of assumption (4) prescribes that u_z does not vary through the thickness.

The linear strain displacement relations are

$$
\epsilon_{xx} = \frac{\partial u_x}{\partial x} \qquad \epsilon_{yy} = \frac{\partial u_y}{\partial y}
$$
$$
\epsilon_{xy} = \frac{1}{2}\left(\frac{\partial u_x}{\partial y} + \frac{\partial u_y}{\partial x}\right) \qquad 4.3.2(b)
$$

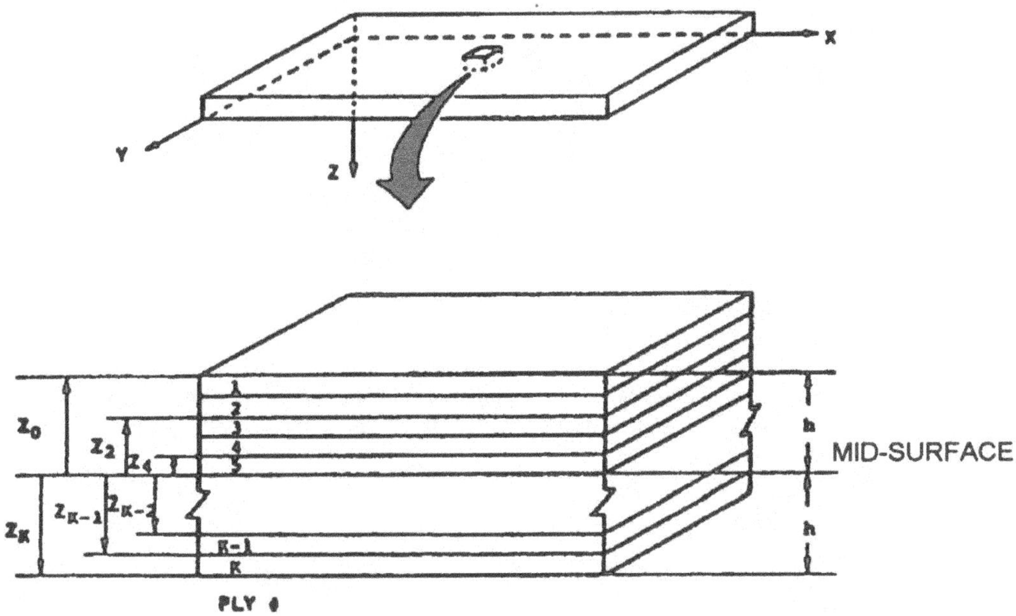

TOTAL LAMINATE THICKNESS 2h

FIGURE 4.3.2(a) *Laminate construction.*

and performing the required differentiations yields

$$\begin{aligned} \epsilon_{xx} &= \epsilon^{\circ}_{xx} + z\kappa_{xx} \\ \epsilon_{yy} &= \epsilon^{\circ}_{yy} + z\kappa_{yy} \\ 2\epsilon_{xy} &= 2\epsilon^{\circ}_{xy} + 2z\kappa_{xy} \end{aligned} \qquad 4.3.2(c)$$

or

$$\{\epsilon_x\} = \{\epsilon^{\circ}\} + z\{\kappa\} \qquad 4.3.2(d)$$

where

$$\{\epsilon^{\circ}\} = \begin{Bmatrix} \dfrac{\partial u^{\circ}_x}{\partial x} \\ \dfrac{\partial u^{\circ}_y}{\partial y} \\ \left[\dfrac{\partial u^{\circ}_x}{\partial y} + \dfrac{\partial u^{\circ}_y}{\partial x}\right] \end{Bmatrix} \qquad 4.3.2(e)$$

and

$$\{\kappa\} = \begin{pmatrix} \kappa_{xx} \\ \kappa_{yy} \\ 2\kappa_{xy} \end{pmatrix} = \begin{pmatrix} -\dfrac{\partial^2 u_z}{\partial x^2} \\ -\dfrac{\partial^2 u_z}{\partial y^2} \\ -2\dfrac{\partial^2 u_z}{\partial x \partial y} \end{pmatrix} \qquad 4.3.2(f)$$

The strain at any point in the plate is defined as the sum of a mid-surface strain $\{\epsilon°\}$, and a curvature $\{\kappa\}$ multiplied by the distance from the mid-surface.

For convenience, stress and moment resultants will be used in place of stresses for the remainder of the development of lamination theory (see Figure 4.3.2(b)). The stress resultants are defined as

$$\{N\} = \begin{pmatrix} N_{xx} \\ N_{yy} \\ N_{xy} \end{pmatrix} = \int_{-h}^{h} \{\sigma_x\}\, dz \qquad 4.3.2(g)$$

and the moment resultants are defined as

$$\{M\} = \begin{pmatrix} M_{xx} \\ M_{yy} \\ M_{xy} \end{pmatrix} = \int_{-h}^{h} \{\sigma_x\}\, z dz \qquad 4.3.2(h)$$

where the integrations are carried out over the plate thickness.

Noting Equations 4.3.1(o) and 4.3.2(c), relations between the stress and moment resultants and the mid-plane strains and curvatures can be written as

$$\{N\} = \int_{-h}^{h} \{\sigma_x\} dz = \int_{-h}^{h} \left[\bar{Q}\right] (\{\epsilon^{\circ}\} + z\,\{\kappa\})\, dz \qquad 4.3.2(i)$$

$$\{M\} = \int_{-h}^{h} \{\sigma_x\}\, z dz = \int_{-h}^{h} \left[\bar{Q}\right] (\{\epsilon^{\circ}\} + z\,\{\kappa\})\, z dz \qquad 4.3.2(j)$$

Since the transformed lamina stiffness matrices are constant within each lamina and the mid-plane strains and curvatures are constant with respect to the z-coordinate, the integrals in Equations 4.3.2(i) and (j) can be replaced by summations.

Introducing three matrices equivalent to the necessary summations, the relations can be written as

$$\{N\} = [A]\{\epsilon^{\circ}\} + [B]\{\kappa\} \qquad 4.3.2(k)$$

$$\{M\} = [B]\{\epsilon^{\circ}\} + [D]\{\kappa\} \qquad 4.3.2(l)$$

where the stiffness matrix is composed of the following 3x3 matrices:

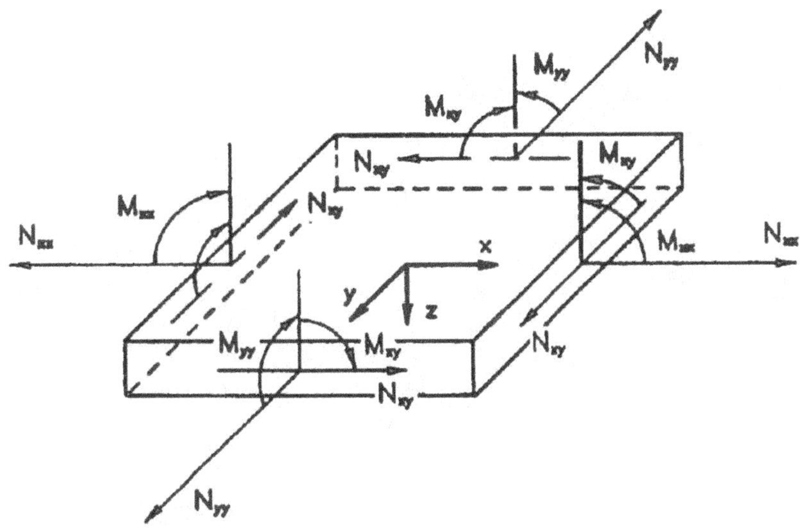

$$\{N\} = \int_{-h}^{h} \{\sigma_x\} dz$$

$$\{M\} = \int_{-h}^{h} \{\sigma_x\} z dz$$

FIGURE 4.3.2(b) *Stress and moment resultants.*

$$[A] = \sum_{i=1}^{N} [\bar{Q}]^i (z_i - z_{i-1})$$

$$[B] = \frac{1}{2} \sum_{i=1}^{N} [\bar{Q}]^i (z_i^2 - z_{i-1}^2) \qquad 4.3.2(m)$$

$$[D] = \frac{1}{3} \sum_{i=1}^{N} [\bar{Q}]^i (z_i^3 - z_{i-1}^3)$$

where N is the total number of plies, z_i is defined in Figure 4.3.2(a) and subscript i denotes a property of the ith ply. Note that $z_i - z_{i-1}$ equals the ply thickness. Here the reduced lamina stiffnesses for the ith ply are found from Equations 4.3.2(k) and (l) using the principal properties and orientation angle for each ply in turn. Thus, the constitutive relations for a laminate have been developed in terms of stress and moment resultants.

Classical lamination theory has been used to predict the internal stress state, stiffness and dimensional stability of laminated composites (e.g., References 4.3.2(a) - (e)). The constitutive law for CLT couples extensional, shear, bending and torsional loads with strains and curvatures. Residual strains or warpage due to differential shrinkage or swelling of plies in a laminate have also been incorporated in lamination theory using an environmental load analogy (See Sections 4.3.3.3 and 4.3.3.4.). The combined influence of various types of loads and moments on laminated plate response can be described using the ABD matrix from Equations 4.3.2(k) and (l). In combined form:

$$\begin{bmatrix} N_x \\ N_y \\ N_{xy} \\ M_x \\ M_y \\ M_{xy} \end{bmatrix} = \begin{bmatrix} A_{11} & A_{12} & A_{16} & B_{11} & B_{12} & B_{16} \\ A_{12} & A_{22} & A_{26} & B_{12} & B_{22} & B_{26} \\ A_{16} & A_{26} & A_{66} & B_{16} & B_{26} & B_{66} \\ B_{11} & B_{12} & B_{16} & D_{11} & D_{12} & D_{16} \\ B_{12} & B_{22} & B_{26} & D_{12} & D_{22} & D_{26} \\ B_{16} & B_{26} & B_{66} & D_{16} & D_{26} & D_{66} \end{bmatrix} \begin{bmatrix} \epsilon_x \\ \epsilon_y \\ \epsilon_{xy} \\ \kappa_x \\ \kappa_y \\ \kappa_{xy} \end{bmatrix} \qquad 4.3.2(n)$$

where N are loads, M are moments, ϵ are strains, κ are curvatures and

A_{ij} = extensional and shear stiffnesses
B_{ij} = extension-bending coupling stiffnesses
D_{ij} = bending and torsional stiffnesses

Several observations regarding lay-up and laminate stacking sequence (LSS) can be made with the help of Equation 4.3.2(n). These include:

(1) The stiffness matrix A_{ij} in Equation 4.3.2(n) is independent of LSS. Inversion of the stiffness matrix [ABD] yields the compliance matrix [A'B'D']. This inversion is necessary in order to calculate strains and curvatures in terms of loads and moments. The inversion results in a relationship between LSS and extension/shear compliances. However, this relationship is eliminated if the laminate is symmetric.

(2) Nonzero values of A_{16} and A_{26} indicates that there is extension/shear coupling (e.g., longitudinal loads will result in both extensional and shear strains). If a laminate is balanced A_{16} and A_{26} become zero, eliminating extension/shear coupling.

(3) Nonzero values of B_{ij} indicates that there is coupling between bending/twisting curvatures and extension/shear loads. Traditionally, these couplings have been suppressed for most applications by choosing an LSS that minimizes the values of B_{ij}. All values of B_{ij} become zero for symmetric laminates. Reasons for designing with symmetric laminates include structural dimensional stability requirements (e.g., buckling, environmental warping), compatibility of structural components at joints and the inability to test for strength allowables of specimens that have significant values of B_{ij}.

(4) In general, the values of D_{ij} are nonzero and strongly dependent on LSS. The average plate bending stiffnesses, torsional rigidity and flexural Poisson's ratio can be calculated per unit width using components of the compliance matrix [A'B'D'], i.e.,

$1/D'_{11}$ = bending stiffness about y-axis
$1/D'_{22}$ = bending stiffness about x-axis
$1/D'_{66}$ = torsional rigidity about x- or y-axis
$-D'_{12}/D'_{11}$ = flexural Poisson's ratio.

The D'_{16} and D'_{26} terms should also be included in calculations relating midplane curvatures to moments except when considering a special class of balanced, unsymmetric laminates.

(5) Nonzero values of D_{16} and D_{26} indicates that there is bending/twisting coupling. These terms will vanish only if a laminate is balanced and if, for each ply oriented at +θ above the laminate midplane, there is an identical ply (in material and thickness) oriented at -θ at an equal distance below the midplane. Such a laminate cannot be symmetric, unless it contains only 0° and 90° plies. Bending/twisting coupling can be minimized by alternating the location of +θ and -θ plies through the LSS (Section 4.6.5.2.2, Recommendation 5).

Additional information on laminate stacking sequence effects is found in Section 4.6.5.

4.3.3 Laminate properties

The relations between the mid-surface strains and curvatures and the membrane stress and moment resultants are used to calculate plate bending and extensional stiffnesses for structural analysis. The effects of orientation variables upon plate properties are also considered. In addition to the mechanical loading conditions treated thus far, the effects of temperature changes upon laminate behavior must be understood. Further, for polymeric matrix composites, high moisture content causes dimensional changes which can be described by effective swelling coefficients.

4.3.3.1 Membrane stresses

Recalling Equations 4.3.2(k) and (l) and noting that for symmetric laminates the [B] matrix is zero, the relations can be rewritten as

$$\begin{pmatrix} N_{xx} \\ N_{yy} \\ N_{xy} \end{pmatrix} = \begin{bmatrix} A_{11} & A_{12} & A_{16} \\ A_{12} & A_{22} & A_{26} \\ A_{16} & A_{26} & A_{66} \end{bmatrix} \begin{pmatrix} \epsilon^{\circ}_{xx} \\ \epsilon^{\circ}_{yy} \\ 2\epsilon^{\circ}_{xy} \end{pmatrix} \qquad 4.3.3.1(a)$$

and

$$\begin{pmatrix} M_{xx} \\ M_{yy} \\ M_{xy} \end{pmatrix} = \begin{bmatrix} D_{11} & D_{12} & D_{16} \\ D_{12} & D_{22} & D_{26} \\ D_{16} & D_{26} & D_{66} \end{bmatrix} \begin{pmatrix} \kappa_{xx} \\ \kappa_{yy} \\ 2\kappa_{xy} \end{pmatrix} \qquad 4.3.3.1(b)$$

Since the extensional and bending behavior are uncoupled, effective laminate elastic constants can be readily determined. Inverting the stress resultant mid-plane strain relations yields

$$\{\epsilon^{\circ}\} = [A]^{-1}\{N\} = [a]\{N\} \qquad 4.3.3.1(c)$$

from which the elastic constants are seen to be

$$E_x = \frac{1}{2ha_{11}} \qquad G_{xy} = \frac{1}{2ha_{66}}$$

$$E_y = \frac{1}{2ha_{22}} \qquad \nu_{xy} = -\frac{a_{12}}{a_{11}} \qquad 4.3.3.1(d)$$

where the divisor 2h corresponds to the laminate thickness.

Note that the [A] matrix is comprised of [Q] matrices from each layer in the laminate. It is obvious that the laminate elastic properties are functions of the angular orientation of the plies. This angular influence is illustrated in Figure 4.3.3.1 for a typical high modulus carbon/epoxy system which has the lamina properties listed in Table 4.3.3.1(a). The laminae are oriented in ±θ pairs in a symmetric, balanced construction, creating what is called an angle-ply laminate.

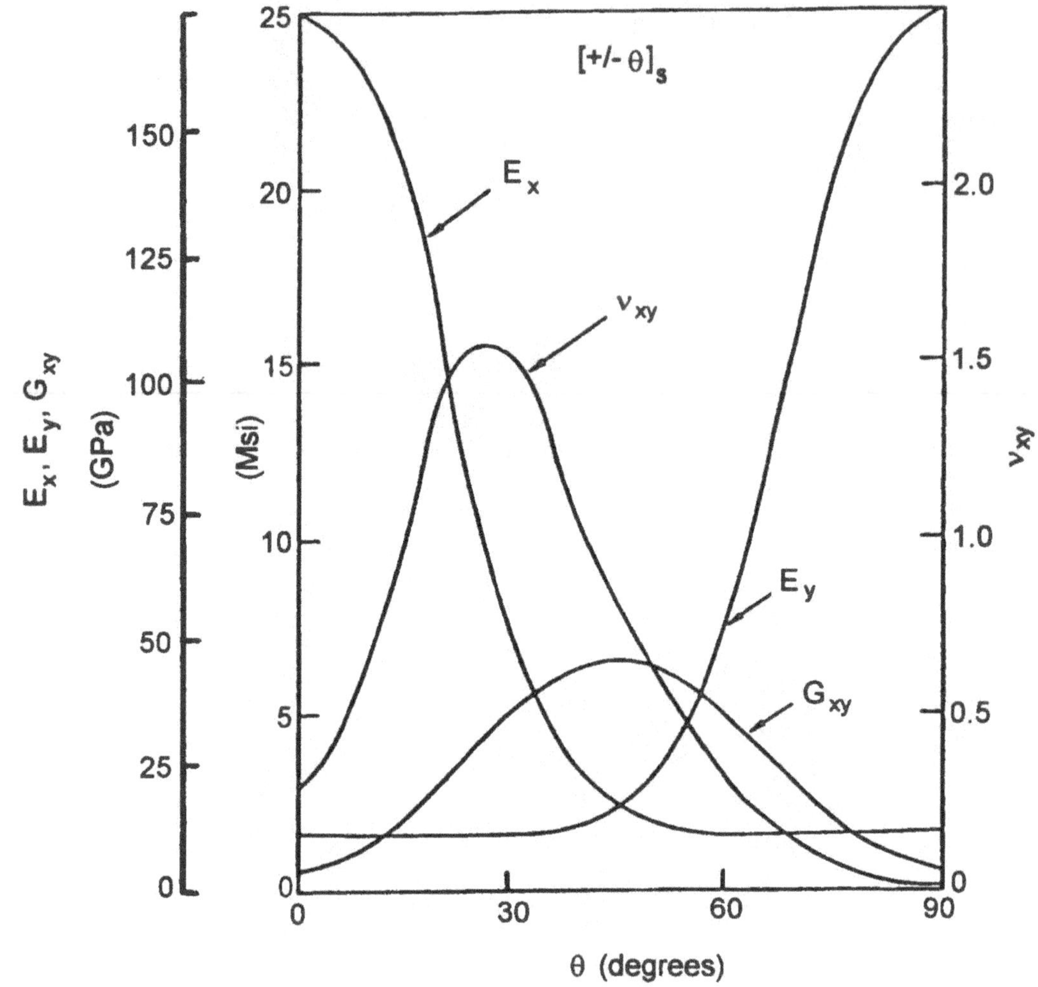

FIGURE 4.3.3.1 *Laminate elastic constants for high modulus carbon/epoxy.*

The variation of shear modulus and Poisson's ratio are noteworthy in Figure 4.3.3.1. The shear modulus is equal to the unidirectional value for 0° and 90° and rises sharply to a maximum at 45°. The peak at 45° can be explained by noting that shear is equivalent to a combined state of tensile and compressive loads oriented at 45°. Thus, the shear loading on a $[\pm45]_s$ laminate is equivalent to tensile and compressive loading on a $[0/90]_s$ laminate. Effectively, the fibers are aligned with the loading and, hence, with the large shear stiffness.

An even more interesting effect is seen in the variation of Poisson's ratio. The peak value in this example is greater than 1.5. In an isotropic material, this would be impossible. In an orthotropic material, the isotropic restriction does not hold and a Poisson's ratio greater than one is valid and realistic. In fact, large Poisson's ratios are typical for laminates constructed from unidirectional materials with the plies oriented at approximately 30°.

TABLE 4.3.3.1(a) *Properties of a high modulus carbon/epoxy lamina.*

E_1 = 25.0 Msi = 172 GPa	α_1 = 0.30x10^{-6} in/in/F° = 0.54x10^{-6} mm/mm/C°
E_2 = 1.7 Msi = 12 GPa	α_2 = 19.5x10^{-6} in/in/F° = 35.1x10^{-6} mm/mm/C°
G_{12} = 0.65 Msi = 4.5 GPa	
ν_{12} = 0.30	
ρ = 0.056 lb/in^3 = 1.55 g/cm^3	
F_1^{tu} = 110 ksi = 760 MPa	F_1^{cu} = 110 ksi = 760 MPa
F_2^{tu} = 4.0 ksi = 28 MPa	F_2^{cu} = 20.0 ksi = 138 MPa
F_{12}^{su} = 9.0 ksi = 62 MPa	
v_f = 0.6	t_ℓ = 0.0052 in = 0.13 mm

Because of the infinite variability of the angular orientation of the individual laminae, one would assume that a laminate having a stiffness which behaves isotropically in the plane of the laminate could be constructed by using many plies having small, equal differences in their orientation. It can be shown that a symmetric, quasi-isotropic laminate can be constructed with as few as six plies as a $[0/\pm 60]_s$ laminate. A general rule for describing a quasi-isotropic laminate states that the angles between the plies are equal to π/N, where N is an integer greater than or equal to 3, and there is an identical number of plies at each orientation in a symmetric laminate. For plies of a given material, all such quasi-isotropic laminates will have the same elastic properties, regardless of the value of N.

A quasi-isotropic laminate has in-plane stiffnesses which follow isotropic relationships

$$E_x = E_y = E_\theta \qquad 4.3.3.1(e)$$

where the subscript θ indicates any arbitrary angle. Additionally,

$$G_{xy} = \frac{E_x}{2(1+\nu_{xy})} \qquad 4.3.3.1(f)$$

There are two items which must be remembered about quasi-isotropic laminates. First and foremost, only the elastic in-plane properties are isotropic; the strength properties, in general, will vary with directions. The second item is that two equal moduli $E_x = E_y$ do not necessarily indicate quasi-isotropy, as demonstrated in Table 4.3.3.1(b). The first two laminates in Table 4.3.3.1(b) are actually the same (a $[0/90]_s$ laminate rotated 45° is a $[\pm 45]_s$ laminate). Note that the extensional moduli of these laminates are not the same and that the shear modulus of each laminate is not related to the extensional modulus and Poisson's ratio. For these laminates, the π/N relation has not been satisfied and they are not quasi-isotropic. The third laminate has plies oriented at 45° to each other but there are not equal numbers of plies at each angle. This laminate is also not quasi-isotropic.

TABLE 4.3.3.1(b) *Elastic properties of laminates.*

	$E_x = E_y$ Msi (GPa)	ν_{xy}	G_{xy} Msi (GPa)
$[0°/90°]_s$	13.4 (92.5)	0.038	0.65 (4.5)
$[\pm45°]_s$	2.38 (16.4)	0.829	6.46 (44.5)
$[0°/90°/\pm45°/-45°/90°/0°]_s$	11.0 (75.6)	0.213	2.59 (17.9)

This discussion of symmetric laminates has centered on membrane behavior. Symmetric laminates can be constructed which are very well behaved in the membrane sense. The bending behavior of symmetric laminates is considerably more complex, primarily due to the arrangement of plies through the thickness of the laminate.

4.3.3.2 Bending

The equations for bending analysis of symmetric laminates has been developed with the extensional analysis. The first complication that arises in the treatment of laminate bending deals with relationships between the extensional (A) and bending (D) elastic properties. In composite laminates, there is no direct relationship between extensional and bending stiffnesses, unlike the case of a homogeneous material where

$$D = \frac{A(2h)^2}{12} \qquad 4.3.3.2$$

In determining the membrane stiffnesses (A), the position of the ply through the thickness of the laminate does not matter (Equation 4.3.2(m)). The relations for the bending stiffnesses are a function of the third power of the distance of the ply from the mid-surface. Therefore, the position of the plies with respect to the mid-surface is critical. The effects of ply position in a unit thickness laminate are shown in Table 4.3.3.2(a).

The three laminates shown in Table 4.3.3.2(a) are all quasi-isotropic. The membrane properties are isotropic and identical for each of the laminates. The bending stiffnesses can be seen to be a strong function of the thickness position of the plies. Additionally, bending stiffness calculations based on homogeneity (Equation 4.3.3.2) do not correspond to lamination theory calculations. Thus, the simple relations between extensional and bending stiffnesses are lost and lamination theory must be used for bending properties. Table 4.3.3.2(a) also demonstrates that quasi-isotropy holds only for extensional stiffnesses.

Another complication apparent in Table 4.3.3.2(a) involves the presence of the bending-twisting coupling terms, D_{16} and D_{26}. The corresponding extensional-shear coupling terms are zero because of the presence of pairs of layers at ±60° orientations. Noting that the bending-twisting terms can be of the same order of magnitude as the principal bending terms, D_{11}, D_{22}, and D_{66}, the bending-twisting effect can be severe. This effect can be reduced by the proper selection of stacking sequence.

TABLE 4.3.3.2a Extensional and bending stiffnesses.

	$[0/\pm60]_s$		$[\pm60/0]_s$		$[60/0/-60]_s$		Homogeneous Laminate	
A_{11}	1.05×10^7	(7.30×10^{10})	1.05×10^7	(7.30×10^{10})	1.05×10^7	(7.30×10^{10})	1.05×10^7	(7.30×10^{10})
A_{12}	3.42×10^6	(2.38×10^{10})	3.42×10^6	(2.38×10^{10})	3.42×10^6	(2.38×10^{10})	3.42×10^6	(2.38×10^{10})
A_{22}	1.05×10^7	(7.30×10^{10})	1.05×10^7	(7.30×10^{10})	1.05×10^7	(7.30×10^{10})	1.05×10^7	(7.30×10^{10})
A_{66}	3.55×10^6	(2.47×10^{10})	3.55×10^6	(2.47×10^{10})	3.55×10^6	(2.47×10^{10})	3.55×10^6	(2.47×10^{10})
D_{11}	1.55×10^6	(1.08×10^{10})	3.36×10^5	(2.34×10^9)	7.42×10^5	(5.16×10^9)	8.75×10^5	(6.09×10^9)
D_{12}	1.50×10^5	(1.04×10^9)	3.92×10^5	(2.73×10^9)	3.12×10^5	(2.17×10^9)	2.85×10^5	(1.98×10^8)
D_{16}	4.74×10^4	(3.30×10^8)	9.50×10^4	(6.61×10^8)	1.42×10^5	(9.88×10^8)	0.0	(0.0)
D_{22}	4.69×10^5	(3.26×10^9)	1.20×10^6	(8.35×10^9)	9.59×10^5	(6.67×10^9)	8.75×10^5	(6.09×10^9)
D_{26}	1.42×10^5	(9.88×10^8)	2.81×10^5	(1.95×10^9)	4.22×10^5	(2.94×10^9)	0.0	(0.0)
D_{66}	1.63×10^5	(1.13×10^9)	4.04×10^5	(2.81×10^9)	3.23×10^5	(2.25×10^9)	2.96×10^5	(2.06×10^9)

Lamina properties are from Table 4.3.3.1(a); unit thickness laminate.

[A] lb/in (N/m)   [D] in-lb (N/m)

Another example that shows how the laminate stacking sequence (LSS) can significantly affect composite behavior is the bending stiffness of a laminated beam with rectangular cross-section (h ≡ laminate thickness). For the purpose of this example, define effective in-plane and bending moduli along the beam axis as

$$E_x = \frac{1}{A'_{11}h} \tag{4.3.3.2(b)}$$

$$E_x^b = \frac{12}{D'_{11}h^3} \tag{4.3.3.2(c)}$$

respectively. The relationship,

$$\Delta = \frac{E_x^b - E_x}{E_x} \times 100 \tag{4.3.3.2(d)}$$

provides a relative measure of the effect of LSS on beam bending stiffness. Bending moduli of laminated beams approach those of homogeneous beams as the number of plies increase provided that there is no preferential stacking of ply orientations through the thickness.

Table 4.3.3.2(b) shows lamination theory predictions of in-plane and effective bending moduli for beams with seven different LSS variations of a 16-ply, carbon/epoxy, quasi-isotropic lay-up.[1] Note that the in-plane moduli are independent of LSS because all lay-ups are symmetric. Bending moduli are shown to vary significantly above or below the in-plane moduli depending on preferential stacking of 0° plies towards the surface or center of the laminate, respectively.

TABLE 4.3.3.2(b) *Stiffness predictions for seven different LSS for 16-ply, quasi-isotropic, carbon/epoxy, laminated beams.*

Stacking Sequence	In-plane Modulus E_x		Bending Modulus E_x^b		Percent Difference Δ
	Msi	GPa	Msi	GPa	%
$[0_2/(\pm45)_2/90_2]_s$	7.67	52.9	12.8	88.2	67
$[0/\pm45/90]_{2s}$	7.67	52.9	10.1	69.6	32
$[\pm45/0_2/\pm45/90_2]_s$	7.67	52.9	7.80	53.8	1.7
$[\pm45/0/90]_{2s}$	7.67	52.9	6.51	44.9	-15
$[(\pm45)_2/0_2/90_2]_s$	7.67	52.9	4.45	30.7	-42
$[(\pm45)_2/90_2/0_2]_s$	7.67	52.9	3.42	23.6	-55
$[90_2/(\pm45)_2/0_2]_s$	7.67	52.9	3.25	22.4	-58

Properties for T300/934 ($V_f = 0.63$): E_{11}= 20.0 Msi (138 GPa), E_{22} = 1.4 Msi (9.7 GPa), G_{12} = 0.65 Msi (4.5 Gpa), ν_{12} = 0.31, Ply Thickness = 0.0056 in. (0.14 mm)

In general, the relationship between effective bending moduli and stacking sequence can be more complex than that shown in Table 4.3.3.2(b). Predictions in the table assumed that the basic lamina moduli were constant (i.e., linear elastic behavior). Depending on material type and the degree of accuracy desired, this assumption may lead to poor predictions. Lamina moduli for graphite/epoxy have been shown to depend on environment and strain level. Since flexure results in a distribution of tension and compression strains through the laminate thickness, nonlinear elastic lamination theory predictions may be more appropriate.

The example from Table 4.3.3.2(b) shows a significant effect of LSS on bending moduli of laminated beams. Similarly, calculations with Equation 4.3.2(n) can be used to indicate that LSS has a strong influence on the bending behavior of laminated plates. However, the bending response of common structures may

[1]The LSS used in Table 4.3.3.2(b) were chosen for illustrative purposes only and do not represent optimal LSS for a given application.

depend more on the resulting moment of inertia, I, for a given geometry than on LSS. This is particularly true for stringer geometries typically used to stiffen composite plates in aerospace structures.

Figure 4.3.3.2 illustrates how structural geometry of a beam section can overshadow the effects of LSS on bending. Web and flange members of each I-beam have LSS indicated in the legend of Figure 4.3.3.2[1] These LSS are the same as those used in Table 4.3.3.2(b). The ordinate axis of the figure indicates a percent difference between laminated and homogeneous beam calculations. As shown in Figure 4.3.3.2, the effect of LSS on the EI of an I-beam diminishes rapidly with increasing web height.

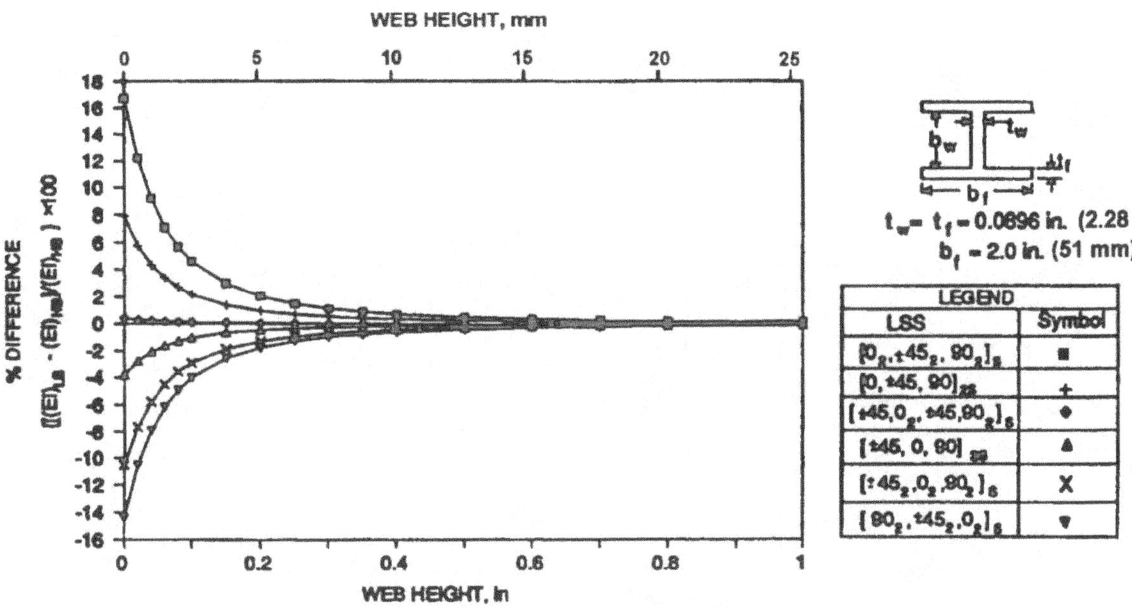

Homogeneous Beam (HB)

$$(EI)_{HB} = E_x\{2[b_f t_f^3/12 + b_f t_f (b_w+t_f)^2/4] + t_w b_w^3/12\}$$

Laminated Beam (LB)

$$(EI)_{LB} = 2[E_x^b b_f t_f^3/12 + E_x b_f t_f (b_w+t_f)^2/4] + E_x t_w b_w^3/12$$

E_x, E_x^b are listed in Table 4.3.3.2 (b).

FIGURE 4.3.3.2 *Laminated and homogeneous EI calculations for an I-beam stringer geometry with variable web height.*

Additional information on laminate stacking sequence effects is found in Section 4.6.5.

4.3.3.3 Thermal expansion

As the use of composite materials becomes more commonplace, they are subjected to increasingly severe mechanical and environmental loading conditions. With the advent of high temperatures in systems, the range of temperatures over which composite systems can be used has increased. The response of laminates to temperature and moisture, as well as to applied loads, must be understood. Previously, laminate

[1]The LSS used in Figure 4.3.3.2 were chosen for illustrative purposes only and do not represent optimal LSS for a given application.

extensional and bending stiffnesses were determined; in this section laminate conductivities and expansion coefficients will be defined.

To determine the laminate thermal expansion coefficients and thermally-induced stresses quantitatively, begin at the ply level. The thermoelastic relations for strain in the principal material directions are

$$\{\epsilon_\ell\} = \{\epsilon_\ell^M\} + \{\alpha_\ell\}\Delta T \qquad 4.3.3.3(a)$$

or

$$\{\epsilon_\ell\} = \{\epsilon_\ell^M\} + \{\epsilon_\ell^T\} \qquad 4.3.3.3(b)$$

where

$\{\epsilon_\ell^M\}$ = strain induced by stress

The change in temperature is represented by ΔT and the vector $\{\alpha_\ell\}$ represents the free thermal expansion coefficients of a ply. The individual components are

$$\{\alpha_\ell\} = \begin{Bmatrix} \alpha_1 \\ \alpha_2 \\ 0 \end{Bmatrix} \qquad 4.3.3.3(c)$$

The thermal strains, $\{\alpha_\ell\}\Delta T$, are the lamina free thermal expansions, which produce no stress in an unconstrained lamina. The thermal expansion coefficients α_1 and α_2 are the effective thermal expansion coefficients α_1^* and α_2^* of the unidirectional composite.

Substituting for the mechanical strain terms in Equation 4.3.3.3(a) and inverting yields

$$\{\sigma_\ell\} = [Q]\{\epsilon_\ell\} - \{\Gamma_\ell\}\Delta T \qquad 4.3.3.3(d)$$

where

$$\{\Gamma_\ell\} = [Q]\{\alpha_\ell\}$$

The components in the thermal stress coefficient vector $\{\Gamma_\ell\}$ are

$$\{\Gamma_\ell\} = \begin{Bmatrix} \dfrac{E_1\alpha_1 + \nu_{12}E_2\alpha_2}{\Delta} \\ \dfrac{E_2\alpha_2 + \nu_{12}E_1\alpha_1}{\Delta} \\ 0 \end{Bmatrix} \qquad 4.3.3.3(e)$$

where

$$\Delta = 1 - \frac{E_2}{E_1}\nu_{12}^2$$

The vector $\{\Gamma_\ell\}\Delta T$ physically represents a correction to the stress vector which results from the full constraint of the free thermal strains in a lamina. Both the thermal expansion vector, $\{\alpha_\ell\}\Delta T$, and the thermal stress vector, $\{\Gamma_\ell\}\Delta T$, can be transformed to arbitrary coordinates using the relations developed for stress and strain transformations, Equations 4.3.1(k) - (n).

With the transformed thermal expansion and stress vectors, the thermal elastic laminate relations can be developed. Following directly from the development of Equations 4.3.2(g) - (l), the membrane relations are:

$$\{N\} = [A]\{\epsilon^\circ\} + [B]\{\kappa\} + \{N^T\} \quad 4.3.3.3(f)$$

where

$$\{N^T\} = -\int_{-h}^{h} \{\Gamma_x\}\Delta T\, dz \quad 4.3.3.3(g)$$

Similarly, the bending relations are

$$\{M\} = [B]\{\epsilon^\circ\} + [D]\{\kappa\} + \{M^T\} \quad 4.3.3.3(h)$$

where

$$\{M^T\} = -\int_{-h}^{h} \{\Gamma_x\}\Delta T\, zdz \quad 4.3.3.3(i)$$

The integral relations for the thermal stress resultant vector $\{N^T\}$ and thermal moment resultant vector $\{M^T\}$ can be evaluated only when the change in temperature through the thickness is known. For the case of uniform temperature change through the thickness of a laminate, the term ΔT is constant and can be factored out of the integral, yielding:

$$\{N^T\} = -\Delta T \sum_{i=1}^{N} \{\Gamma_x\}^i (z_i - z_{i-1}) \quad 4.3.3.3(j)$$

$$\{M^T\} = -\frac{1}{2}\Delta T \sum_{i=1}^{N} \{\Gamma_x\}^i \left(z_i^2 - z_{i-1}^2\right) \quad 4.3.3.3(k)$$

With Equations 4.3.3.3(f) - (i), it is possible to determine effective laminate coefficients of thermal expansion and thermal curvature. These quantities are the extension and curvature changes resulting from a uniform temperature distribution.

Noting that for free thermal effects $\{N\} = \{M\} = 0$, and defining a free thermal expansion vector as

$$\{\alpha_x\} = \{\epsilon^\circ\}\frac{1}{\Delta T} \quad 4.3.3.3(l)$$

and a free curvature vector as

$$\{\delta_x\} = \{\kappa\}\frac{1}{\Delta T} \quad 4.3.3.3(m)$$

Equations 4.3.3.3(f) - (i) can be solved. After suitable matrix manipulations, the following expressions for thermal expansion and thermal curvature for symmetric laminates are found:

$$\{\alpha_x\} = -\frac{1}{\Delta T}[A]^{-1}\{N^T\} \quad 4.3.3.3(n)$$

$$\{\delta_x\} = -\frac{1}{\Delta T}[D]^{-1}\{N^T\} \quad 4.3.3.3(o)$$

If the relation for $\{M^T\}$ in Equation 4.3.3.3(i) is examined, symmetry eliminates the $\{M^T\}$ vector. Therefore $\{\delta_x\} = 0$ and no curvatures occur due to uniform temperature changes in symmetric laminates.

The variation of the longitudinal thermal expansion coefficient for a symmetric angle-ply laminate is shown in Figure 4.3.3.3 to illustrate the effect of lamina orientation. At 0° the term α_x is simply the axial lamina coefficient of thermal expansion, and at 90°, α_x equals the lamina transverse thermal expansion coefficient. An interesting feature of the curve is the large negative value of α_x in the region of 30°. Referring to Figure 4.3.3.1, the value of Poisson's ratio also behaves peculiarly in the region of 30°. The odd variation

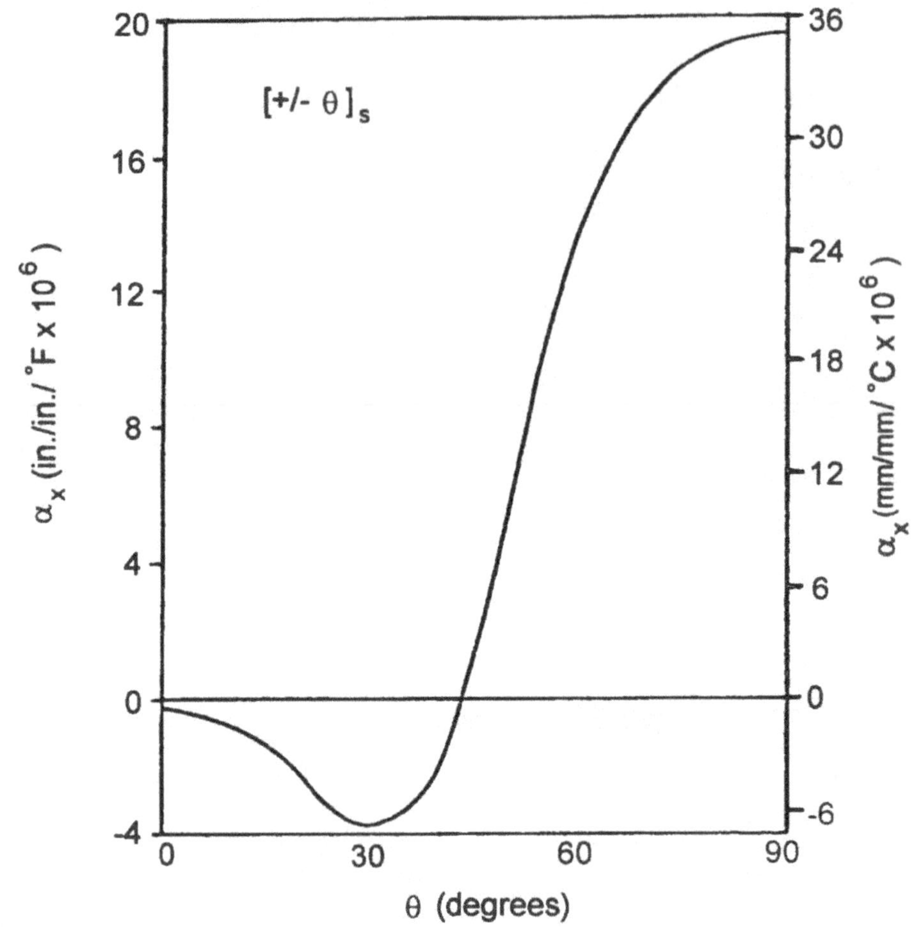

FIGURE 4.3.3.3 *Thermal expansion coefficients for high modulus carbon/epoxy.*

of both the coefficient and Poisson's ratio stems from the magnitude and sign of the shear-extensional coupling present in the individual laminae.

Previously, classes of laminates were shown to have isotropic stiffnesses in the plane of the laminate. Similarly, laminates can be specified which are isotropic in thermal expansion within the plane of the laminate. The requirements for thermal expansion isotropy are considerably less restrictive than those for elastic constants. In fact, any laminate which has two identical, orthogonal thermal expansion coefficients and a zero shear thermal expansion coefficient is isotropic in thermal expansion. Therefore, $[0/90]_s$ and $[\pm 45]_s$ laminates are isotropic in thermal expansion even though they are not quasi-isotropic for elastic stiffnesses.

Laminates which are isotropic in thermal expansion have thermal expansions of the form:

$$\alpha_x = \begin{Bmatrix} \alpha_x \\ \alpha_y \\ \alpha_{xy} \end{Bmatrix} = \begin{Bmatrix} \alpha^* \\ \alpha^* \\ 0 \end{Bmatrix} \qquad 4.3.3.3(p)$$

where the term α^* can be shown to be a function of lamina properties only, as follows:

$$\alpha^* = \alpha_1 + \frac{(\alpha_2 - \alpha_1)(1 + \nu_{12})}{1 + 2\nu_{12} + \frac{E_1}{E_2}} \qquad 4.3.3.3(q)$$

Thus, all laminates of a given ply material, which are isotropic in thermal expansion, have identical thermal expansion coefficients.

4.3.3.4 Moisture expansion

The term hygroelastic refers to the phenomenon in resin matrix composites when the matrix absorbs and desorbs moisture from and to the environment. The primary effect of moisture is a volumetric change in the laminae. When a lamina absorbs moisture, it expands, and when moisture is lost, the lamina contracts. Thus, the effect is very similar to thermal expansion.

In a lamina, a free moisture expansion vector can be defined as

$$\{\epsilon_\ell\} = \{\beta_\ell\}\Delta c \qquad 4.3.3.4(a)$$

where

$$\{\beta_\ell\} = \begin{pmatrix} \beta_1 \\ \beta_2 \\ 0 \end{pmatrix} \qquad 4.3.3.4(b)$$

and Δc is the change in specific moisture. Noting that the relations 4.3.3.4(a) and (b) are identical to thermal expansion with $\{\beta_\ell\}$ substituted for $\{\alpha_\ell\}$ and Δc for ΔT, it can easily be seen that all the relations developed for thermal effects can be used for moisture effects.

4.3.3.5 Conductivity

The conductivity (thermal or moisture) of a laminate in the direction normal to the surface is equal to the transverse conductivity of a unidirectional fiber composite. This follows from the fact that normal conductivity for all plies is identical and unaffected by ply orientation.

In-plane conductivities will be required for certain problems involving spatial variations of temperature and moisture. For a given uniform state of moisture in a laminate, the effective thermal conductivities in the x and y directions can be obtained by methods entirely analogous to those used for stiffnesses in Section 4.3.2:

$$\mu_x = \frac{1}{2h} \sum_{i=1}^{N} (\mu_1 m^2 + \mu_2 n^2)\, t_\ell^i \qquad 4.3.3.5$$

where

μ_1 = conductivity in the fiber direction
μ_2 = conductivity transverse to the fibers
m = $\cos \theta^i$
n = $\sin \theta^i$
θ^i = orientation of ply i
t_ℓ^i = thickness of ply i
N = the number of plies
$2h$ = laminate thickness

The results apply to both symmetric and unsymmetric laminates. The results for moisture conductivity are identical.

4.3.4 Thermal and hygroscopic analysis

The distribution of temperature and moisture through the thickness of a laminate influences the behavior of that laminate. The mathematical descriptions of these two phenomena are identical and the physical effects are similar. Some of these aspects have already been discussed in Sections 4.2.2.3 - 4.2.2.4 and 4.3.3.3 - 4.3.3.5.

A free lamina undergoes stress-free deformation due to temperature change or moisture swelling. In a laminate, stress-free deformation is constrained by adjacent layers producing internal stresses. In addition to these stresses, temperature and moisture content also affect the properties of the material. These effects are primarily related to matrix-dominated strength properties.

The principal strength-degrading effect is related to a change in the glass transition temperature of the matrix material. As moisture is absorbed, the temperature at which the matrix changes from a glassy state to a viscous state decreases. Thus, the elevated-temperature strength properties decrease with increasing moisture content. Limited data suggest that this process is reversible. When the moisture content of the composite is decreased, the glass transition temperature increases and the original strength properties return.

The same considerations also apply for a temperature rise. The matrix, and therefore the lamina, lose strength and stiffness when the temperature rises. Again, this effect is primarily important for the matrix-dominated properties such as E_2, G_{12}, F_2^{tu}, F_2^{cu}, and F_{12}^{su}.

The differential equation governing time-dependent moisture sorption of an orthotropic homogeneous material is given by

$$D_1 \frac{\partial^2 c}{\partial x_1^2} + D_2 \frac{\partial^2 c}{\partial x_2^2} + D_3 \frac{\partial^2 c}{\partial x_3^2} = \frac{\partial c}{\partial t} \qquad 4.3.4(a)$$

where

t = time
x_1, x_2, x_3 = coordinates in principal material directions
c = specific moisture concentration
D_1, D_2, D_3 = moisture diffusivity coefficients

Equation 4.3.4(a) is based on Fick's law of moisture diffusion. The equation is analogous to the equation governing time dependent heat conduction with temperature ϕ replacing concentration c and thermal conductivities μ_1, μ_2, and μ_3 replacing the moisture diffusivities. For a transversely isotropic lamina with x_1 in the fiber direction, x_2 in the transverse direction, and $x_3 = z$ in the direction normal to the lamina,

$$D_2 = D_3 \qquad 4.3.4(b)$$

These quantities are analogous to the thermal conductivities of a unidirectional fiber composite and have been discussed in Section 4.2.2.4.

An important special case is one-dimensional diffusion or conduction through the thickness of a lamina. In this case, Equation 4.3.4(a) reduces to

$$D_3 \frac{\partial^2 c}{\partial z^2} = \frac{\partial c}{\partial t} \qquad 4.3.4(c)$$

This equation also applies to moisture diffusion or thermal conduction through a laminate, in the direction normal to its laminae planes, since all laminae are homogeneous in the z direction with equal diffusion coefficients, $D_3 = D_z$.

Equation 4.3.4(c) is applicable to the important problem of time-dependent moisture diffusion through a laminate where the two faces are in different moisture environments. After a sufficiently long time has elapsed, the concentration reaches a time-independent state. In this state, since c is no longer time-dependent, Equation 4.3.4(c) simplifies to

$$\frac{d^2c}{dz^2} = 0 \qquad 4.3.4(d)$$

The specific moisture concentration is a linear function of z and, if the laminate faces are in environments with constant saturation concentrations, c_1 and c_2, then

$$c = \frac{1}{2}\left[(c_2 - c_1) z/h + c_2 + c_1\right] \qquad 4.3.4(e)$$

where the laminate thickness is 2h and z originates at the mid-surface. In the case where $c_1 = c_2$, Equation 4.3.4(e) reduces to $c = c_1$ = constant as would be expected.

The above discussion of moisture conduction also applies to heat conduction.

Solutions to the time-dependent problem are readily available and considerable work has been performed in the area of moisture sorption (Reference 4.3.4). The most interesting feature of the solutions relates to the magnitude of the coefficient D_z. This coefficient is a measure of how fast moisture diffusion can occur. In typical epoxy matrix systems, D_z is of the order of 10^{-8} (in^2/s, cm^2/s) to 10^{-10} (in^2/s, cm^2/s). The diffusion coefficient is sufficiently small that full saturation of a resin matrix composite may require months or years even when subjected to 100% relative humidity.

The approach typically taken for design purposes is to assume a worst case. If the material is assumed to be fully saturated, it is possible to compute reduced allowable strengths. This is a conservative approach, since typical service environments do not generate full saturation. This approach is used since it allows for inclusion of moisture effects in a relatively simple fashion. It is to be expected that as the design data base and analytical methodologies mature, more physically realistic methods will be developed.

For heat conduction, the time required to achieve the stationary, or time-independent, state is extremely small. Therefore, the transient time-dependent state is generally of little practical importance for laminates.

4.3.4.1 Symmetric laminates

The laminate stacking sequence (LSS) can be chosen to control the effects of environment on stiffness and dimensional stability. When considering the special case of constant temperature and moisture content distributions in symmetric laminates, the effect of environment on in-plane stiffness relates to the relative percentages of chosen ply orientations. For example, LSS dominated by 0° plies will have longitudinal moduli that are nearly independent of environment. Note that increasing the environmental resistance of one laminate in-plane modulus may decrease another.

Bending and torsional stiffnesses depend on both LSS and environment. Preferential stacking of outer ply groups having relatively high extensional or shear moduli will also promote high bending or torsional stiffness, respectively. As with in-plane moduli, the higher the bending or torsional stiffness the better the corresponding environmental resistance. When optimizing environmental resistance, compromises between longitudinal bending, transverse bending and torsion need to be made due to competing relationships with LSS.

Unsymmetric temperature and moisture content distributions will affect the components of the stiffness matrix [ABD] differently, depending on LSS. In general, coupling components which were zero for symmetric laminates having symmetric temperature and moisture content distributions become nonzero for an unsymmetric environmental state. This effect can be minor or significant depending on LSS, material type, panel thickness and the severity of temperature/moisture content gradients.

Environmentally-induced panel warpage will occur in symmetric laminates when conditions yield an unsymmetric residual stress distribution about the laminate midplane. This may occur during the cure process due to uneven heating or crystallization through the laminate thickness. Unsymmetric temperature and moisture content distributions can also lead to panel warpage in symmetric laminates. This is due to the unsymmetric shrinkage or swelling through the laminate thickness.

4.3.4.2 Unsymmetric laminates

The in-plane thermal and moisture expansion of unsymmetric laminated plates subjected to any environmental condition (i.e., constant, symmetric and unsymmetric temperature and moisture content distributions) is dependent on LSS (e.g., Reference 4.3.4.2(a)). In general, environmentally induced panel warpage occurs with unsymmetric laminates.

Panel warpage in unsymmetric laminates depends on LSS and changes as a function of temperature and moisture content. Zero warpage will occur in unsymmetric laminates only when temperature and moisture content distributions result in either zero or symmetric residual stress distributions. Equilibrium environmental states that result in zero residual stresses are referred to as stress-free conditions (see Reference 4.3.2(e).

Since unsymmetric LSS warp as a function of temperature and moisture content, their use in engineering structures has generally been avoided. The warped shape of a given unsymmetric laminate has been found to depend on LSS and ratios of thickness to in-plane dimensions (e.g., References 4.3.4.2(b) and (c)). Relatively thin laminates tend to take a cylindrical shape rather than the saddle shape predicted by classical lamination theory. This effect has been accurately modeled using a geometrically nonlinear theory.

Additional information on laminate stacking sequence effects is found in Section 4.6.5.

4.3.5 Laminate stress analysis

The physical properties defined in Section 4.3.3 enable any laminate to be represented by an equivalent homogeneous anisotropic plate or shell element for structural analysis. The results of such analyses will be the definition of stress resultants, bending moments, temperature, and moisture content at any point on the surface which defines the plate. With this definition of the local values of state variables, a laminate analysis can be performed to determine the state of stress in each lamina to assess margins for each critical design condition.

4.3.5.1 Stresses due to mechanical loads

To determine stresses in the individual plies, the laminate mid-plane strain and curvature vectors are used. Writing the laminate constitutive relations

$$\begin{Bmatrix} N \\ --- \\ M \end{Bmatrix} = \begin{bmatrix} A & | & B \\ --- & | & --- \\ B & | & D \end{bmatrix} \begin{Bmatrix} \epsilon^\circ \\ --- \\ \kappa \end{Bmatrix} \qquad 4.3.5.1(a)$$

a simple inversion will yield the required relations for $\{\epsilon^\circ\}$ and $\{\kappa\}$. Thus

$$\begin{Bmatrix} \epsilon^\circ \\ --- \\ \kappa \end{Bmatrix} = \begin{bmatrix} A & | & B \\ --- & | & --- \\ B & | & D \end{bmatrix}^{-1} \begin{Bmatrix} N \\ --- \\ M \end{Bmatrix} \qquad 4.3.5.1(b)$$

Given the strain and curvature vectors, the total strain in the laminate can be written as

$$\{\epsilon_x\} = \{\epsilon^\circ\} + z\,\{\kappa\} \qquad 4.3.2(d)$$

The strains at any point through the laminate thickness are now given as the superposition of the mid-plane strains and the curvatures multiplied by the distance from the mid-plane. The strain field at the center of ply i in a laminate is

$$\{\epsilon_x\}^i = \{\epsilon^\circ\} + \frac{1}{2}\{\kappa\}(z^i + z^{i-1}) \qquad 4.3.5.1(c)$$

where the term

$$\frac{1}{2}(z^i + z^{i-1})$$

corresponds to the distance from the mid-plane to the center of ply i. It is possible to define curvature induced strains at a point through the laminate thickness simply by specifying the distance from the mid-plane to the point in question.

The strains defined in Equation 4.3.5.1(c) correspond to the arbitrary laminate coordinate system. These strains can be transformed into the principal material coordinates for this ply using the transformations developed previously (Equation 4.3.1(m)). Thus

$$\{\epsilon_\ell\}^i = \left[\theta^i\right]^{-1}\{\epsilon_x\}^i \qquad 4.3.5.1(d)$$

where the superscript i indicates which layer and, therefore, which angle of orientation to use.

With the strains in the principal material coordinates defined, stresses in the same coordinates are written by using the lamina reduced stiffness matrix (Equation 4.3.1(h)).

$$\{\sigma_\ell\}^i = \left[Q^i\right]\{\epsilon_\ell\}^i \qquad 4.3.5.1(e)$$

Again, the stiffness matrix used must correspond to the correct ply, as each ply may be a different material.

The stresses in the principal material coordinates can be determined without the use of principal material strains. Using the strains defined in the laminate coordinates (Equation 4.3.5.1(c)) and the transformed lamina stiffness matrix (Equations 4.3.1(o,q,r)), stresses in the laminate coordinate system can be written as

$$\{\sigma_x\}^i = \left[\bar{Q}^i\right]\{\epsilon_x\}^i \qquad 4.3.5.1(f)$$

and these stresses are then transformed to the principal material coordinates using the relations 4.3.1(m). Thus

$$\{\sigma_\ell\}^i = \left[\theta^i\right]^{-1}\{\sigma_x\}^i \qquad 4.3.5.1(g)$$

By reviewing these relations, it can be seen that, for the case of symmetric laminates and membrane loading, the curvature vector is zero. This implies that the laminate coordinate strains are identical in each ply and equal to the mid-plane strains. The differing angular orientation of the various plies will promote different stress and strain fields in the principal material coordinates of each ply.

4.3.5.2 Stresses due to temperature and moisture

In Section 4.3.3.3, equations for the thermoelastic response of composite laminates were developed. It was indicated that thermal loading in laminates can cause stresses even when the laminate is allowed to expand freely. The stresses are induced because of a mismatch in thermal expansion coefficients between plies oriented in different directions. Either the mechanical stresses of the preceding section or the thermomechanical stresses can be used to evaluate laminate strength.

To determine the magnitude of thermally induced stresses, the thermoelastic constitutive relations (Equations 4.3.3.3(f) - (i)) are required. Noting that free thermal stress effects require that $\{N\} = \{M\} = 0$, these relations are written as

$$\begin{Bmatrix} 0 \\ --- \\ 0 \end{Bmatrix} = \begin{bmatrix} A & | & B \\ --- & | & --- \\ B & | & D \end{bmatrix} \begin{Bmatrix} \epsilon^{o} \\ --- \\ \kappa \end{Bmatrix} + \begin{Bmatrix} N^{T} \\ --- \\ M^{T} \end{Bmatrix} \qquad 4.3.5.2(a)$$

Inverting these relations yields the free thermal strain and curvature vectors for the laminate. Proceeding as before, the strain field in any ply is written as

$$\{\epsilon_x\}^i = \{\epsilon^{\circ}\} + \frac{1}{2}(z + z^{i-1})\{\kappa\} \qquad 4.3.5.2(b)$$

Stresses in the laminate coordinates are

$$\{\sigma_x\}^i = \left[\bar{Q}^{i}\right]\{\epsilon_x\}^i - \{\Gamma_x\}^i \Delta T^{i} \qquad 4.3.5.2(c)$$

which can then be transformed to the principal material coordinates. Thus

$$\{\sigma_\ell\} = \left[\theta^i\right]^{-1}\{\sigma_x\}^i \qquad 4.3.5.2(d)$$

The stresses can also be found by transforming the strains directly to principal material coordinates and then finding the principal material coordinate stresses.

For uniform temperature fields in symmetric laminates, the coupling matrix, $[B]$, and the thermal moment resultant vector, $\{M^T\}$, vanish and:

$$\{\epsilon^{\circ}\} = \{\alpha_x\}\Delta T \qquad 4.3.5.2(e)$$

and

$$\{\kappa\} = 0 \qquad 4.3.5.2(f)$$

In this case, the strains in the laminate coordinates are identical in each ply with the value

$$\{\epsilon_x\}^i = \{\epsilon^{\circ}\} = \{\alpha_x\}\Delta T \qquad 4.3.5.2(g)$$

and the stresses in the principal material coordinates are

$$\{\sigma_x\}^i = \left[\bar{Q}^{i}\right](\{\alpha\} - \{\alpha_x\}^i)\,\Delta T \qquad 4.3.5.2(h)$$

These relations indicate that the stresses induced by the free thermal expansion of a laminate are related to the differences between the laminate and ply thermal expansion vectors. Therefore, the stresses are proportional to the difference between the amount the ply would freely expand and the amount the laminate will allow it to expand.

A further simplification can be found if the laminate under investigation is isotropic in thermal expansion. It can be shown that, for this class of laminates subjected to a uniform temperature change, the stresses in the principal material coordinates are identical in every ply. The stress vector is

$$\{\sigma_\ell\} = \frac{E_{11}(\alpha_{22} - \alpha_{11})\Delta T}{1 + 2\nu_{12} + \dfrac{E_{11}}{E_{22}}}\begin{Bmatrix} 1 \\ -1 \\ 0 \end{Bmatrix} \qquad 4.3.5.2(i)$$

where it can be seen that the transverse direction stress is equal and opposite to the fiber direction stress.

Similar developments can be generated for moisture-induced stresses. All of the results of this section apply when moisture swelling coefficients, $\{\beta_\ell\}$, are substituted for thermal expansion coefficients, $\{\alpha_\ell\}$.

4.3.5.3 Netting analysis

Another approach to the calculation of ply stresses is sometimes used for membrane loading of laminates. This procedure is netting analysis and, as the name implies, treats the laminate as a net. All loads are carried in the fibers while the matrix material serves only to hold the geometric position of the fibers.

Since only fibers are assumed to load in this model, stress-strain relations in the principal material directions can be written as

$$\sigma_{11} = E_1\, \epsilon_{11} \qquad \text{4.3.5.3(a)}$$

or

$$\epsilon_{11} = \frac{1}{E_1}\, \sigma_{11} \qquad \text{4.3.5.3(b)}$$

and

$$E_2 = G_{12} = \sigma_{22} = \sigma_{12} = 0 \qquad \text{4.3.5.3(c)}$$

The laminate stiffnesses predicted with a netting analysis will be smaller than those predicted using lamination theory, due to the exclusion of the transverse and shear stiffnesses. This effect is demonstrated in Table 4.3.5.3 for a quasi-isotropic laminate comprised of high-modulus graphite/epoxy. The stiffness properties predicted using a netting analysis are approximately 10% smaller than lamination theory predictions. Experimental work has consistently shown that lamination theory predictions are more realistic than netting analysis predictions.

TABLE 4.3.5.3 *Laminate elastic constants.*

Analysis	E_x Msi (GPa)	E_y Msi (GPa)	G_{xy} Msi (GPa)	γ_{xy}
Lamination Theory	9.42 (64.9)	9.42 (64.9)	3.55 (24.5)	0.325
Netting Analysis	8.33 (57.4)	8.33 (57.4)	3.13 (21.6)	0.333

Although the stiffness predictions using netting analysis are of limited value, the analysis can be used as an approximation of the response of a composite with matrix damage. It may be considered as a worst case analysis and is frequently used to predict ultimate strengths of composite laminates.

4.3.5.4 Interlaminar stresses

The analytical procedures which have been developed can be used to predict stresses within each lamina of a laminate. The stresses predicted are planar due to the assumed state of plane stress. There are cases where the assumption of plane stress is not valid and a three-dimensional stress analysis is required.

An example of such a case exists at certain free edges in laminates where stress free boundary conditions must be imposed.

4.3.5.5 Nonlinear stress analysis

All the preceding material in this chapter has related to laminae which behave in a linear elastic fashion. Composites can behave in a non-linear manner due to internal damage or non-linear behavior of the matrix material. Matrix nonlinearity or micro-cracking can result in laminae which have non-linear stress-strain curves for transverse stress or axial shear stress. When this situation exists, the elastic laminate stress analysis of Section 4.3.5.1 must be replaced by a nonlinear analysis. A convenient procedure for the non-linear analysis is presented in Reference 4.3.5.5.

4.3.6 Summary

- When laminae are at an angle to the laminate reference axes, the lamina stiffness relations described in Section 4.2 must be transformed into the laminate coordinate system to perform laminate stress-strain analysis.

- Stresses and strains are related in the principal lamina material directions by 6 x 6 symmetric compliance $[S]$ and stiffness $[Q]$ matrices.

- The transformation of stresses and strains from the principal lamina material direction to the laminate coordinate system is accomplished by following the rules for transformation of tensor components (equations 4.3.1(k) and 4.3.1(m)).

- Lamination theory makes the same simplifications as classical thin plate theory for isotropic materials. Therefore, the procedures used to calculate stresses and deformations are dependent on the fact that laminate thickness is considerably smaller than the laminate's in-plane dimensions.

- The strain at a y point in a laminate is defined as the sum of the mid-surface strain (ϵ), and the product of the curvature (κ) and the distance from the mid surface (z).

- Laminate load (N) and moment (M) resultants are related to mid-plane strains and curvatures as described by the $[A]$, $[B]$, and $[D]$ 3 x 3 stiffness matrices (Equations 4.3.2 (k) - (m)).

- Two-dimensional lamination theory can generally be used to predict stresses within each lamina of a laminate. The planar stresses are predicted based on an assumption of plane stress. In cases where interlaminar stresses exist, three-dimensional stress analysis is required.

- In symmetric laminates, bending-extensional coupling is eliminated by a symmetric stacking sequence whereby $[B] = 0$.

- Since they are susceptible to warping as a result of processing and usage conditions, use of unsymmetric laminates in composite structures should generally be avoided for both design and manufacture.

4.4 LAMINATE STRENGTH AND FAILURE

Methods of stress analysis of laminates subjected to mechanical loads, temperature changes, and moisture absorption were presented in Section 4.3.5. The results of such a stress analysis can be used to assess the strength of a laminate. As a result of the complexity of the structure of a composite laminate, several modes of failure are possible, and it is desirable for the failure mode as well as the failure stress or strain to be predicted. The analytical problem is to define the failure surface for the laminate in either stress or strain space.

Laminate failure may be calculated by applying stress or strain limits at the laminate level or, alternatively, at the ply level. Ply level stresses or strains are the more frequently used approach to laminate strength. The average stresses in a given ply may be used to calculate either an onset of damage, which is frequently called "first ply failure", or a critical failure which is regarded as ultimate strength. In the former case, subsequent damages leading to laminate failure are then calculated. This calculation of subsequent damage is sometimes performed using the "sequential ply failure" methodology, and sometimes performed using "netting" analysis. These approaches are discussed subsequently. Four factors should be considered in assessing the validity of using ply level stresses for failure calculation. The first is the question of which tests (or analyses) should be used to define the ply strength values. In particular, it must be recognized that a crack parallel to the fibers may result in failure of a transverse tensile test specimen of a unidirectional composite, while the same crack may have an insignificant effect in a laminate test. The second factor is the assumption that local failures within a ply are contained within the ply and are determined solely by the stress/strain state in that ply. There is evidence that the former assumption is not valid under fatigue loading, during which a crack within one ply may well propagate into adjacent plies. In this case, the ply-by-ply model may not be the best analytical approach. Furthermore, matrix cracking within one ply is not determined uniquely by the stresses and strains within that ply but is influenced by the orientation of adjacent layers as well as by the ply thickness (Reference 4.4). The third factor is the existence of residual thermal stresses, usually of unknown magnitude, resulting from the fabrication process. The fourth factor is that it does not cover the possibility of delaminations which can occur, particularly at free edges. Thus, the analysis is limited to in-plane failures.

4.4.1 Sequential ply failure approach

4.4.1.1 Initial ply

To predict the onset of damage, consider stresses remote from the edges in a laminate which is loaded by in-plane forces and/or bending moments. If there is no external bending, if the membrane forces are constant along the edges, and if the laminate is balanced and symmetric, the stresses in the ith layers are constant and planar. With reference to the material axes of the laminae, fiber direction x_1 and transverse direction x_2, the stresses in the ith ply are written σ^i_{11}, σ^i_{22}, and σ^i_{12}. Failure is assumed to occur when the selected semi-empirical failure criteria involving these calculated stresses or the associated strains are satisfied. Numerous criteria have been proposed for calculation of onset of damage. These may be grouped into two broad categories - mode-based and purely empirical. Mode based criteria treat each identifiable physical failure mode, such as fiber-direction tensile failure and matrix-dominated transverse failure, separately. A purely empirical criterion generally consists of a polynomial combination of the three stress or strain components in a ply. Such criteria attempt to combine the effects of several different failure mechanisms into one function and may, therefore, be less representative than physically based criteria. All criteria rely on test data at the ply level to set parameters and are therefore at least partially empirical in nature.

The selection of appropriate criteria can be a controversial issue and the validity of any criterion is best determined by comparison with test data. As a consequence, different criteria may be best for different materials. Two mode-based failure criteria are presented here as examples: the maximum strain criteria and the failure criteria proposed by Hashin. It is important, however, for the engineer to consider the material, the application, and the test data in choosing and utilizing a failure criterion.

The maximum strain criteria may be written as

$$\begin{aligned} \epsilon_{11}^{cu} &\le \epsilon_{11}^{i} \le \epsilon_{11}^{tu} \\ \epsilon_{22}^{cu} &\le \epsilon_{22}^{i} \le \epsilon_{22}^{tu} \\ |\,\epsilon_{12}^{i}\,| &\le \epsilon_{12}^{su} \end{aligned} \qquad 4.4.1.1(a)$$

For given loading conditions, the strains in each ply are compared to these criteria. Whichever strain reaches its limiting value first indicates the failure mode and first ply to fail for those loading conditions. The limiting strains, ϵ_{11}^{tu}, ϵ_{11}^{cu}, etc., are the specified maximum strains to be permitted in any ply. Generally, these quantities are specified as some statistical measure of experimental data obtained by uniaxial loading of a unidirectional laminate. For example, in the case of axial strain, ϵ_{11}, a B-basis strain allowable from unidirectional tests can be used. Other limits may also be imposed. For example, in the case of shear strain, something equivalent to a "yield" strain may be used in place of the ultimate shear strain.

The failure criteria proposed by Hashin (Reference 4.4.1.1(a)) may be written as:

Fiber modes

Tensile

$$\left(\frac{\sigma_{11}}{F_1^{tu}}\right)^2 + \left(\frac{\sigma_{12}}{F_{12}^{su}}\right)^2 = 1 \qquad 4.2.4(f)$$

Compressive

$$\left(\frac{\sigma_{11}}{F_1^{cu}}\right)^2 = 1 \qquad 4.2.4(g)$$

Matrix modes

Tensile

$$\left(\frac{\sigma_{22}}{F_2^{tu}}\right)^2 + \left(\frac{\sigma_{12}}{F_{12}^{su}}\right)^2 = 1 \qquad 4.2.4(h)$$

Compressive

$$\left(\frac{\sigma_{22}}{2F_{23}^{su}}\right)^2 + \left[\left(\frac{F_2^{cu}}{2F_{23}^{su}}\right)^2 - 1\right]\left(\frac{\sigma_{22}}{F_2^{cu}}\right) + \left(\frac{\sigma_{12}}{F_{12}^{su}}\right)^2 = 1 \qquad 4.2.4(i)$$

It should be noted that some users of these criteria add a shear term to equation 4.2.4(g) to reflect the case in which shear mode instability contributes to the compressive failure mechanism (Reference 4.4.1.1(b)). In that case, equation 4.2.4(g) is replaced by:

$$\left(\frac{\sigma_{11}}{F_1^{cu}}\right)^2 + \left(\frac{\sigma_{12}}{F_{12}^{su}}\right)^2 = 1 \qquad 4.4.1.1(b)$$

The limiting stresses in the criteria, F_1^{cu}, F_{12}^{su}, etc., are the specified maximum stresses to be permitted in any ply. As with the case of strains, statistical data from unidirectional tests are generally used to define these quantities. However, as an example of the care required, it should be noted that the stress which produces failure of a 90° coupon in tension is not necessarily a critical stress level for a ply in a multidirectional laminate. One may wish to use, instead, the stress level at which crack density in a ply reduces the effective stiffness by a specified amount. Such a stress level could be determined by either a fracture mechanics analysis or testing of a crossply laminate (Reference 4.4).

In an onset of damage approach, the selected failure criteria are used for each layer of the laminate. The layer for which the criteria are satisfied for the lowest external load set will define the loading which produces the initial laminate damage. The layer which fails and the nature of the failure (i.e., fiber failure or cracking along the fibers) are identified. This is generally called first-ply failure. When the first ply failure is the result of fiber breakage, the resulting ply crack will introduce stress concentrations into the adjacent plies. In this case, it is reasonable to consider that first ply failure is equivalent to laminate failure. A different criterion exists when the first ply failure results from matrix cracking and/or fiber/matrix interface separations. Here it is reasonable to consider that the load-carrying capacity of the ply will be changed significantly when there is a substantial amount of matrix mode damage. Treatment of this case is discussed in the following section.

Additional concerns to be addressed in considering the initial failure or onset of damage include bending, edge stresses, and residual thermal stresses. Bending occurs when there are external bending and/or twisting moments or when the laminate is not symmetric. In these cases the stresses σ_{11}^i, σ_{22}^i, and σ_{12}^i in a layer are symmetric in x_3. Consequently, the stresses assume their maximum and minimum values at the layer interfaces. The failure criteria must be examined at these locations for each layer. Different approaches utilize the maximum value or the average value in such cases.

The evaluation of onset of failure as a result of the edge stresses is much more complicated as a result of the sharp gradients (indicated by analytical singularities) in these stresses. Numerical methods cannot uncover the nature of such stress singularity, but there are analytical treatments (e.g. Reference 4.4.1) which can. The implication of such edge stress fields for failure of the laminate is difficult to assess. This situation is reminiscent of fracture mechanics in the sense that stresses at a crack tip are theoretically infinite. Fracture mechanics copes with this difficulty with a criterion for crack propagation based on the amount of energy required to open a crack, or equivalently, the value of the stress intensity factor. Similar considerations may apply for laminate edge singularities. This situation in composite materials is more complicated since a crack initiating at the edge will propagate between anisotropic layers. It appears, therefore, that at the present time the problem of edge failure must be relegated to experimentation, or approximate analysis.

In the calculation of first-ply failure, consideration must also be given to residual thermal stresses. The rationale for including residual thermal stresses in the analysis is obvious. The stresses exist after processing. Therefore, they can be expected to influence the occurrence of first-ply failure. However, matrix materials exhibit viscoelastic, or time-dependent, effects, and it may be that the magnitude of the residual stresses will be reduced through a process of stress relaxation. Additionally, the processing stresses may be reduced through the formation of transverse matrix microcracks. The question of whether to include residual stresses in the analysis is complicated by difficulties in measuring these stresses in a laminate and by difficulties in observing first-ply failure during a laminate test. It is common practice to neglect the residual thermal stresses in the calculation of ply failure. Data to support this approach do not appear to be available. However, at the present time, damage tolerance requirements limit allowable strain levels in polymeric matrix laminates to 3000 to 4000 $\mu\epsilon$. This criterion becomes the dominant design restriction and obviates, temporarily, the need to resolve the effects of residual thermal stresses.

4.4.1.2 Subsequent failures

Often laminates have substantial strength remaining after the first ply has experienced a failure, particularly if that first failure is a matrix-dominated failure. A conservative approach for analyzing subsequent failure is to assume that the contribution of that first failed ply is reduced to zero. If failure occurs in the fiber-dominated mode, this may be regarded, as discussed earlier, as ultimate laminate failure. If not, then the stiffness in the fiber direction E_L is reduced to zero. If failure occurs in the matrix-dominated mode, the elastic properties E_T and G_L are reduced to zero. The analysis is then repeated until all plies have failed. Generally, the progressive failures of interest are initial and subsequent failures in the matrix mode. In that case, the basic assumptions for netting analysis result where the ultimate load is defined by E_T and G_L vanishing in all laminae. The basic issues involved in modeling post-first-ply behavior are described in Reference 4.4.2. For some materials and/or for some properties, matrix mode failures may not have an important effect. However, for some properties, such as thermal expansion coefficients, ply cracking may have a significant effect.

4.4.2 Fiber failure approach (laminate level failure)

In composites laminates, there are two characteristic stress or strain levels which can be considered in the evaluation of strength. One is the stress or strain state at which a non-catastrophic first-ply failure can occur and the other is the maximum static stress or strain state which the laminate can carry. In those cases where the material exhibits minimal micro-cracking, or where the application is such that effects of micro-cracking need not be considered, a failure criterion based only upon fiber failure may be used. A common practice in the aerospace industry is to use a failure criterion based only upon fiber strain allowables, for which fiber failure in any lamina is considered laminate ultimate failure. Hence, failure is a single event rather than the result of a process.

Perhaps the most common example of this laminate level failure criterion is a modification of the maximum strain criterion. The same assumptions of no external bending, membrane forces constant along the edges, and a balanced and symmetric laminate, are initially used. The basic lamina failure envelope is the same as the conventional maximum-strain envelope for tension- and compression-dominated loads, but introduces truncations in the tension-compression (shear) quadrants as shown in Figure 4.4.2. A critical assumption in this criterion is that the laminate behavior is fiber-dominated meaning that there are fibers in sufficient multiple directions such that strains are limited by the presence of the fibers to inhibit matrix cracking. In many practical applications, this typically translates into having fibers in (at least) each of four directions relative to the primary loads: 0°, 90°, and ±45°. Furthermore, plies are not "clustered" (that is, several plies of the same orientation are not layed together) in order to inhibit matrix macrocracking. With these assumptions, the first translation of the maximum strain criterion to the laminate level is a limiting of the strain in the transverse direction, θ_{90}, to the fiber direction limiting strain to reflect the fact that such "well-designed" laminates with fibers in multiple directions restrict strains in any in-plane direction. Alternatively, if there is reason to believe that matrix cracking will be structurally significant, the 90° strain cutoff based on fiber direction strain could be replaced by an empirically established tensile limit reflecting a matrix-dominated mode. This limit was originally expressed as a constant strain limit. However, if such a limit is based upon the case of a constant 90° stress in a ply, this would result in a sloped line in the strain plane with the slope related to the Poisson's ratio of the unidirectional lamina:

$$\alpha = \tan^{-1}\left(\nu_{LT}^{\text{lamina}}\right) \qquad \text{4.4.2(a)}$$

Such a cutoff is parallel to the uniaxial load line shown in Figure 4.4.2. It should be further noted that possible limitations due to lamina level shear strains are inoperative due to the assumption that the fibers in multiple directions restrict such strains to values below their failure values.

Many users recognize a need to truncate the maximum strain predictions in the tension-compression quadrants. While the particular truncations vary, perhaps the most widely used version is that shown in Figure 4.4.2. These truncations were originally based on data obtained for shear loading of such fiber-dominated

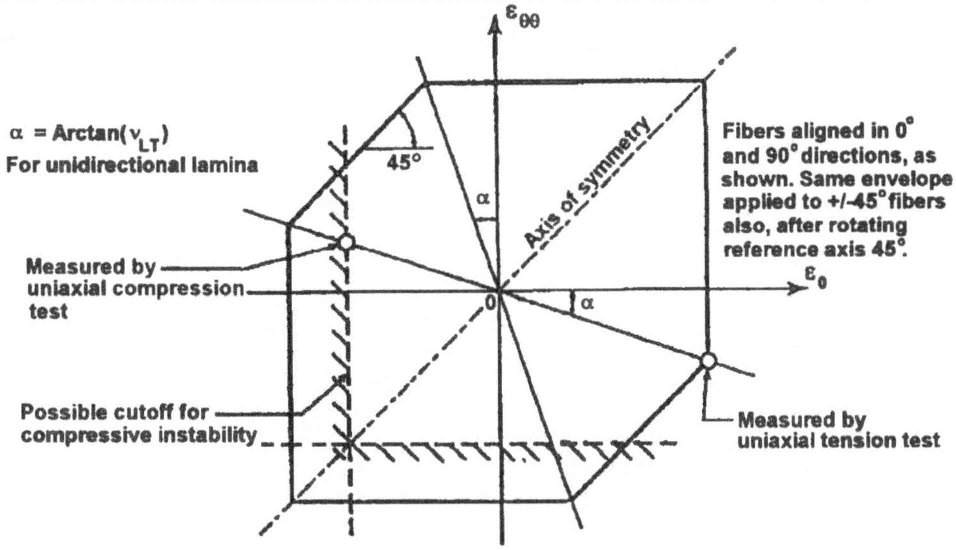

FIGURE 4.4.2 *Illustration of laminate level failure approach.*

laminates. These data lie in the second and fourth quadrants. The 45° cutoffs represent the locus of constant shear strain. These two symmetric truncations are located by finding the intersections of the limiting uniaxial strain lines with the lines representing pure uniaxial stress conditions in fiber directions in 0° and 90° unidirectional plies. At this point, the axial strain now becomes more critical than the shear. The endpoints of the truncations are therefore found by drawing lines through the origin with angles from the relative axes of a which account for the unidirectional ply Poisson's ratio:

$$\alpha = \tan^{-1}\left(\nu_{LT}^{lamina}\right) \qquad 4.4.2(b)$$

thereby yielding the desired pure uniaxial state of stress in the fiber direction. The intersection of these two lines with the greater of the two pure uniaxial stress conditions in the unidirectional plies locates the endpoint of each cutoff. It is always necessary that the cutoff be located by the higher of the uniaxial strengths since, otherwise, the cutoff would undercut the measured uniaxial strain to failure at the other end. This procedure results in the same failure diagram for all fiber-dominated laminates. It should be emphasized that this procedure requires the use of the Poisson's ratio of the unidirectional ply even when the laminate contains fabric plies.

This failure model, as represented in Figure 4.4.2, has been developed from experience with fiber-reinforced polymer matrix composites used on subsonic aircraft, particularly with carbon/epoxy materials, for which the lamina n_{TL} is approximately 0. It should not be applied to other composites, such as whisker-reinforced metal-matrix materials. Figure 4.4.2 addresses only fiber-dominated failures because, for the fiber polymer composites used on subsonic aircraft, the microcracking in the matrix has not been found to cause reductions in the static strength of laminates, particularly if the operating strain level has been restricted by the presence of bolt holes or provision for damage tolerance and repairs. However, with the advent of new composite materials, cured at much higher temperature to withstand operation at supersonic speeds, this approach may no longer be appropriate. The residual stresses developed during cool-down after cure will be far higher, because of the greater difference between the cure temperature and the minimum operating temperature.

This set of truncations together at the laminate level with the original maximum strain criterion results in the following operative set of equations applied *at the laminate level* with respect to axes oriented along and normal to each fiber direction in the laminate

$$\epsilon_{11}^{cu} \le \epsilon_{11}^{i} \le \epsilon_{11}^{tu}$$

$$\epsilon_{11}^{cu} \le \epsilon_{22}^{i} \le \epsilon_{11}^{tu} \qquad 4.4.2(c)$$

$$|\epsilon_{11}^{i} - \epsilon_{22}^{i}| \le \left(1 + \nu_{LT}^{lamina}\right) |\epsilon_{11}^{tu} \text{ or } \epsilon_{11}^{cu}|^{*}$$

* whichever is greater

However, it is important to note that these equations can only be applied in the context of a fiber-dominated laminate as previously described. It should further be noted that the limits on the transverse strain in each ply, ϵ_{22}^{i}, are set by the fibers in plies transverse to the ply under consideration and thus cannot characterize matrix cracking. This must be carefully taken into account if hybrid laminates are utilized. Furthermore, as previously discussed, if matrix cracking is considered to be structurally significant, a stress or strain cutoff must be added based on empirical observation. In this case, an assessment of the effects of the matrix cracks on subsequent properties of the laminate must be made.

As noted in Section 4.4.1, bending occurs when there are external bending and/or twisting moments or when the laminate is not symmetric. In these cases, as with other failure criteria, it is necessary to take into account the fact that the laminate level strains vary through the thickness.

4.4.3 Laminate design

Design charts in the form of "carpet plots" are valuable for selection of the appropriate laminate. Figure 4.4.3 presents a representative carpet plot for the axial tensile strength of laminates having various proportions of plies oriented at 0°, ±45°, and 90°. Appropriate strength data suitable for preliminary design can be found for various materials in References 4.4.3(a) and (b).

The development of laminate stacking sequence (LSS) optimization routines for strength-critical designs is a difficult task. Such a scheme must account for competing failure mechanisms that depend on material, load type (e.g., tension versus compression), environment (e.g., temperature and moisture content) and history (e.g., fatigue and creep). In addition, the load transfer must be adequately modeled to account for component geometry and edge effects. Even for a simple uniaxial load condition, the relationship between LSS and strength can be complex. Some qualitative rules currently exist for optimizing LSS for strength but they have been developed for a limited number of materials and load cases.

Relationships between LSS and laminate strength depend on several considerations. The initiation and growth of local matrix failures are known to depend on LSS. As these failures occur, internal stress distributions also depend on LSS strength through local stiffness and dimensional stability considerations. For example, delamination divides a base laminate into sublaminates having LSS that are generally unsymmetric. Reduced stiffness due to edge delaminations, causes load redistribution and can decrease the effective tensile strength of laminates. Likewise, local instability of sublaminates also causes load redistribution which can lower the effective compressive strength of laminates. As a result, both laminate and sublaminate LSS affect laminate strength.

Shear stress distributions play a significant role in determining the mechanical behavior and response of multi-directional laminates. As was the case for ply transverse tensile strength, ply shear strength depends on LSS. Laminates with homogeneous LSS have been found to yield higher in-situ ply shear strengths than those with ply orientations clumped in groups (Reference 4.4.3(b)). An inherent flaw density and interlaminar stresses appear to be major factors affecting the distribution of ply shear strengths in a LSS.

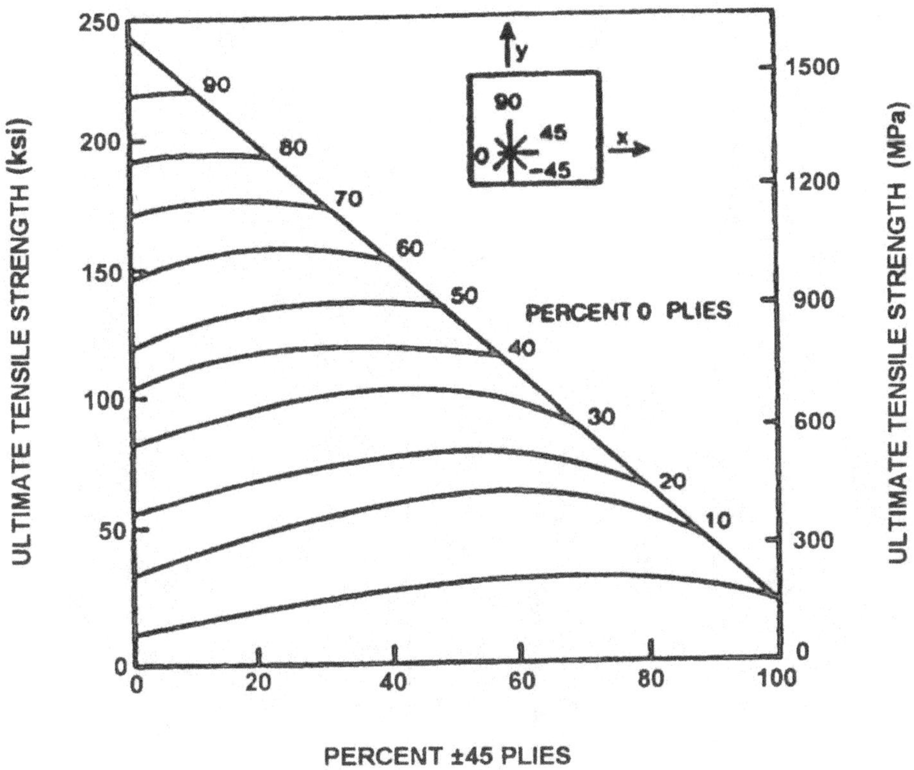

FIGURE 4.4.3 *Tensile strength of $[0_i/\pm45_j/90_k]_s$ family of high strength carbon/epoxy laminates (Reference 4.4.3).*

As was the case for bending stiffness, bending strength in composite laminates is strongly dependent on LSS. Failure mechanisms characteristic of tension, shear, and compression load conditions may all combine to affect bending strength. Table 4.3.3.2(b) showed that preferential stacking of plies in outer layers of the LSS increased bending stiffness. The bending strength performance of undamaged laminates may show similar trends; however, surface damage due to impact or other in-service phenomena would cause severe degradation to such laminates.

Additional information on laminate stacking sequence effects is found in Section 4.6.5.

4.4.4 Stress concentrations

The presence of a hole or other discontinuity in a structure introduces local stress concentrations. These high local stresses can result in initial localized failure. The analysis of failure due to cracking, or fracture, which can result in this situation is complicated for composite materials because of material heterogeneity at the microscale and in a layer-to-layer basis. Effective in-plane laminate stiffnesses, E_x, E_y, and G_{xy}, may be calculated for any laminate by using the methods presented in Section 4.3.3. With these properties specified, a balanced symmetric laminate may be regarded as a homogeneous orthotropic plate, for structural analysis. Orthotropic elasticity theory may be used for the evaluation of stresses around a hole in such a plate (Reference 4.4.4(a)). Examples of the resulting stress concentrations are shown in Figure 4.4.4(a) for carbon/epoxy laminates. The laminae orientation combinations influence both the magnitude and the shape of the stress variation near the hole. The high stresses at the edge of the hole may initiate fracture.

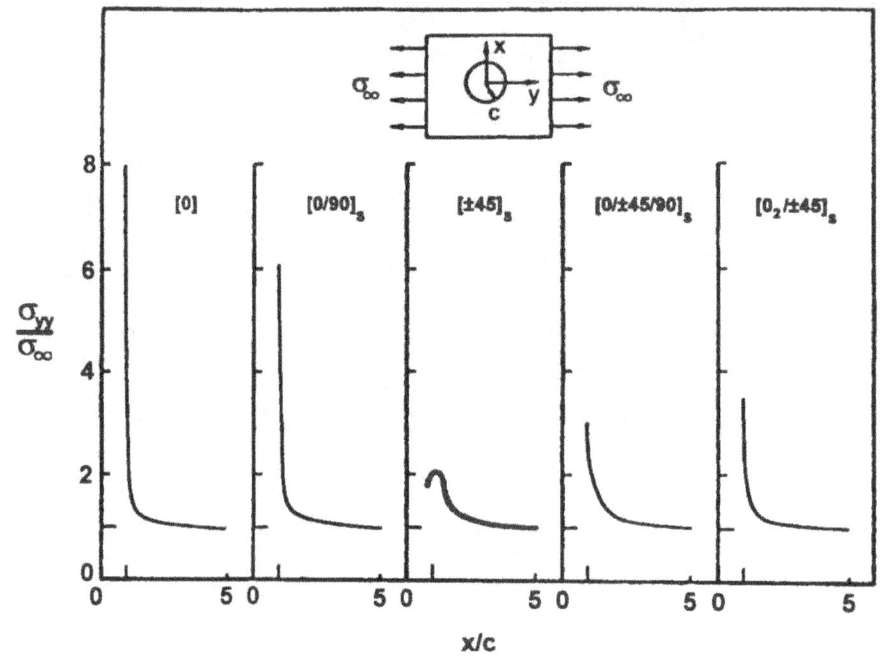

FIGURE 4.4.4(a) *Stress concentration factors for a circular hole in a homogeneous, orthotropic, infinite plate.*

If the laminate fails as a brittle material, fracture will be initiated when the maximum tensile stress at the edge of the hole equals the strength of the unnotched material. In a tensile coupon with a hole, as shown in Figure 4.4.4(a), failure will occur at the minimum cross-section. The failure will initiate at the edge of the hole, where the stress concentration is a maximum.

Consider the stress concentration factor in a finite width isotropic plate with a central circular hole. Stress distribution for this configuration are shown in Figure 4.4.4(b) for various ratios of hole diameter, a, to plate width, W. The basic stress concentration factor for this problem is the ratio of the axial stress at the edge of the hole ($x = a/2$, $y = 0$) to the applied axial stress, σ_∞. For small holes in an isotropic plate, this factor is three. The average stress at the minimum section, σ_n, is higher than the applied stress, σ_∞, and is given by the following relationship:

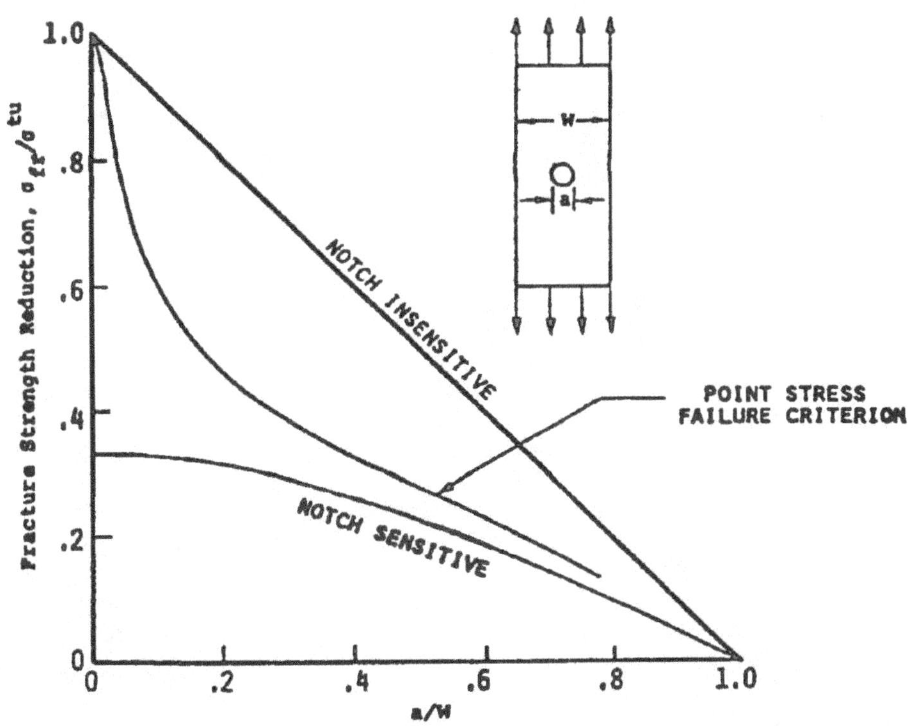

FIGURE 4.4.4(b) *Approximate failure theories.*

$$\sigma_n = \frac{\sigma_\infty}{(1 - \frac{a}{W})} \tag{4.4.4(a)}$$

The net section stress concentration factor, k_n, is the ratio of the maximum stress to this average stress.

$$k_n = \frac{\sigma(\frac{a}{2},0)}{\dfrac{\sigma_\infty}{(1 - \frac{a}{W})}} \tag{4.4.4(b)}$$

Laminate fracture for the elastic-brittle case will occur at stress σ_{fr}:

$$\sigma_{fr} = F^{tu}/k_n \tag{4.4.4(c)}$$

A material which fails in this fashion is denoted a notch-sensitive material. In contrast, a ductile, or notch-insensitive, material will yield locally to alleviate the stress concentration effect.

Various matrix damage effects are expected to occur at the maximum stress locations. This localized damage reduces the material stiffness and diminishes and spreads the stress concentration effects. Semi-empirical methods have been proposed to account for this reduction in the stress concentration.

The "point stress theory" (Reference 4.4.4(b)) proposes that the elastic stress distribution curve, e.g., Figure 4.4.4(a), be used, but that the stress concentration be evaluated at a distance, d_o, from the edge of the hole. The numerator of Equation 4.4.4(b) is evaluated at $x = a/2 + d_o$. The characteristic length, d_o, must be evaluated experimentally. The "average stress theory" (Reference 4.4.4(b)) takes a similar approach by proposing that the elastic stress distribution be averaged over a distance, a_o, to obtain the stress concentration.

$$k_n = \frac{\int_{a/2}^{(a/2)+a_o} \sigma_y dx}{\int_{a/2}^{W/2} \sigma_y dx} \tag{4.4.4(d)}$$

Again, the characteristic dimension, a , must be found experimentally. For both methods, the resulting stress concentration is used in Equation 4.4.4(c) to define the fracture stress. Representative results are plotted in Figure 4.4.4(b) to illustrate the differences associated with different types of material behavior.

The relationship between tensile strength and laminate stacking sequence (LSS) for laminates with holes, cutouts, and through-penetrations (i.e., a damage tolerance consideration) is complex (see Reference 4.4.4(c) - (g)). Certain combinations of ply splitting and delamination that occur at the tip of a notch can enhance residual strength by effectively reducing the stress concentration. Delaminations which uncouple plies, allowing individual plies to fail without fiber breaks, reduce the residual strength. Most existing analysis methods for predicting notched tensile strength are based on parameters determined by some notched laminate tests (e.g., characteristic dimension, fracture energy parameter). The effects of LSS on failure is included in the test parameter. Future analysis development that simulates progressive damage accumulation will provide a more efficient approach for studying the effects of LSS. Additional information on laminate stacking sequence effects is found in Section 4.6.5.

4.4.5 Delamination

The formation and growth of delaminations is generally related to LSS. Delaminations can have varying effects on tensile strength performance, depending on delamination location and the specific property of interest. Most studies performed to date have considered specimens with significant free edge surface area where interlaminar stresses are known to concentrate. Although all structures have some free edges, it is important to realize the limits of analysis and tests performed with specimen geometries. For example, the magnitude of interlaminar tensile stresses, which are crucial to edge delamination, approach zero for plate width to thickness ratios of 30 and greater (Ref. 4.4.5(a)).

As shown in Figure 4.4.5, laminated specimens prone to edge delamination have been shown to exhibit generally lower strength (ultimate stress level) when loaded in uniaxial tension (e.g., References 4.6.5.5.1.1(b), 4.4.5(b) - (f)). The reduction in strength has been directly tied to a drop in stiffness with increased edge delamination area for laminates exhibiting stable delamination growth (References 4.6.5.5.1(b), 4.4.5 (b) - (e)). The onset of edge delamination has been shown to relate to tensile strength for laminates exhibiting unstable delamination growth coupled with matrix cracks (Reference 4.4.5(f)).

The reduced laminate stiffness due to edge delamination can affect the measured tensile strength in two distinct ways (e.g., Reference 4.4.5(e)). If all plies remain loaded after delamination, the ultimate laminate strain has been found to equal the critical strain of primary load bearing plies. In these cases laminate strength drops in proportion to the apparent axial modulus. However, if off-axis plies cease to carry loads because they have been isolated by an interconnected network of matrix cracks and delamination, a local strain concentration can form. When this occurs, the global laminate strain for failure can be less than the critical strain of primary load bearing plies.

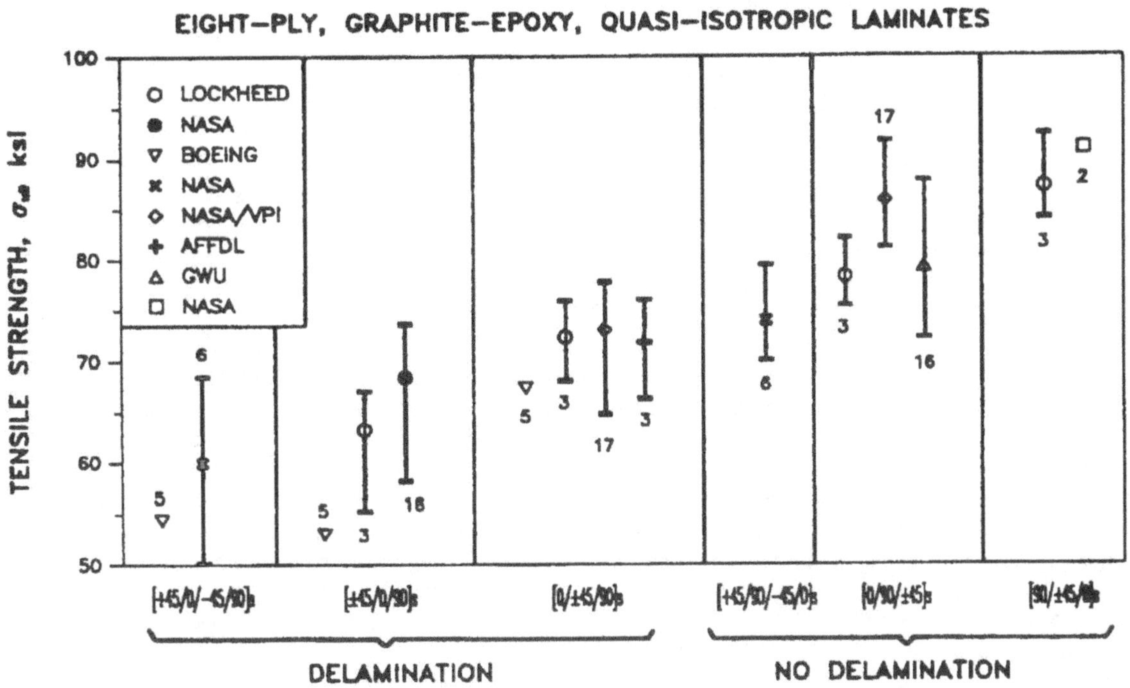

FIGURE 4.4.5 *Unnotched tensile strength variation with LSS (from Reference 4.4.5(d)).*

Free edge delaminations split a laminate into sublaminates, each of which continue to carry tensile loads. The apparent modulus of this laminate depends on delamination length and the sublaminate moduli which may be calculated using lamination theory. These moduli will depend on LSS if unsymmetric sublaminates with strong extension/bending couplings are involved (References 4.4.5(g) and (h)). A simple rule-of-mixtures approach has been used to accurately calculate apparent moduli for edge delamination (References 4.4.5(e), (g) and (h)).

Local coupling between intralaminar matrix cracks and delaminations can cause complete or partial ply isolation. Note that complete ply isolation cannot occur unless associated damage extends the full laminate width. When this occurs, the apparent laminate stiffness and strain concentration can be calculated in a modified rule-of-mixtures approach which discounts isolated ply groups (References 4.6.5.5.1.1(c) and 4.4.5(e)). A local area of reduced stiffness also causes strain concentration (Reference 4.4.5(i)). The strain concentration depends on both the local reduced stiffness and global laminate stiffness. For example, hard laminates with strong anisotropy, such as lay-ups dominated by 0° plies and loaded uniaxially, will have large strain concentration factors. Consequently, hard laminates will be less tolerant of local damage than relatively soft laminates (e.g., quasi-isotropic).

When high interlaminar shear stresses are present, coupled edge delamination and matrix crack growth are possible and may lead to catastrophic failure. Edge delamination behavior of laminates commonly used in design (e.g., quasi-isotropic laminates) become dominated by interlaminar shear stresses when subjected to off-axis loading. Note that for this problem the laminate lay-up is generally unbalanced relative to the loading axis. The measured tensile strength coincides with the onset of edge delamination for such laminates (Reference 4.4.5(f)). As a result, failure criteria that account for interlaminar stresses are needed to predict the tensile strength.

The use of a suitable analysis method is recommended to evaluate edge effects in composite materials (e.g., References 4.4.6.1.1(f), 4.4.5(d), (g), (h), (j) - (l)). Applied mechanical loads and environmental effects should be included in the free edge analysis. Two approaches have been successfully applied to quantify free edge stresses and predict edge delamination: (1) a fracture mechanics based method using strain energy release rates (References 4.6.5.5.1.1(f), 4.4.5(d), (g), (h), (j)), and (2) a strength of materials based approach using an average stress failure criterion (References 4.4.5(k) and (l)).

The combined use of resin interlayers between the plies in a laminate and specimen edge polishing have been found to be effective methods for suppressing edge delamination (References 4.4.5(f)). Materials with high interlaminar toughness have an inherent resistance to delamination. Other methods that have been used to suppress edge delamination include resin interlayer strips at critical interfaces along the edge of laminates (References 4.4.5(m)), termination of critical plies offset from the edge (Reference 4.4.5(n)), hybridization (References 4.4.5(o) and (p)), and serrated edges (Reference 4.4.5(p)).

Most of the above discussion on the effects of delamination suggest a decrease in tensile properties. This is generally true for unnotched specimen geometries prone to edge delamination. Isolated delaminations that occur away from the edge of a laminate (e.g., manufacturing defects) and are not coupled with matrix cracks have been shown to have little effects on tensile strength (Reference 4.4.5(r)). Theoretically, such delaminations do not result in local reduced laminate stiffness when loaded in tension due to compatibility considerations. Multiple delaminations located away from the edge of a laminate have been shown to cause a small reduction in tensile strength (Reference 4.4.5(r)). This was explained by coupling between delaminations and other matrix damage (e.g., ply splits) that occurred during loading, resulting in partial ply isolation and local reduced stiffness. Most of the discussion in this section is related to free edge effects (Section 4.6.3) and laminate stacking sequence effects (Section 4.6.5).

4.4.5.1 Compression

Delaminations generally have a stronger affect on compressive strength than on tensile strength. As a result, the potential for delamination should always be considered when selecting a suitable LSS. The effect of delamination occurring due to manufacturing defects and/or in-service events such as impact needs to be included in this evaluation. For example, the best LSS for avoiding edge delamination in specimen geometries may not be best for suppressing the effects of delaminations occurring in structures due to impact.

Delamination breaks the laminate into sublaminates, each having associated stiffness, stability, and strength characteristics. Sublaminates are usually unsymmetric and, therefore, all of the sublaminate stiffnesses will depend on LSS. As shown in Figure 4.4.6.2, stability and local compressive performance of sublaminate ply groups ultimately determines catastrophic failure.

Compressive failure in composite laminates having delaminations is strongly tied to the stability of sublaminate plates. Since delaminations may occur at many different interfaces in a laminate, sublaminates LSS will generally not be balanced and symmetric. As discussed earlier, the bending/extension couplings characteristic of such LSS reduce buckling loads. The sublaminate boundary conditions and shape are also crucial to the relationship between LSS and stability.

Several methods exist for predicting sublaminate stability in composite laminates (e.g., References 4.4.5.1(a) - (e)). These models differ in assumed bending stiffness, boundary conditions, and sublaminate shape. Experimental data bases are needed to determine which assumption is appropriate for a given problem. The in-plane and out-of-plane stress redistribution due to a buckled sublaminate is crucial to compressive strength.

Environment can play a significant role in delamination growth and load redistribution if the environmental resistance of combined sublaminate stiffnesses are significantly different than those of the base laminate. The combined effects of environment and LSS on laminate dimensional stability were covered in earlier sections. The stability of unsymmetric sublaminates is expected to relate to warpage. The warp depends on both LSS and environmental conditions. Warp may be treated as an imperfection in stability analysis.

The initiation of free edge delamination in compressively loaded laminates can be predicted using methods similar to those used for tension (e.g., Reference 4.4.5.1(a)). Once initiated, delamination growth depends on sublaminate stability. An adequate sublaminate stability analysis model must, therefore, be coupled with the growth model (e.g., References 4.4.5.1(b) and (c)). Delamination growth can be stable or unstable, depending on sublaminate LSS, delamination geometry, structural geometry, and boundary conditions. Growth of multiple delaminations, characteristic of impact damage, is currently not well understood.

4.4.6 Damage and failure modes

4.4.6.1 Tension

Tensile rupture of laminates with multidirectional plies normally involves a series of pre-catastrophic failure events, including both matrix damage and localized fiber breaks. Catastrophic failure is expected whenever the longitudinal tensile strength of any ply in a laminate is exceeded; however, laminates can separate without fiber failure by coupling various forms of matrix damage. Example laminates that can fail due to matrix damage include those with less than three distinct ply orientations (angle-ply laminates loaded in the 0° direction). Recommendation 2 in Section 4.6.5.2.1 is intended to avoid the low strengths associated with catastrophic failures occurring without fiber breaks.

Figure 4.4.6.1 shows the various failure mechanisms that can occur at micro and lamina dimensional scales for a multidirectional laminate loaded in tension. Depending on load conditions and material properties, matrix failure (e.g., transverse matrix cracks, delamination) or isolated fiber breaks occur at stress levels less than the static strength. Load redistributes around local failures until a critical level of damage is reached, upon which catastrophic fiber failure occurs. Resin is of secondary importance through its effect on resistance to matrix damage accumulation and local load transfer (i.e., near matrix damage and isolated fiber breaks). The LSS also plays a secondary role by affecting damage accumulation and load transfer.

Critical micro failure mechanisms shown in Figure 4.4.6.1 include localized fiber failure and fiber/matrix interfacial cracking. These mechanisms occur mostly in plies aligned with a major axis of tensile stress. The laminate stress levels at which these failures occur depend on load redistribution due to the characteristic damage state in adjacent plies. A limited number of fiber breaks are tolerated within a lamina before the entire ply fails, which can trigger catastrophic laminate failure.

Matrix failure mechanisms at the lamina scale for laminates with multidirectional plies are also shown in Figure 4.4.6.1. Intralaminar matrix cracks align parallel to the fiber direction and span the thickness of a ply or group of plies stacked with the same orientation. These have also been referred to as transply cracks or ply splits depending on whether a crack orients at an angle or parallel to the tensile load axis, respectively.

Interlaminar matrix failure, often referred to as delamination, can form near free edges or at intersections between intralaminar cracks. Delaminations form due to excessive interlaminar normal and shear stresses. The accumulation of intralaminar and interlaminar matrix failures depends strongly on LSS.

4.4.6.1.1 Matrix cracks

Matrix cracks occur in plies of laminated composites due to combined mechanical and environmental stresses. These transverse cracks align with fibers and, when fully formed, span the thickness of individual plies or ply groups stacked together in the same orientation. Matrix cracks redistribute local stress in multidirectional laminates, allowing a crack density to develop in the ply of ply group as a function of load and environmental history. These cracks can also form prior to service exposure due to processing.

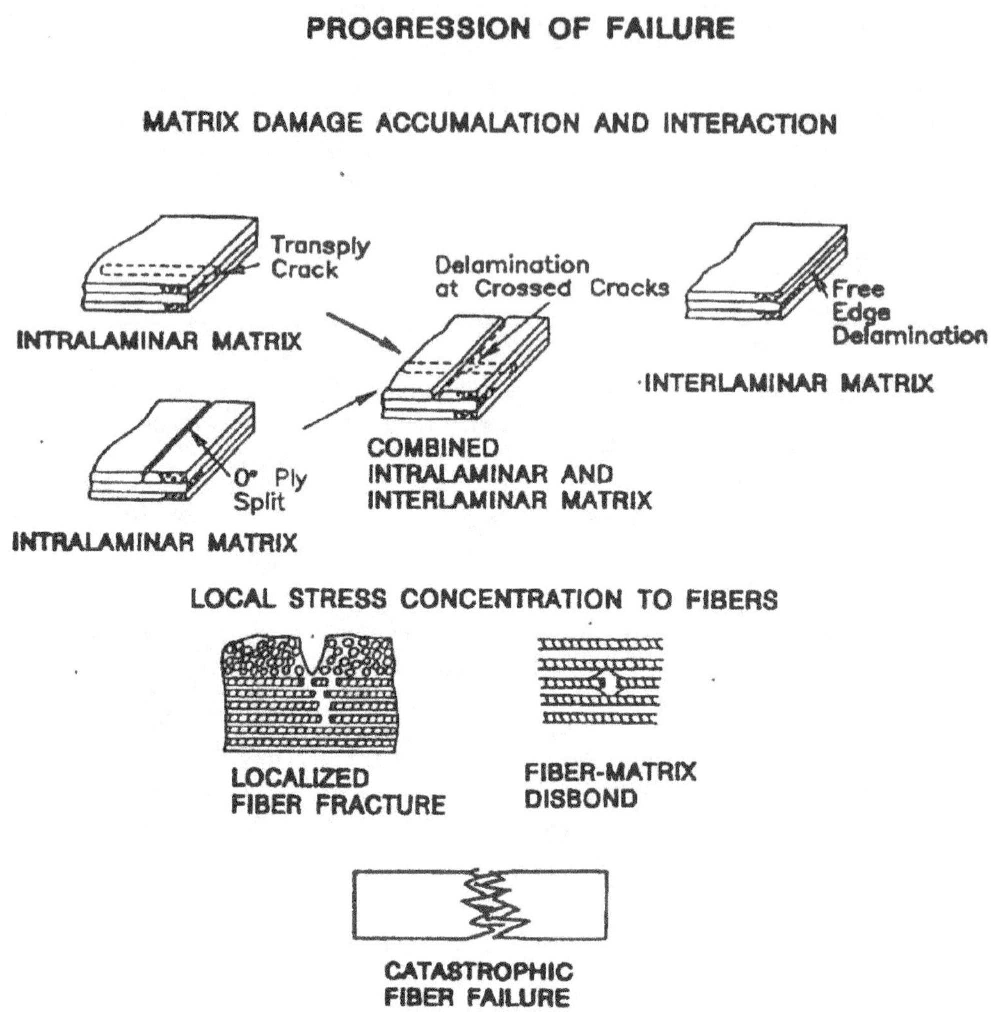

FIGURE 4.4.6.1 *Failure mechanisms for laminates loaded in tension.*

Studies with coupons loaded in uniaxial tension have shown that initial fiber failures found in 0° plies occur near intralaminar matrix cracks in neighboring off-axis plies (Ref. 4.4.6.1.1(a)). When matrix cracks span a single off-axis ply, the stress concentration in a neighboring 0° ply is generally small and localized over a small portion of the neighboring ply thickness. This has been found to influence the location of laminate failure, but has little effect on tensile strength (Refs. 4.4.6.1.1(b) and 4.4.6.1.1(c)).

Intralaminar matrix cracks normally span the full thickness of multiple off-axis plies that have been stacked together. The associated stress concentration in a neighboring ply increases with the thickness of a cracked group of stacked plies. The stress concentration in a 0° ply due to matrix cracks in a large group of stacked 90° plies was found to significantly decrease laminate tensile strength (Refs. 4.4.6.1.1(c) and 4.4.6.1.1(d)). This is one of the reasons for Recommendation 3, Section 4.6.5.2.1.

Even when strength is not altered by the presence of matrix cracks, it is important to understand the mechanics of matrix cracking for composite materials used in aerospace applications. For example, matrix cracks can play a fundamental role in the generation of delaminations. The increased surface area due to a network of matrix cracks can also alter physical properties such as composite thermal expansion, liquid permeability, and oxidative stability.

Residual stresses, that develop due to differences in thermal and moisture expansion properties of constituents, affect the formation of matrix cracks. In general, tensile residual stress develops in the transverse-fiber directions of lamina when multidirectional polymer matrix composites are exposed to temperatures below the residual stress free temperature. This occurs during a temperature drop because unconstrained shrinkage of tape lamina is much greater in transverse-fiber directions than in fiber directions. As moisture is absorbed into a laminate, matrix swelling counteracts thermal shrinkage, decreasing the transverse-fiber tensile stress.

The critical stress or strain causing the onset of matrix cracking in plies of a laminate has been referred to as in situ transverse lamina strength. This strength is not a material constant since it depends on LSS. Experiments and analysis have shown that in situ strength increases as the thickness of plies grouped together with the same orientation decreases (e.g., Refs. 4.4.6.1.1(e) - (i)). These studies have also shown that neighboring plies can impose differing constraints on matrix crack formation, depending on fiber orientation. Many materials currently used in the aerospace industry have resin-rich interlaminar layers (RIL). The magnitude of the in situ strengthening effect decreases if a RIL with significant thickness exists between plies (Ref. 4.4.6.1.1(j)). Relatively soft RIL eliminate some of the constraint imposed by neighboring plies.

4.4.6.2 Compression

Compressive strength is ultimately related to the local response of individual ply groups. Assuming no matrix damage exists due to impact or previous load history, the local stability and strength of plies aligned with the axis of loading will determine final failure. The location of load-carrying plies relative to the laminate surface can play a role in this instance. The short wavelength buckling load is reduced when critical plies are located in outer layers of the laminate stacking sequence. When matrix damage does exist, the combined local response of individual ply groups affects the compression strength. The stability and load redistribution within individual ply groups or sublaminates is crucial to the local response.

Figure 4.4.6.2 shows three different types of local compressive failure mechanisms. These mechanisms were observed to occur as a function of θ for $(\pm\theta)_s$ type laminates (References 4.4.6.2(a) and (b)). When delamination occurs, all three of the local failure modes may combine to determine the compressive strength of a laminate stacking sequence. (Additional information on the effects of the laminate stacking sequence is found in Section 4.6.5.) In-plane matrix shearing and matrix compression failures were observed for $(\pm\theta)_s$ type laminates with $15° \leq \theta \leq 50°$ and $60° \leq \theta \leq 90°$, respectively. The shear mode of fiber microbuckling is most commonly observed for composites. This mode was shown to initiate compressive failure for $(\pm\theta)_s$ type laminates with $0° \leq \theta \leq 10°$. Depending on matrix and fiber combination, final local failures for such laminates involved some combination of fiber failure (shear, kinking, or bending) and matrix splitting or yielding (References 4.4.6.2(c) and (d)).

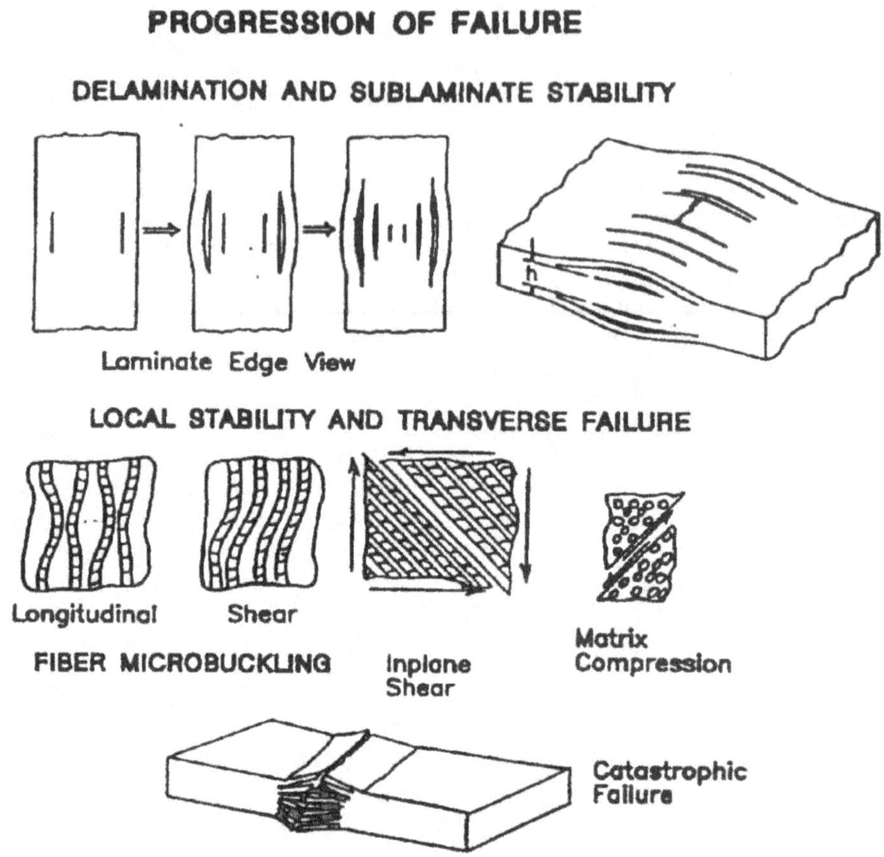

FIGURE 4.4.6.2 *Failure mechanisms for laminates loaded in compression.*

4.4.7 Summary

- Ply level stresses are commonly used to predict first ply and subsequent ply failures leading up to laminate failure. Once a ply has failed, its contribution to laminate strength and stress is conservatively reduced. Typically, in-plane failure criteria are applied only to lamina fiber loading conditions; in-plane matrix-dominated static failure criteria should not be used since it will generally lead to overly conservative failure predictions.

- Under static loading conditions, composites are particularly notch-sensitive as a function of layup and more specifically stacking sequence.

4.5 COMPLEX LOADS

4.5.1 Biaxial in-plane loads

4.5.2 Out-of-plane loads

4.6 LAMINA TO LAMINATE CONSIDERATIONS

4.6.1 Residual stresses and strains

Residual curing stresses and strains have virtually no effect on fiber-dominated laminate properties. However, residual stresses in the resin can be greater than the mechanical stresses needed to cause failure. Neglecting these residual stresses therefore may be nonconservative. The residual stresses may be high enough that resin microcracking may occur before any mechanical load is applied. Consequently, the principle of superposition may not be applicable as the mechanical loading may result in nonlinear behavior. As an example, typical epoxy matrix residual strains at the microlevel, resulting from cool down after curing at 350°F (180°C), may be approximately 25 to 100% of the laminate failure strain.

4.6.2 Thickness effects

Much of the difference in properties found when comparing laminates with different thicknesses can be explained by the residual stresses developed during processing. Internal stresses developed during processing may produce voids, delaminations, and microcracks or cause residual stresses in the laminate that may affect material properties. Excessive porosity, generally caused by poor processing, or environmental effects due to temperature and moisture conditions may also degrade the material and affect its behavior.

Variations in material properties between thick laminate test data from different sources, for laminates having the same thickness, can generally be attributed to differences in processing. Such variations can be minimized by optimizing the cure cycle and by proper process control.

The residual stresses may be caused by non-isothermal conditions present during the solidification phase. Different layers of the laminate will undergo different degrees of volume contraction at any given time during the process cycle. This gives rise to a self-equilibrating force system producing tension stresses in the center and compression stresses in the surface layers of the laminate as reported by Manson and Seferis (Reference 4.6.2(a)). Thickness effects observed in composite laminates are primarily due to this phenomena.

In thermosetting materials, these through-the-thickness stress gradients can be virtually eliminated by modeling the total process, including cool-down, so isothermal conditions are present near the resin gelation point and are maintained for a sufficient period of time. In some high-temperature processing materials where rapid cooling is required, significant thermal stresses may build up in the laminate.

In their work, the authors in Reference 4.6.2(a) present a method to experimentally determine and analyze the internal stresses developed during processing of a composite laminate. This method consists of laying up a certain number of plies, separated by a release ply that can be removed after processing. The internal stresses in the laminate can then be analyzed by considering the deformations of the individual sublaminates.

In summary, variations in material properties in laminated composites are mostly the result of thermally induced residual stresses, although environmental effects and process parameters other than temperature may affect test data. True thickness effects are caused by temperature gradients across the thickness of the laminate. These effects may be minimized by mathematical modeling of the total process and can be virtually eliminated in thermosetting materials. Advance process models such as ROAST, described in Reference 4.6.2(b), may be used to optimize the process parameters.

4.6.3 Edge effects

Consideration of edge effects in laminated composites is necessary due to behavior not observed in homogeneous solids. A complex stress state exists between the layers of different orientation at the free edge of a laminate, such as along a straight edge or around the perimeter of a hole. Where a fiber in a laminate has been subjected to thermal or mechanical strain, the end of the cut fibers must transfer the load to adjacent fibers. If these adjacent fibers have a different orientation, they will present a locally stiffer path and accept the load. The matrix is the only mechanism for this load transfer. The stresses due to this load, namely interlaminar stresses, can be sufficient to cause local microcracking and edge delamination. These interlaminar stresses, in general, include normal (peel stress σ_z) and shear components (τ_{yz}, τ_{xz}) and are only present in a small region near the free edge. A typical interlaminar stress distribution is shown in Figure 4.6.3. The high gradients of these stresses depend on differences in Poisson's ratio and in-plane shear stiffness that exist between the laminae groups in a laminate. The same kinds of stresses are induced by residual thermal stresses due to cool-down after cure at elevated temperatures.

Failure often occurs as a result of delamination at the locations of high interlaminar stresses because of low interlaminar strength. The effects of free edge stress are sufficient to reduce the strength of certain coupons in both static and fatigue tests significantly. This premature failure makes coupon data difficult to apply to large components because of the local effects of the free edge failure mode. Classical laminate theory which assumes a state of plane stress is incapable of predicting the edge stresses. However, determination of such stresses by higher order plate theory or finite element analysis is practical. Therefore, consideration of edge interlaminar stresses in a laminate design is feasible. The gradients of this stress can be reduced by such measures as 1) changing the laminate stacking order, 2) minimizing the mismatch of the Poisson's ratio, the coefficient of mutual influence, and coefficients of thermal and moisture expansion between adjacent laminae, and 3) by inserting an inner layer which has a lower shear modulus and a finite thickness between laminae, thus allowing greater local strain to occur (Reference 4.6.3(a)).

Edge effects may be analyzed by fracture mechanics, strength of materials, or other methods (References 4.6.3(a) - (d)). These methods can be used to provide a guideline for designers to select the laminate configuration and material system best suited for a particular application.

Very little work has been performed to date on free edge effects for load conditions other than uniaxial tension or compression. Some analysis results indicate that in-plane shear, out-of-plane shear/bending, in-plane bending, twisting moments, and combined loading yield a higher magnitude of interlaminar stress relative to those associated with axial load conditions (Reference 4.6.3(f)). For example, out-of-plane shear due to bending causes free edge interlaminar stresses that are an order of magnitude higher than that caused by axial tension. For more information on delaminations and free edge effects, see Section 4.4.5. Information on the laminate stacking sequence effects is found in Section 4.6.5.

4.6.4 Effects of transverse tensile properties in unidirectional tape

The transverse strength properties play only a minor role in establishing cross-plied laminate strengths. It is, however, well-known that the effective "in-situ" transverse strength of transverse plies is much greater than the strength measured on the lamina. This effect has been handled by post-first ply failure analysis methods.

In-plane shear tests on laminae exhibit relatively high strains to failure (4 -5%). The much lower transverse tensile strains to failure (1/2%) indicate a marked notch sensitivity that is suppressed in cross-plied laminates. The initial cracks that fail laminae are arrested by fibers in other directions; thus laminae with microcracks are still effective. Most laminae develop cracks due to residual thermal stresses and continue to function.

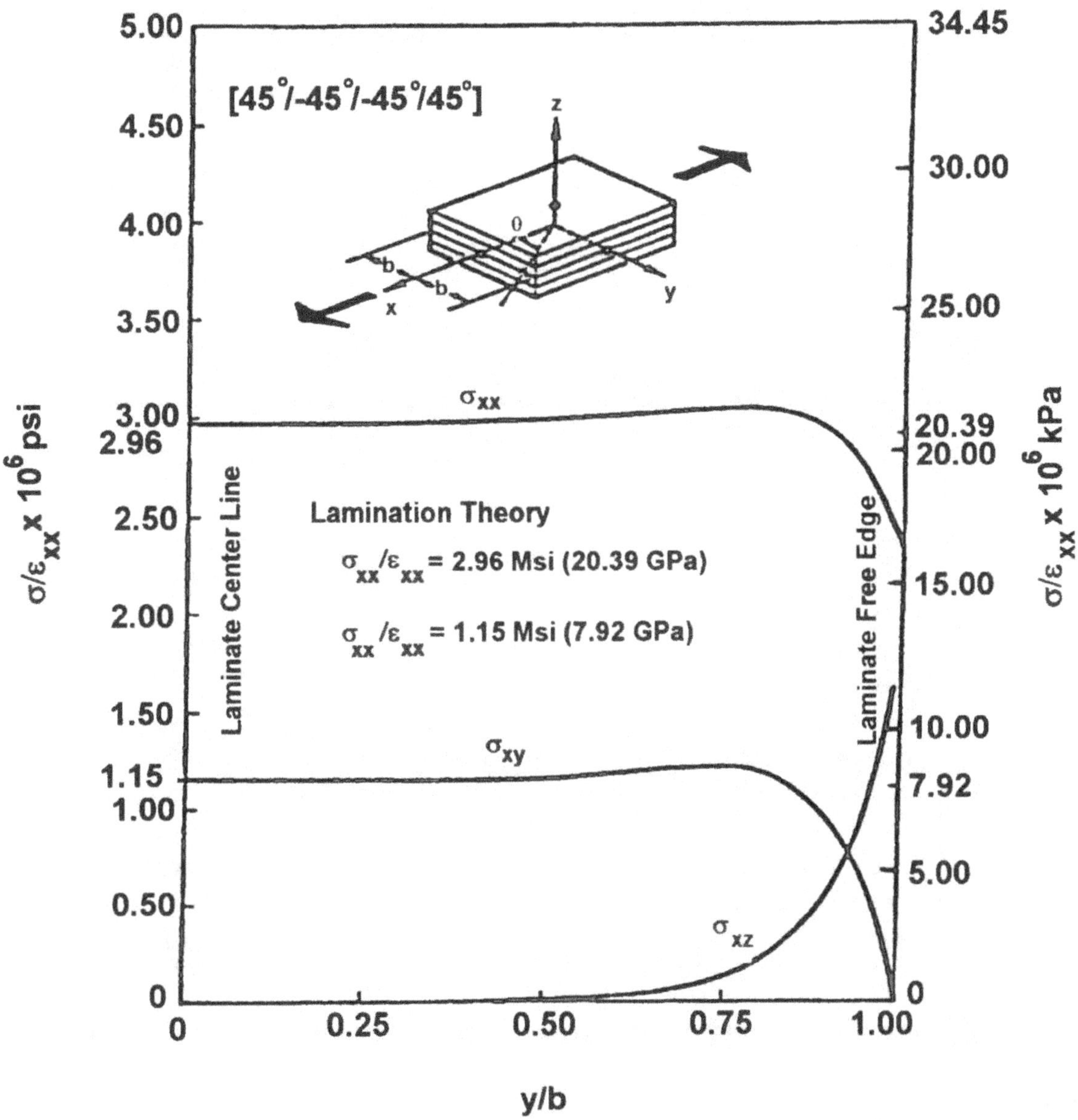

FIGURE 4.6.3 *Interlaminar stresses normalized with respect to the applied strain (reproduced by permission from Reference 4.6.3(e))*

4.6.5 Laminate stacking sequence effects

4.6.5.1 Introduction

Stacking sequence describes the distribution of ply orientations through the laminate thickness. As the number of plies with chosen orientations increase, more stacking sequences are possible. For example, a symmetric 8-ply laminate with four different ply orientations has 24 different stacking sequences. This presents a predicament when attempting to optimize composite performance as a function of stacking sequence.

Laminated composite structural properties such as stiffness, dimensional stability, and strength have all been found to depend on laminate stacking sequence (LSS). Generally, each property has a different relationship with LSS. Therefore, the choice of LSS for a particular design application may involve a compromise. Design optimization requires verified analysis methods and an existing materials database. The development of verified analysis methods for predicting stiffness and stability of laminated composites is more mature than that for predicting strength.

Some simplified design guidelines for LSS are provided in Section 4.6.5.2. These guidelines are generally conservative; however, they limit design optimization, and may even be misleading for some special cases. As a result, a comment on the reason for each guideline is included in the discussion. Verified analysis methods should be used to help judge the effects of LSS whenever possible.

Additional discussion of stacking sequence effects on particular topics are provided in the sections noted in Table 4.6.5.1.

TABLE 4.6.5.1 *Additional discussions on stacking sequence effects.*

Topic	Section	Page
Bending	4.3.3.2, 4.4.3	4-42, 4-62
Buckling	4.7.1.8	4-85
Compression after impact	4.11.1.4	4-107
Delamination	4.4.5	4-66
Free edge effects	4.6.3	4-74
Hygroscopic analysis	4.3.4	4-50
Lamination theory	4.3.2	4-33
Notched strength	4.4.4	4-63
Ply shear strength	4.4.3	4-62
Thermal analysis	4.3.4	4-50
Vibration	4.12.2	4-115

4.6.5.2 Design guidelines

Laminate design starts by selecting the number of plies and ply angles required for a given application. Once the number of plies and ply angles are selected, a LSS is chosen. A LSS is considered heterogeneous when there is preferential stacking of specific ply orientations in different locations through the thickness of the laminate. Thick laminates with heterogeneous LSS are created by clumping plies of similar orientation. A LSS is said to be homogeneous if ply angles are evenly distributed through the laminate thickness. The ability to generate homogeneous LSS depends on the number of plies and ply angles. For example, it is impossible to create a homogeneous LSS for a four-ply laminate consisting of four different ply angles.

The following LSS guidelines are based on past experience from test and analysis. Guidelines are lumped under two categories; (1) strong recommendation, and (2) recommendation. Despite this classification, exceptions to the guidelines should be considered based on an engineering evaluation of the specific application.

4.6.5.2.1 Strong recommendations

1. Homogeneous LSS are recommended for strength controlled designs (In other words, thoroughly intersperse ply orientations throughout the LSS).

   Comment: *Heterogeneous laminates should be avoided for strength-critical designs unless analysis and test data is available that indicates a clear advantage. In cases where heterogeneous laminates cannot be avoided (e.g., minimum gage laminates), it is generally best to stack primary load-carrying plies toward the laminate core. The best way to view possible strength problems with heterogeneous LSS is to consider the behavior of individual sublaminates (i.e., groups of plies separated by delaminations) that may be created during manufacturing or service exposure. This will be discussed later in greater detail.*

   Heterogeneous LSS can yield optimum stiffness or stability performance; however, the effects on all other aspects of the design (e.g., strength, damage tolerance, and durability) should be considered before ignoring Recommendation 1. For example, interlaminar stress distributions are affected by variations in the in-plane stress field around the periphery of holes and cutouts and the "effective" LSS (i.e., ply orientations relative to a tangent to the edge). Since it is difficult to optimize for a single lay-up in this case, the best solution is to make the LSS as homogeneous as possible.

2. A LSS should have at least four distinct ply angles (e.g., 0°, ±θ°, 90°) with a minimum of 10% of the plies oriented at each angle. Ply angles should be selected such that fibers are oriented with principal load axes.

   Comment: *This rule is intended to avoid the matrix-dominated behavior (e.g., nonlinear effects and creep) of laminates not having fibers aligned with principal load axes. Such behavior can lead to low strengths and dimensional stability problems.*

3. Minimize groupings of plies with the same orientation. For tape plies, stack no more than four plies of the same orientation together (i.e., limit stacked ply group thicknesses ≤0.03 in. (0.8 mm)). In addition, stacked ply group thicknesses with orientations perpendicular to a free edge should be limited to ≤0.015 in.(0.38 mm).

   Comment: *This guideline is used for laminate strength-critical designs. For example, it will help avoid the shear-out failure mode in bolted joints. It also considers relationships between stacked ply group thickness, matrix cracking (i.e., transverse tension and shear ply failures) and delamination.*

   In general, ply group thickness should be limited based on details of the design problem (e.g., loads, free edges, etc.) and material properties (e.g., interlaminar toughness). Note that the absolute level of ply group thickness identified in this guideline is based on past experience. It should be confirmed with tests for specific materials and design considerations.

4. If possible, LSS should be balanced and symmetric about the midplane. If this is not possible due to other requirements, locate the asymmetry or imbalance as near to the laminate midplane as possible. A LSS is considered symmetric if plies positioned at an equal distance above and below the midplane are identical (i.e., material, thickness, and orientation). Balanced is defined as having equal numbers of +θ and -θ plies, where θ is measured from the primary load direction.

   Comment: *This guideline is used to avoid shear/extension couplings and dimensional stability problems (e.g., warpage which affects component manufacturing tolerances). The extension/bending coupling of unsymmetric laminates can reduce buckling loads. Note that some* *coupling may be desired for certain applications (e.g., shear/extension coupling has been used for aeroelastic tailoring).*

4.6.5.2.2 Recommendations

5. Alternate +θ and -θ plies through the LSS except for the closest ply either side of the symmetry plane. A +θ/-θ pair of plies should be located as closely as possible while still meeting the other guidelines.

   Comment: *This guideline minimizes the effect of bending/twisting coupling, which is strongest when angle plies are separated near the surface of a laminate. Modifications to this rule may promote more efficient stiffness and stability controlled designs.*

6. Shield primary load carrying plies from exposed surfaces.

   Comment: *The LSS for laminates primarily loaded in tension or compression in the 0° direction should start with angle and transverse plies. Tensile strength, microbuckling resistance, impact damage tolerance and crippling strength can all increase by shielding the main load bearing plies from the laminate surface. With primary load fibers buried, exterior scratches or surface ply delamination will not have a critical effect on strength. For laminates loaded primarily in shear, consideration should be given to locating +45° and -45° plies away from the surface. For cases in which an element is shielded by other structures (e.g., shear webs), it may not be necessary to stack primary load carrying plies away from the surface.*

7. Avoid LSS that create high interlaminar tension stresses (σ_z) at free edges. Analyses to predict free edge stresses and delamination strain levels are recommended to help select LSS.

   Comment: *Composite materials tend to have a relatively low resistance to mode I delamination growth. Edge delamination, followed by sublaminate buckling can cause premature failure under compressive loads. Edge delamination occurring under tensile loads can also effectively reduce stiffness and lower the load carrying capability. Since delaminations occurring at the core of the laminate can have the strongest effect on strength, avoid locating tape plies with fibers oriented perpendicular to a free edge at the laminate midplane.*

8. Minimize the Poisson's ratio mismatch between adjacent laminates that are cocured or bonded.

   Comment: *Excessive property mismatches between cobonded elements (e.g., skin and stringer flange) can result in delamination problems. In the absence of more sophisticated analysis tools, a general rule of thumb is*

$$|\nu_{xy}\,(\text{laminate 1}) - \nu_{xy}\,(\text{laminate 2})| < 0.1. \qquad 4.6.5.2.1$$

As opposed to static strength, composites are not particularly notch-sensitive in fatigue; hole wear is often used as the governing criterion constituting fatigue failure of composites loaded in bearing.

4.6.6 Lamina-to-laminate statistics

4.6.7 Summary

- Laminate properties such as strength, stiffness, stability, and damage resistance and damage tolerance have been found to have some dependency upon laminate stacking sequence (LSS). Each property can have a different relationship with LSS. Thus, each given design application may involve a compromise relative to LSS determination.

- Homogeneous LSS are recommended for strength controlled designs (in other words, thoroughly intersperse ply orientations throughout the LSS).

- A LSS should have at least four distinct ply angles (e.g., 0°, ±θ°, 90°) with a minimum of 10% of the plies oriented at each angle. Ply angles should be selected such that fibers are oriented with principal load axes.

- Minimize groupings of plies with the same orientation. For tape plies, stack no more than four plies of the same orientation together (i.e., limit stacked ply group thicknesses <0.03 in. (0.8 mm)). In addition, stacked ply group thicknesses with orientations perpendicular to a free edge should be limited to ≤ 0.015 in. (0.38 mm).

- If possible, LSS should be balanced and symmetric about the midplane. If this is not possible due to other requirements, locate the asymmetry or imbalance as near to the laminate midplane as possible. A LSS is considered symmetric if plies positioned at an equal distance above and below the midplane are identical (i.e., material, thickness, and orientation). Balanced is defined as having equal numbers of $+\theta$ and $-\theta$ plies, where θ is measured from the primary load direction.

4.7 COMPRESSIVE BUCKLING AND CRIPPLING

4.7.1 Plate buckling and crippling

4.7.1.1 Introduction

Rectangular flat plates are readily found in numerous aerospace structures in the form of unstiffened panels and panels between stiffeners of a stiffened panel, and as elements of a stiffener. Closed form classical buckling solutions available in the literature are limited to orthotropic plates with certain assumed boundary conditions. These boundary conditions may be fixed, simply supported, or free. For expediency, the engineer may wish to assume the most appropriate boundary conditions and obtain a quick solution rather than resort to using a buckling computer program such as Reference 4.7.1.1(a). However, the closed form solutions of laminated orthotropic plates are appropriate only when the lay-ups are symmetrical and balanced. Symmetrical implies identical corresponding plies about the plate mid-surface. Balanced refers to having a minus θ ply for every plus θ ply on each side of the mid-surface. Symmetrical and balanced laminated plates have B_{ij} terms vanish and the D_{16} and D_{26} terms virtually vanish. However, the balanced plies (±θ) should be adjacent; otherwise the D_{16} and D_{26} terms could become significant and invalidate the use of the orthotropic analysis. The buckling solutions could be significantly nonconservative for thin unbalanced or unsymmetric plates (see Reference 4.7.1.1(b)). Note that not all closed form solutions give direct answers; sometimes the equations must be minimized with respect to certain parameters as will be shown later.

The behavior of flat plates in compression involves initial buckling, postbuckling out-of-plane displacements, and crippling (ultimate postbuckling failure). Only at crippling does permanent damage occur, usually some form of delamination due to interlaminar tensile or shear stresses.

Nomenclature used to describe the buckling behavior of composite plates in Section 4.7.1 is given in Table 4.7.1.1.

4.7.1.2 Initial buckling

Initial buckling is defined to occur at a load that results in incipient out-of-plane displacements. The classical equations are elastic, and finite transverse shear stiffness effects are neglected. (Reference 4.7.1.2). The buckling of certain plate geometries, however, can be influenced by the finite shear stiffness effects as shown in Section 4.7.1.8.

4.7.1.3 Uniaxial loading - long plate with all sides simply supported

The case of a long plate (a/b > 4) with all sides simply supported (SS) and loaded uniaxially is shown in Figure 4.7.1.3(a) and described by Equation 4.7.1.3.

$$N_{x,cl}^{cr} = \frac{2\Pi^2}{b^2}\left[(D_{11}\,D_{22})^{1/2} + D_{12} + 2\,D_{66}\right] \qquad 4.7.1.3$$

Equation 4.7.1.3 is the most frequently used plate buckling equation. It can be shown by the use of the STAGS computer program (Reference 4.7.1.1(a)) that this equation is also valid for fixed boundary conditions (FF) on the loaded edges, which is important since all testing is performed with fixed boundary conditions on the loaded edges to prevent local brooming. Comprehensive testing has shown these equation to be valid except for very narrow plates. Figure 4.7.1.3(b) shows the comparisons between experiment and classical theory from References 4.7.1.3(a) and (b), where the test results are plotted as $N_{x,i}^{cr} / N_{x,cl}^{cr}$ versus the b/t ratios. Notice the discrepancy becomes worse at the low b/t ratios (narrow plates). Thus the equation should be used with caution at b/t ratios less than 35. In Figure 4.7.1.3(c) the same experimental data has been normalized by the buckling load prediction which includes the effects of transverse shear ($N_{x,w}^{cr}$) from References 4.7.1.3(c) and (d)). Note that most available computer buckling programs will not account for this transverse shear effect.

TABLE 4.7.1.1 *Buckling and crippling symbols.*

SYMBOL	DEFINITION
a	length
b	width
B_{ij}	stiffness coupling terms of laminated plate
D_{ij}	flexural/twisting stiffness terms of laminated plate
$F_{x,cl}^{cr}$	classical orthotropic longitudinal compressive buckling stress
$F_{x,i}^{cr}$	initial longitudinal compressive buckling stress from test
F_{x}^{cc}	longitudinal crippling stress from test
F_{x}^{cu}	longitudinal ultimate compressive stress of laminate
$N_{x,cl}^{cr}$, $N_{y,cl}^{cr}$	classical orthotropic longitudinal and transverse compressive uniform buckling loads, respectively
$N_{x,i}^{cr}$	initial longitudinal uniform buckling load from test
$N_{x,w}^{cr}$	longitudinal compressive uniform buckling load based on anisotropic theory, including transverse shear effects
N_x, N_y	longitudinal and transverse applied uniform loads, respectively, on a plate
$P_{x,i}^{cr}$	total longitudinal initial buckling load form test
$P_{x,i}^{cc}$	total longitudinal crippling load from test
t	thickness

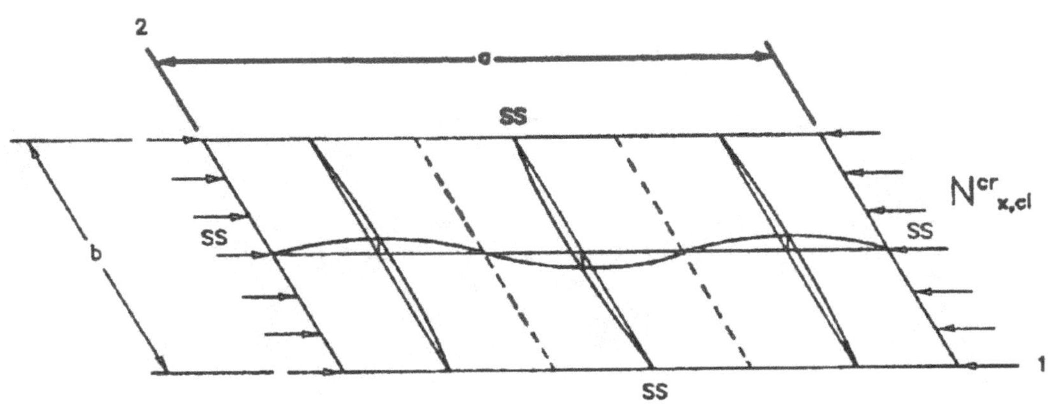

Figure 4.7.1.3(a) *Uniaxial loading - long plate (a/b > 4) with all sides simply supported (SS).*

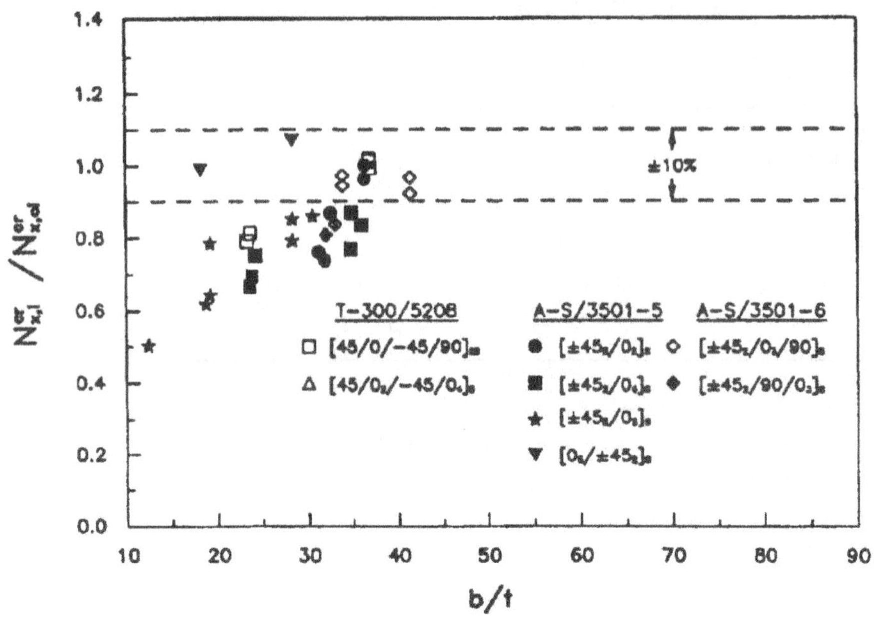

FIGURE 4.7.1.3(b) *Predicted classical buckling loads compared to experimental data (Reference 4.7.1.3(b)).*

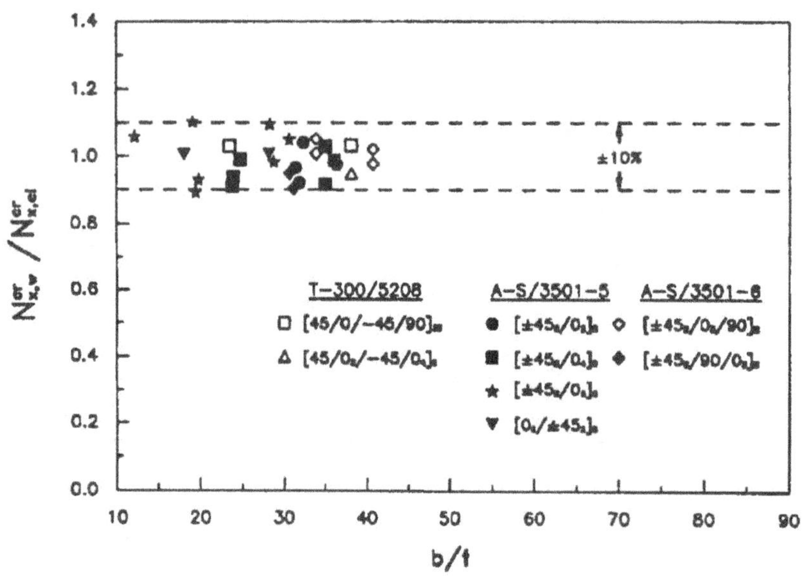

FIGURE 4.7.1.3(c) *Predicted buckling loads of the current theory compared to experimental data (Reference 4.7.1.3(b)).*

4.7.1.4 Uniaxial loading - long plate with all sides fixed

The case of a long plate (a/b > 4) with all sides fixed (FF) and loaded uniaxially is shown in Figure 4.7.1.4 and described by Equation 4.7.1.4.

$$N_{x,cl}^{cr} = \frac{\Pi^2}{b^2}\left[4.6\,(D_{11}\,D_{22})^{1/2} + 2.67\,D_{12} + 5.33\,D_{66}\right] \qquad 4.7.1.4$$

This equation has not had the comprehensive experimental study as has Equation 4.7.1.3. However, by conjecture the effect of transverse shear for narrow plates would be quite similar to that found for plates with all edges simply supported.

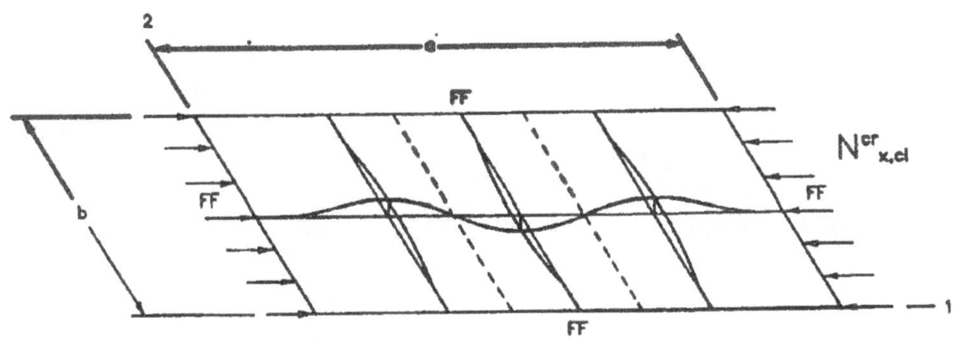

FIGURE 4.7.1.4 *Uniaxial loading - long plate (a/b > 4) with all sides fixed.*

4.7.1.5 Uniaxial loading - long plate with three sides simply supported and one unloaded edge free

Figure 4.7.1.5 shows the case of a long plate (a/b > 4) with three sides simply supported and the remaining unloaded edge free. This plate is uniaxially loaded. This loading situation is described by Equation 4.7.1.5.

$$N_{x,cl}^{cr} = \frac{12\,D_{66}}{b^2} + \frac{\Pi^2\,D_{11}}{a^2} \qquad 4.7.1.5$$

where b/t must be greater than 20 because of transverse shear effects in narrow plates as discussed in Section 4.7.1.3.

4.7.1.6 Uniaxial and biaxial loading - plate with all sides simply supported

Biaxial and uniaxial loading of a simply supported plate is shown in Figure 4.7.1.6, where $1 < a/b < \infty$. The following classical orthotropic buckling equation must be minimized with respect to the longitudinal and transverse half-waves numbers, m and n:

$$N_{x,cl}^{cr} = \frac{\Pi^2}{b^2}\,\frac{D_{11}m^4\,(b/a)^4 + 2(D_{12} + 2D_{66})\,m^2n^2\,(b/a)^4 + D_{22}\,n^4}{m^2\,(b/a)^2 + \phi n^2}\text{, min} \qquad 4.7.1.6(a)$$

where

$$\phi = N_y/N_x \qquad 4.7.1.6(b)$$

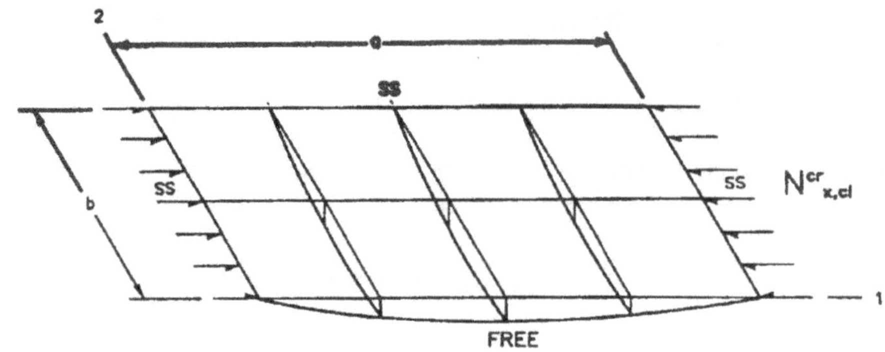

FIGURE 4.7.1.5 *Uniaxial loading - long plate with three sides simply supported and one unloaded edge free.*

which is the ratio of applied transverse to longitudinal loading. Accordingly, the corresponding transverse buckling load is

$$N_{y,cl}^{cr} = \phi N_{x,cl}^{cr} \qquad 4.7.1.6(c)$$

For uniaxial loading, let $\phi = 0$.

4.7.1.7 Uniaxial loading - plate with loaded edges simply supported and unloaded edges fixed

The case of a uniaxially loaded plate ($1 < a/b < \infty$) with the loaded sides simply supported (SS) and the unloaded sides fixed (FF) can also be considered. For this case, the following classical orthotropic buckling equation must be minimized with respect to the longitudinal half-wave number, m:

$$N_{x,cl}^{cr} = \frac{\Pi^2}{b^2}\left\{D_{11}m^2(b/a)^2 + 2.67\,D_{12} + 5.33\left[D_{22}(a/b)^2 + (1/m)^2 + D_{66}\right]\right\} \qquad 4.7.1.7$$

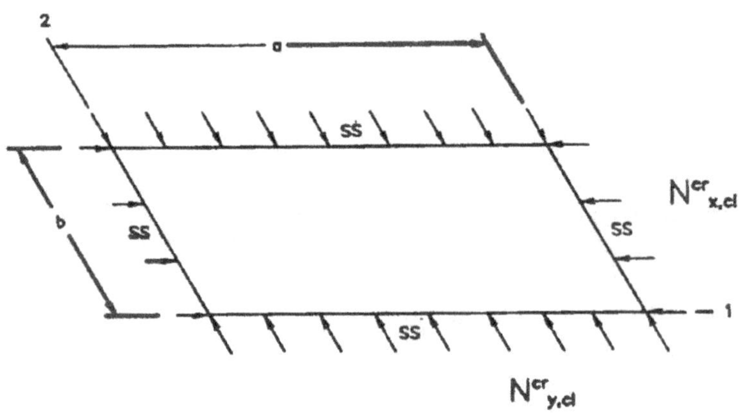

FIGURE 4.7.1.6 *Uniaxial and biaxial loading - plate with all sides simply supported.*

4.7.1.8 Stacking sequence effects in buckling

Methods to accurately predict the stability of laminated plates have been documented (e.g., References 4.7.1.8(a)-(c)). Laminated plate stability can be strongly affected by LSS. However, factors such as plate geometry, boundary conditions and load type each contribute to the relationship between LSS and plate stability. As a result, general rules that define the best LSS for plate stability do not exist. Instead, such relationships must be established for specific structure and loading types. Three examples that illustrate this point will be shown in this section. Two different analysis methods were used in these examples. The first, utilized design equations from Reference 4.7.1.8(c) and bending stiffnesses as calculated using lamination theory. This method assumed the plate bending behavior to be "specially orthotropic" (D_{16} and D_{26} terms were set equal to zero). The second method was a Boeing computer program called LEOTHA (an enhanced version of OTHA, Reference 4.7.1.8(a) which uses the Galerkin method to solve equations for buckling. This method allowed nonzero D_{16} and D_{26} terms.

Figures 4.7.1.8(a), (b), and (c) show plate buckling predictions for the seven LSS used in an earlier example (see Table 4.3.3.2(b)).[1] All plates were assumed to have simply-supported boundary conditions on the four edges. Figures 4.7.1.8(a) and (b) are rectangular plates loaded by uniaxial compression in long and short directions, respectively. Figure 4.7.1.8(c) shows shear buckling predictions for a square plate. Horizontal dashed lines on Figure 4.7.1.8(a) - (c) represent the results obtained when using the DOD/NASA design equations and assuming no LSS effect (i.e., a homogeneous orthotropic plate). The homogeneous plate assumption results in a buckling load that is roughly an average of the predictions for all LSS shown in the figures.

The highest buckling loads for rectangular plates loaded in the long direction occur with preferential clumping of ±45° plies toward the surface layers (Figure 4.7.1.8(a)). Such is not the case for rectangular plates loaded in the short direction, where preferential stacking of 0° plies yield the highest buckling loads (4.7.1.9(b)). Note that predictions using the homogeneous plate assumption can be conservative or nonconservative depending on LSS. The DOD/NASA equations compare well with LEOTHA for conditions shown in Figures 4.7.1.8(a) and (b).

The highest buckling loads for square plates loaded in shear occur with preferential clumping of ±45° plies toward the surface layers (4.7.1.9(c)). Predictions using LEOTHA are different for positive and negative shear due to the relative positions of +45° and -45° plies. Predictions from DOD/NASA equations were generally lower than those of LEOTHA for positive shear loads. The opposite was true for negative shear loads. Differences may be attributed to the influence of D_{16} and D_{26} terms which were not included in the DOD/NASA design equations.

As with bending, structural geometry can overshadow the effects of LSS on stability (see the discussion pertaining to Figure 4.3.3.2). For example, the Euler buckling load of a laminated I-section used as a column is more strongly dependent on geometrical dimensions than on LSS of web and flanges. In fact, the effects of LSS on Euler buckling load diminishes sharply with increasing web height.

Design for local buckling and crippling of composite plates has typically relied on empirical data (e.g., Reference 4.7.1.8(c)). Local buckling and crippling have been found to relate to LSS. The lowest values for local buckling and crippling under uniaxial compression occurred with preferential stacking of 0° plies towards the outside surface of a laminate. Hence, when considering an I-section, Euler buckling loads may be independent of LSS while local buckling and crippling can relate to LSS.

---

[1]The LSS used in Figures 4.7.1.9(a), (b), and (c) were chosen for illustrative purposes only and do not represent optimal LSS for a given application.

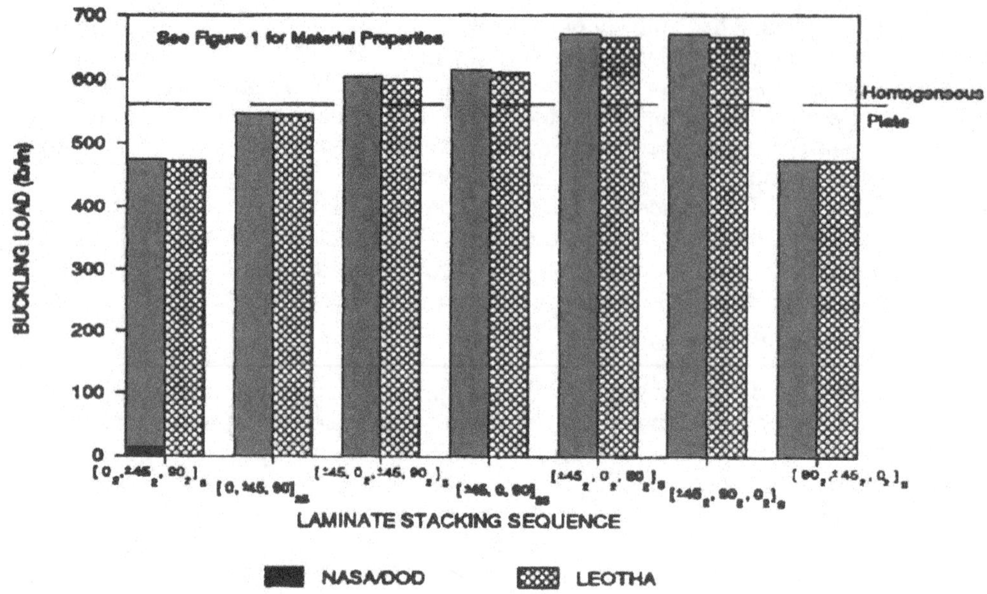

FIGURE 4.7.1.8(a) *Buckling analysis of 4 sides simply-supported, 24 in. by 6 in., laminated plates loaded in the long direction.*

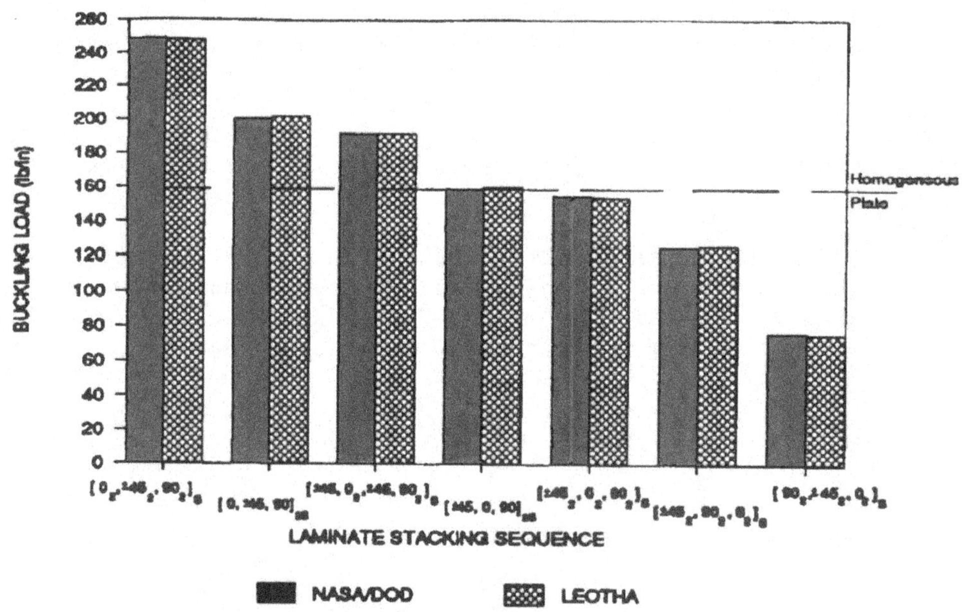

FIGURE 4.7.1.8(b) *Buckling analysis of 4 sides simply-supported, 6 in. by 24 in., laminated plates loaded in the short direction.*

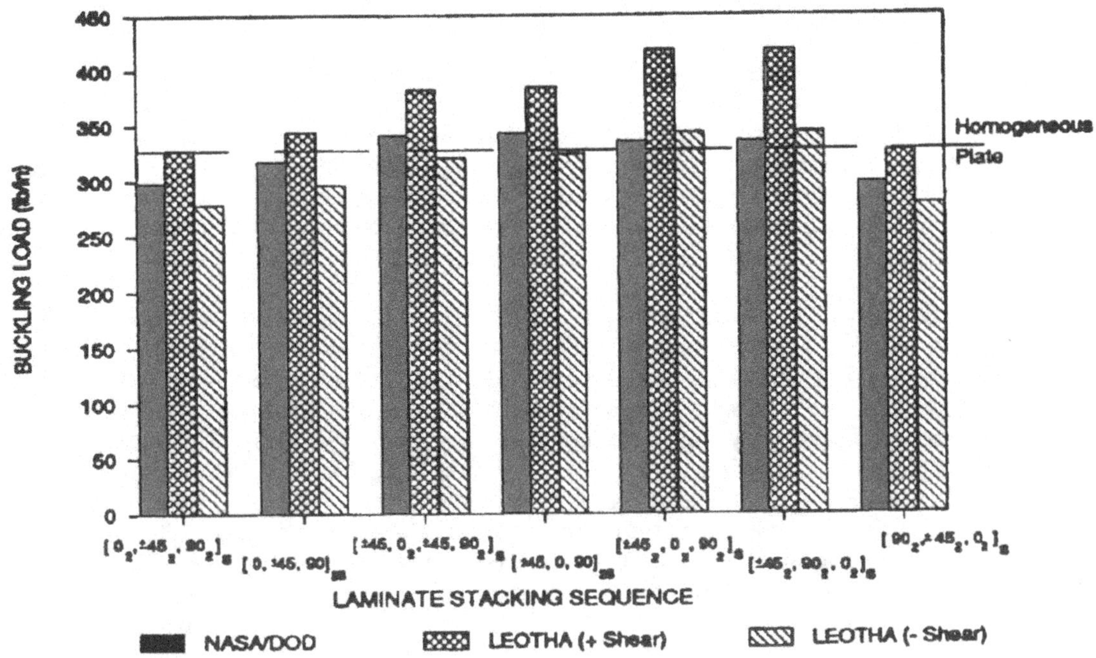

FIGURE 4.7.1.8(c) *Buckling analysis of 4 sides simply-supported, 12 in. by 12 in., laminated plates loaded in shear.*

The effects of LSS on the stability of a stiffened panel is more complex. Assuming no local buckling and crippling, stiffener stability will not depend directly on LSS. However, post-buckling behavior of the skin and load redistribution to the stringer is strongly affected by the skin's LSS. As a result, overall stiffened panel stability can be influenced by the skin's LSS.

Basic information on laminate stacking sequence effects is found in Section 4.6.5.

4.7.2 Compression postbuckling and crippling

Wide exploitation of advanced composites in stability critical structural designs depends to a large degree on the ability of composites to support loads well beyond the initial buckling level. Unquestionably, the high stiffness-to-weight ratio of composites renders them potentially attractive up to initial buckling. However, since postbuckling design has been established over several decades for certain types of conventional metallic alloy construction, it should be anticipated that composites demonstrate a similar capability. Hence, this section addresses this vitally important issue as it pertains to the design of structural compression members.

Postbuckling. Postbuckling is the ability of a compression member or stiffened panel to carry loads well in excess of the initial buckling load. The "postbuckling range" may be considered to exist between the initial buckling load and some higher load representing failure, e.g., delamination at the free edge of a compression member or the disbonding of a stiffener from the panel in a stiffened panel. When stiffened panels are loaded in compression, load is shared between skin and stiffeners in proportion to their respective stiffnesses. At initial buckling, the tangent stiffness of the skin is reduced sharply and as a result, a greater portion of the total load will be carried by the stiffeners. For an isotropic material with linear elastic behavior prior to initial buckling, the tangent stiffness at buckling is reduced to one half of its initial value. For composite panels, tangent stiffnesses are a function of material properties and lay-up. Local buckling of one or more of the plate elements comprising a stiffener will similarly reduce the in-plane stiffnesses of the affected elements and will cause the load to shift to the unbuckled portions of the stiffener. The upper limit of the postbuckling range is sometimes referred to as "local crippling" or simply "crippling".

Crippling. Compression crippling is a failure in which the cross section of a stiffener is loaded in compression and becomes distorted in its own plane without translation or rotation of the entire column taking place. Typical deflected shapes seen in crippling tests of angles and channel section stiffeners are shown in Figure 4.7.2(a). Angles or cruciforms loaded in compression are commonly used as crippling specimens for the "one-edge-free" case. Channels or simply supported compression panels are normally used for the "no-edge-free" case, in which the center channel segment is approximately simply supported with "no-edge-free".

The postbuckling behavior of composite plates presented here is derived from the empirical graphite tape data obtained from References 4.7.2(a) through (h). Relatively narrow plates, with simply supported unloaded edges or one-edge-free and fixed loading edges were tested and analyzed. The simply supported unloaded edges were simulated by the use of steel V-blocks mounted on the compression test fixture. Specifically, the plates with both unloaded edges simply supported are defined as "no-edge-free". Plates with one unloaded edge simply supported and the other free are defined as "one-edge-free". A typical no-edge-free test in progress with the specimen in the postbuckling range is shown in Figure 4.7.2(b). In addition, a typical one-edge-free test where crippling of the specimen has occurred is shown in Figure 4.7.2(c). Typical load-displacement curves of no-edge-free and one-edge-free tests are shown in Figures 4.7.2(d) and 4.7.2(e), respectively. Figure 4.7.2(d) clearly shows the reduction in stiffness at initial buckling as indicated by the change in slope of the load deflection curve at that point. A convenient plot that exemplifies the postbuckling strength of the no-edge-free composite plates is shown in Figure 4.7.2(f). The value for F_{11}^{cu} is the ultimate compressive strength of the particular laminate. A typical failed test specimen is shown in Figure 4.7.2(g). Figure 4.7.2(h) illustrates the postbuckling strengths of one-edge-free plates. Note that all the empirical data presented involved the testing of high strength carbon/epoxy tape. Other material systems or other forms of carbon/epoxy composites may yield different results.

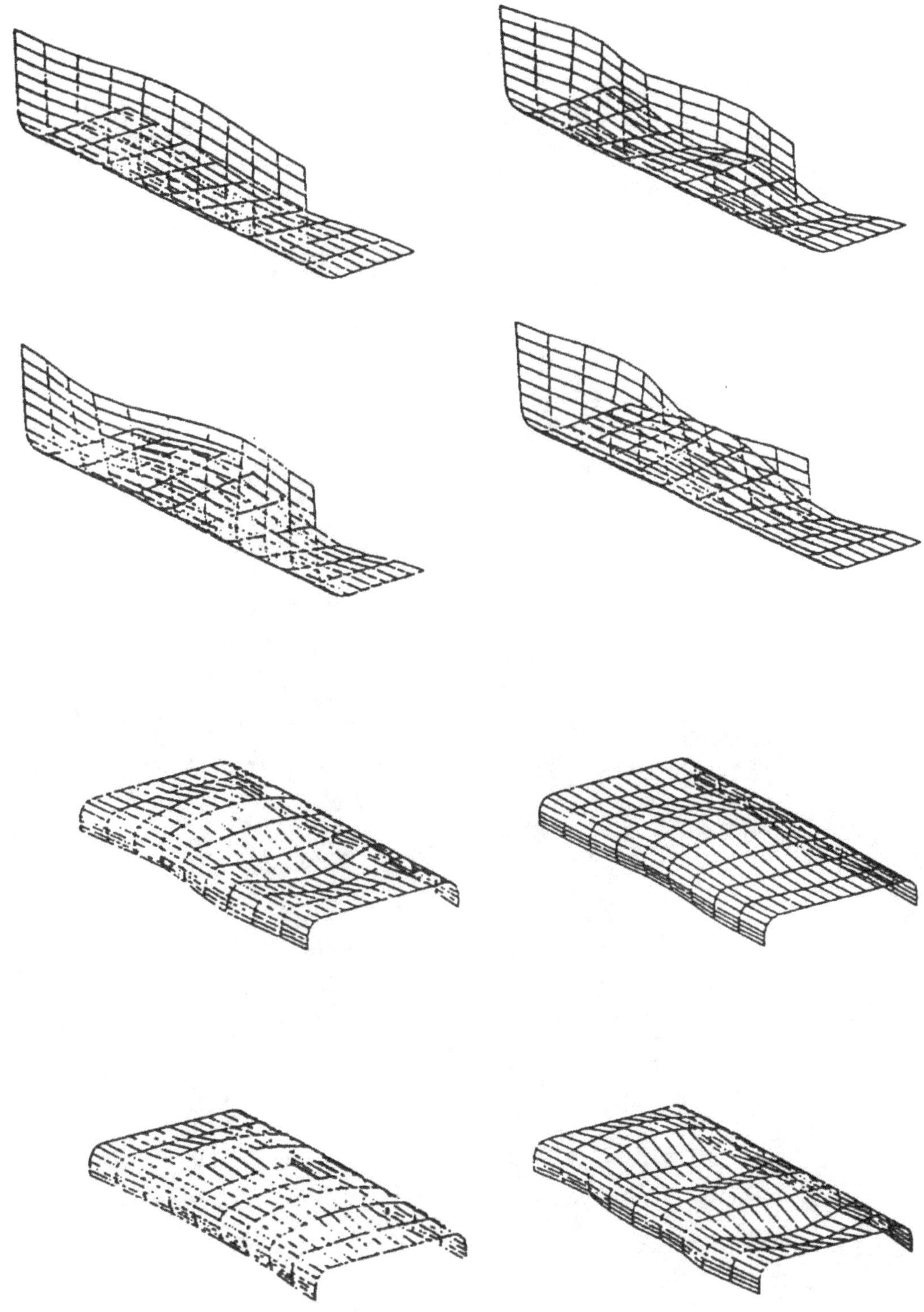

FIGURE 4.7.2(a) *Typical crippling shapes.*

FIGURE 4.7.2(b) *No-edge-free carbon/epoxy test.*

FIGURE 4.7.2(c) *One-edge-free carbon/epoxy postbuckling test at crippling.*

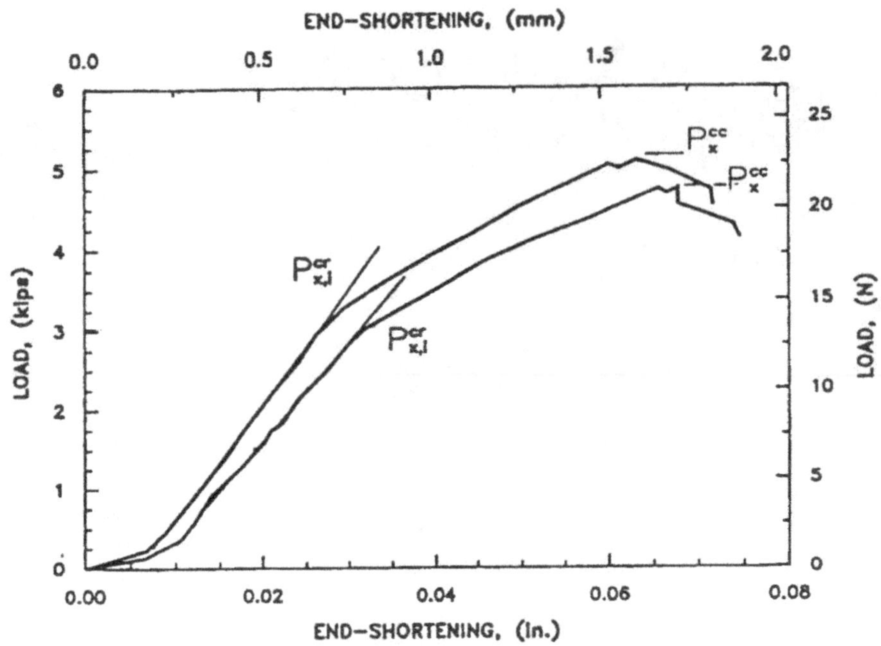

FIGURE 4.7.2(d) *No-edge-free plate. Crippling tests - AS/3501-6 [±45/90/0$_3$]$_s$ - b/t ≈ 32.*

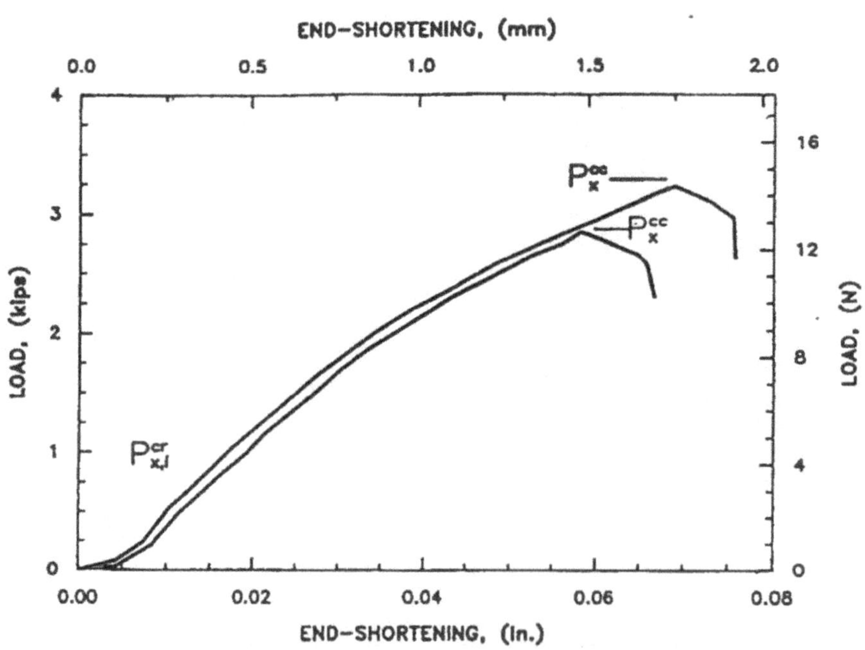

FIGURE 4.7.2(e) *One-edge-free plate. Crippling tests - AS/3501-6 [±45/90/0$_3$]$_s$ - b/t ≈ 30.*

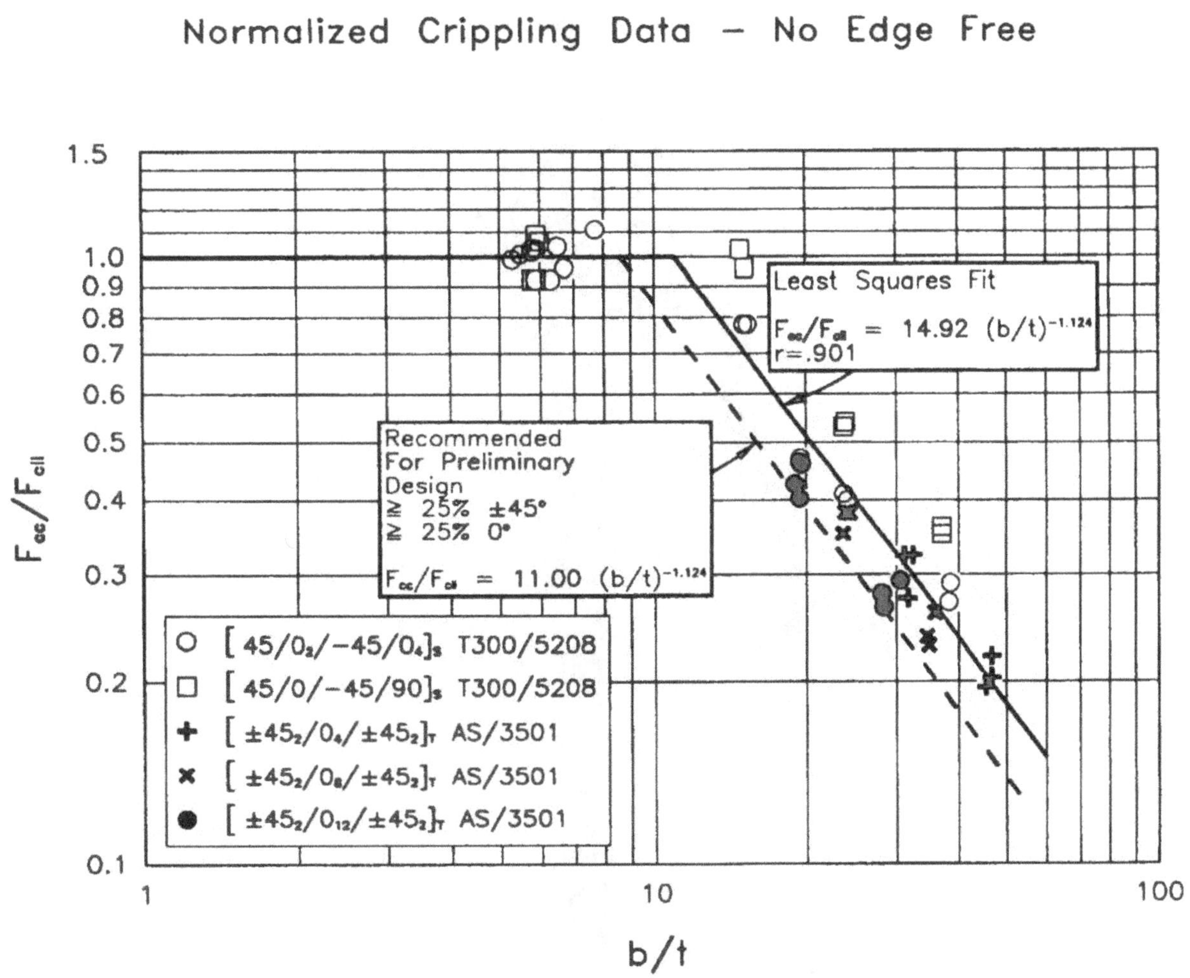

FIGURE 4.7.2(f) *Normalized crippling data - no-edge-free.*

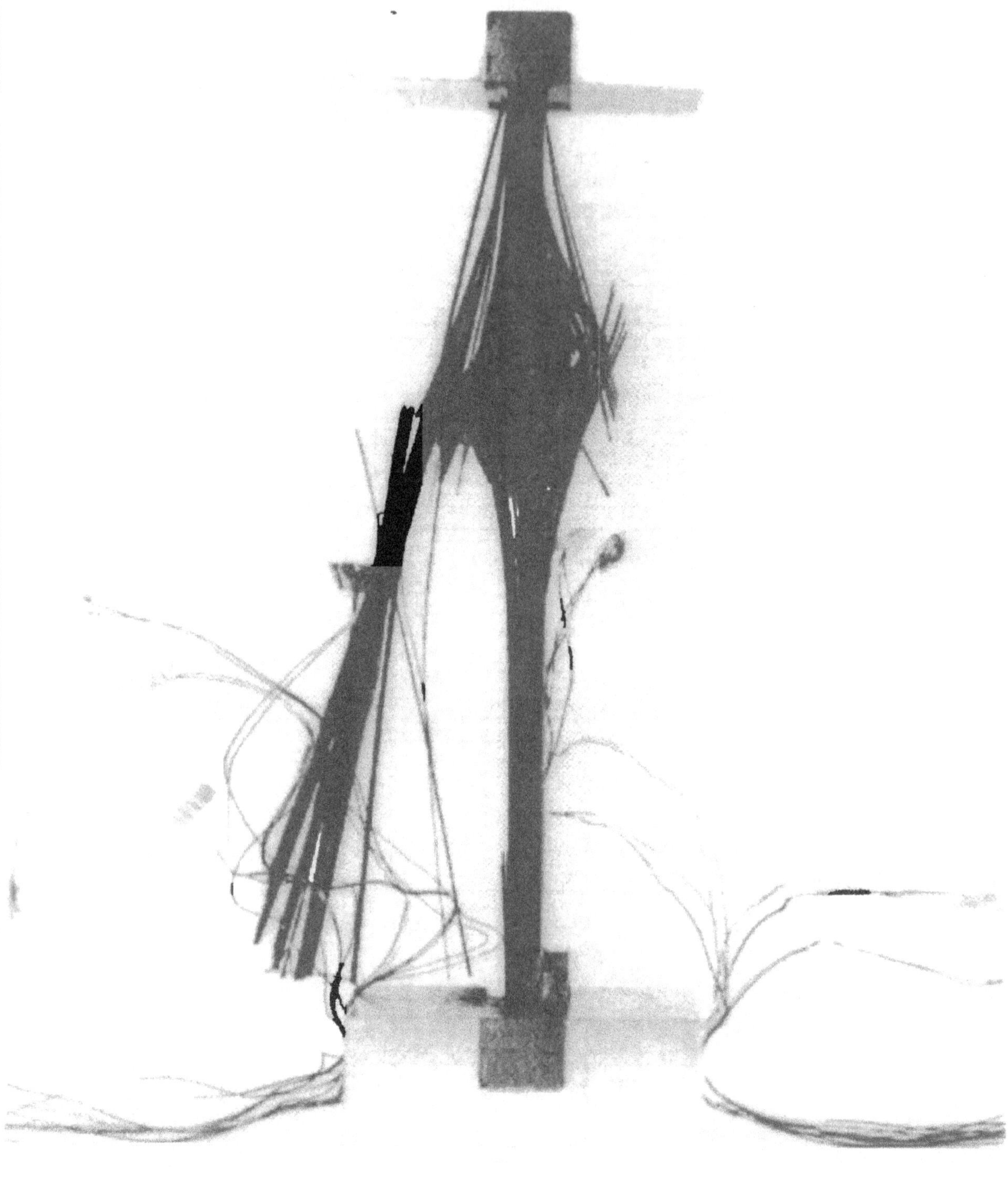

FIGURE 4.7.2(g) *Typical carbon/epoxy failed ultimate compression specimen.*

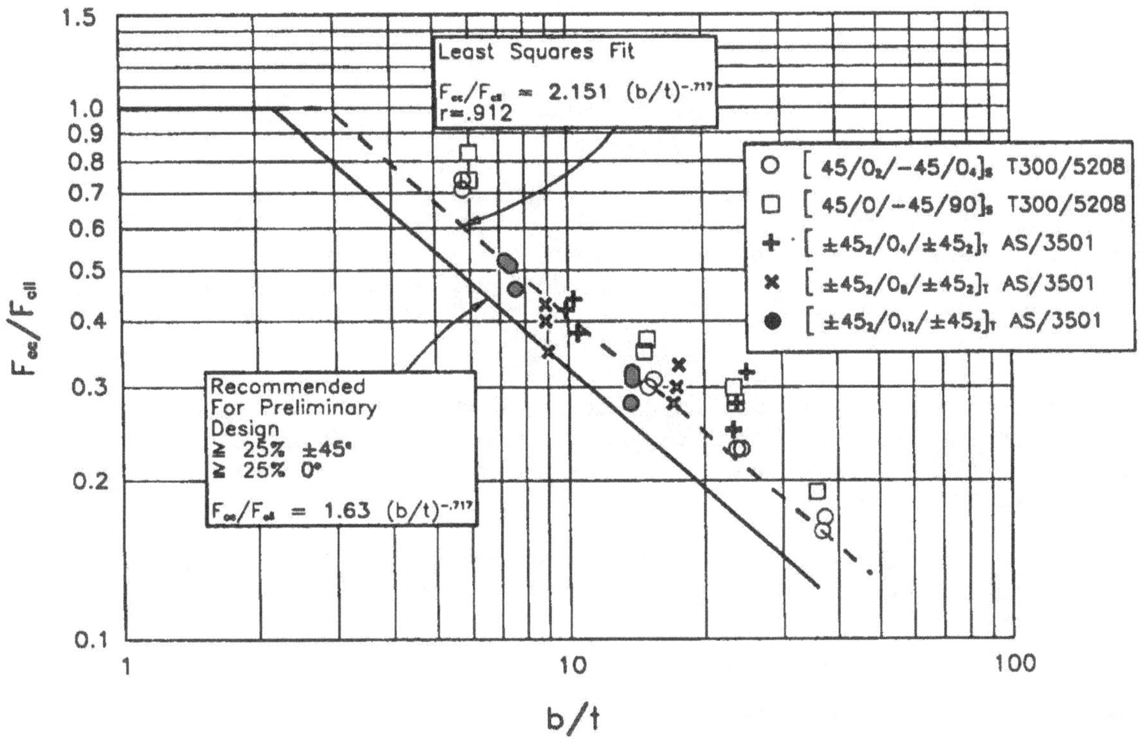

FIGURE 4.7.2(h) *Normalized crippling data - one-edge free.*

4.7.2.1 Analytical models

As stated in Section 4.7.1.2, initial buckling is more accurately determined by including the effects of transverse shear and material nonlinearity as is done in References 4.7.1.3(c) and (d). Transverse shear effects become especially important for thick laminates ($b/t < 20$). Stress-strain curves for laminates with a high percentage of ±45° plies may show significant material nonlinearity prior to initial buckling. These effects are equally important, of course, for plates loaded in the postbuckling range. Some examples of test results vs. the theory of these references are shown in Figures 4.7.2.1(a) and (b). Unfortunately, most of the computer programs available today are based on linear elastic theory and do not include transverse shear effects. Consequently, experimental data must be obtained to correct for these and other deficiencies in the analytical models.

The theoretical buckling loads for orthotropic one-edge-free and no-edge-free plates are given by:

$$N_x^{cr}\ (OEF) = \frac{12 D_{66}}{b^2} + \frac{\pi^2 D_{11}}{L^2}$$

$$N_x^{cr}\ (NEF) = \frac{2\pi^2}{b^2}\left[\sqrt{D_{11}D_{22}} + D_{12} + 2D_{66}\right] \qquad 4.7.2.1(a)$$

These expressions do not include the bending-twisting terms D_{16} and D_{26}. These terms are present in all laminates that contain angle plies but, except in laminates having very few plies, their effect on the initial buckling load is generally not significant. Hence, the above equations are accurate for most practical laminates that are balanced and symmetrical about their mid-surface. The reader is referred to studies performed by Nemeth (Reference 4.7.2.1(a)) for additional information on the buckling of anisotropic plates and the effect of the various parameters on the buckling loads.

The Euler term in the first of the above equations is generally found to be negligible and, therefore, initial buckling of a one-edge-free plate is largely resisted by the torsional stiffness (D_{66}) of the laminate. This explains why higher initial buckling loads may be obtained for a given lay-up when the ±45° plies are on the outside surfaces of the plate.

For laminates that are only slightly unbalanced or unsymmetrical, approximate values for the initial buckling load may be obtained by substituting "equivalent" bending stiffnesses $\bar{D}_{ij}$ in place of D_{ij} in the buckling equations, where

$$[\bar{D}] = [D] - [B][A]^{-1}[B] \qquad 4.7.2.1(b)$$

The analysis of panels loaded in the postbuckling range becomes a geometrically nonlinear problem and, therefore, "conventional" plate buckling programs or other linear analysis codes cannot be used to accurately predict the crippling strength of composite plates. One example is shown in Figure 4.7.2.1(c), which shows experimental crippling curves and theoretical buckling curves for a quasi-isotropic T300/5208 laminate. (The AS/3501 and T300/5208 carbon/epoxy crippling data was taken from References 4.7.2(b) - (e)). The theoretical buckling curves shown in Figure 4.7.2.1(c) are very conservative at high b/t values and very unconservative at low b/t values. This may be explained by the fact that thin plates buckle at low strain levels and may thus be loaded well into the postbuckling range. On the other hand, neglecting transverse shear effects will cause strength predictions at low b/t ratios to be unconservative. The analysis of laminated plates is further complicated by the fact that high interlaminar stresses in the corners or at the free edge of the plate may trigger a premature failure.

As it would not be practical during preliminary design to conduct nonlinear analyses for a large number of lay-ups and b/t ratios, a better approach may be to use semi-empirical data to correct initial buckling predictions.

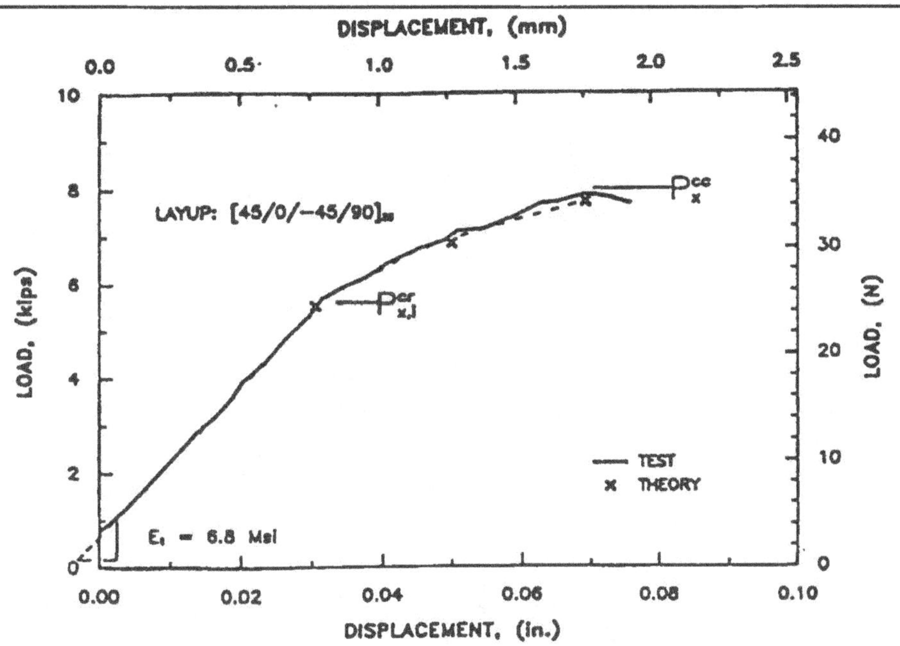

FIGURE 4.7.2.1(a) *Comparison of theory in References 4.7.1.3(b) and (c) with experiments for postbuckling curves and crippling strengths.*

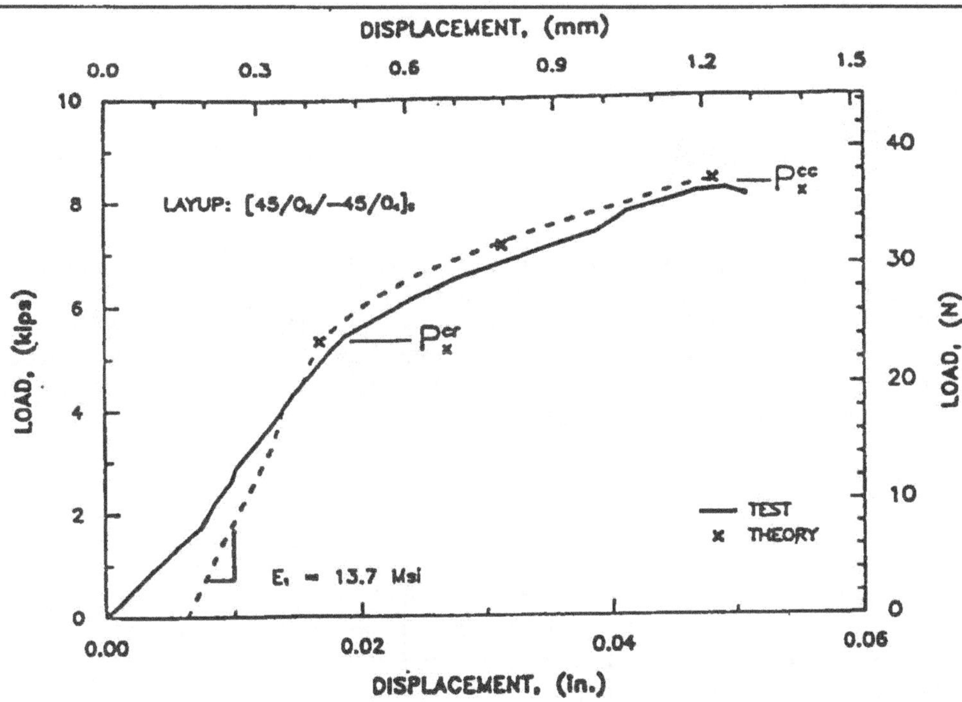

FIGURE 4.7.2.1(b) *Comparison of theory in References 4.7.1.3(b) and (c) with experiments for postbuckling curves and crippling strengths.*

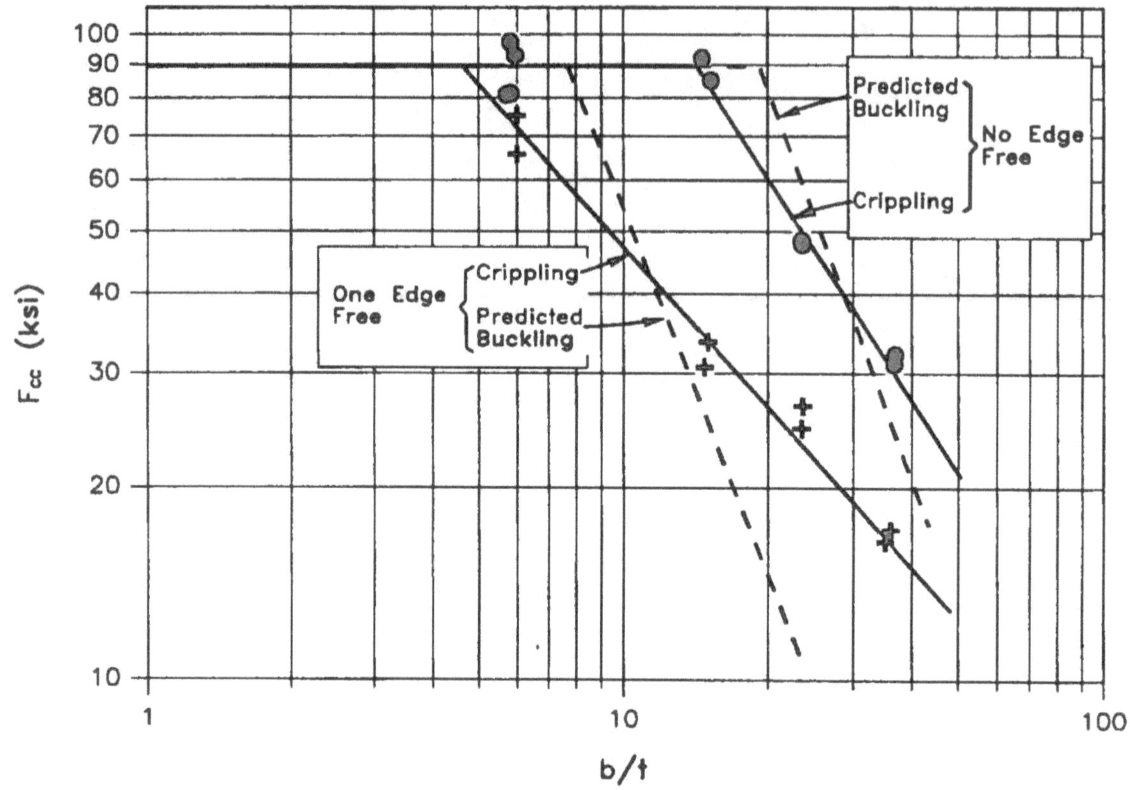

FIGURE 4.7.2.1(c) *Comparison of predicted buckling and crippling curves.*

4.7.2.2 Fatigue effects

Postbuckling fatigue may be permitted under certain circumstances without jeopardizing the structural integrity of the plate (References 4.7.2(b), 4.7.2(g), and 4.7.2(h)). Significant conclusions identified in Reference 4.7.2(i) stated: "Composite panels demonstrated a high fatigue threshold relative to the initial skin buckling loads. Composite panels showed a greater sensitivity to shear dominated fatigue loading as compared with compression dominated fatigue loading. The fatigue failure mode in composite panels was separation between the cocured stiffener and skin."

4.7.2.3 Crippling curve determination

Non-dimensional crippling curves are used to determine the crippling strength of the one-edge and no-edge-free composite elements. Different normalization techniques have been suggested for composites, most of which are modifications of those currently used in the aircraft industry for metallic structures. Perhaps, the most obvious change in the analysis and presentation of crippling data is the proposed use of the ultimate compression strength, F^{cu} to normalize the crippling strength, F^{cc}, for composites, instead of the material yield stress, F^{cy} commonly used for metallic elements.

Crippling curves for carbon/epoxy one- and no-edge-free plates are presented in References 4.7.2(e) in terms of the non-dimensional parameters F^{cc}/F^{cu} and $(b/t) * \left[F^{cu}/(E_x * E_y)^{1/2}\right]^{1/2}$. The latter parameter was chosen to reflect the orthotropic nature of composites. Test data for the one-edge-free plate elements were found to be in excellent agreement with the expected behavior, when the data were presented in terms of these non-dimensional parameters, but test results for the no-edge-free elements fell below the expected values.

A shortcoming in the methodology presented in Reference 4.7.2(e) is that the curves are non-dimensionalized on the basis of laminate extensional modulus only. The plate bending stiffnesses play an important role in determining the initial buckling and crippling loads of the element. Unlike in metallic plates, however, there exists no direct relationship between the extensional and bending stiffnesses of a composite plate and, therefore, laminates with equal in-plane stiffnesses may buckle at different load levels if their stacking sequences are not identical. Tests conducted by Lockheed and McDonnell Douglas under their respective Independent Research and Development (IRAD) programs have confirmed that more accurate buckling and crippling predictions may be obtained when the curves are defined in terms of the non-dimensional parameters

$$\frac{F^{cc}}{F^{cu}}\,\frac{E_x}{\bar{E}} \quad \text{and} \quad \frac{b}{t}\,\frac{\bar{E}}{E_x}\sqrt{\frac{F_{cu}}{\sqrt{E_x E_y}}} \qquad 4.7.2.3(a)$$

in which

$$\bar{E} = \frac{12D_{11}}{t^3}\left(1 - \nu_{xy}\nu_{yx}\right) \qquad 4.7.2.3(b)$$

is an effective modulus accounting for stacking sequence effects through the bending stiffness term D_{11}.

4.7.2.4 Stiffener crippling strength determination

The commonly used procedure for predicting the crippling strength of a metallic stiffener, composed of several one-edge and no-edge-free elements, is to compute the weighted sum of the crippling strengths of the individual elements:

$$F_{ST}^{cc} = \frac{\sum_{i=1}^{N} F_i^{cc}\cdot b_i\cdot t_i}{\sum_{i=1}^{N} b_i \cdot t_i} \qquad 4.7.2.4$$

Test results appear to indicate that the same procedure can be successfully applied to composite stiffeners of uniform thickness if the element crippling strengths are determined with the aid of the non-dimensional parameters in Equation 4.7.2.3. Lockheed tests involved crippling of angles and channels made from thermoplastic (IM8/HTA) and thermoset (IM7/5250-4) materials. Tests results for one- and no-edge-free plates are presented in Figures 4.7.2.4(a) and 4.7.2.4(b). McDonnell Douglas also reported that, using this approach, predictions for carbon/epoxy stiffeners and AV-8B forward fuselage longerons have shown excellent correlation with test results.

Optimum design of stiffened panels made of composite materials may require the use of stiffeners of non-uniform thickness. Typical examples of frequently used stiffener configurations are shown in Figure 4.7.2.4(c). Insufficient experimental data currently exist to accurately predict the crippling strength of such stiffeners. At the juncture of two plate elements of different thickness, the thicker element will provide additional restraint to the thinner element. As a result, both the buckling and crippling strength of the thinner element will be increased while that of the thicker one will be decreased. The net effect could be an increase or decrease of the allowable stiffener stress depending on which of these two elements is more critical and thus is driving the buckling process. Equation 4.7.2.4 may be used to predict stiffener crippling but appropriate adjustments should be made to the crippling strength of the affected elements if that strength was based on data obtained from uniform thickness test specimens.

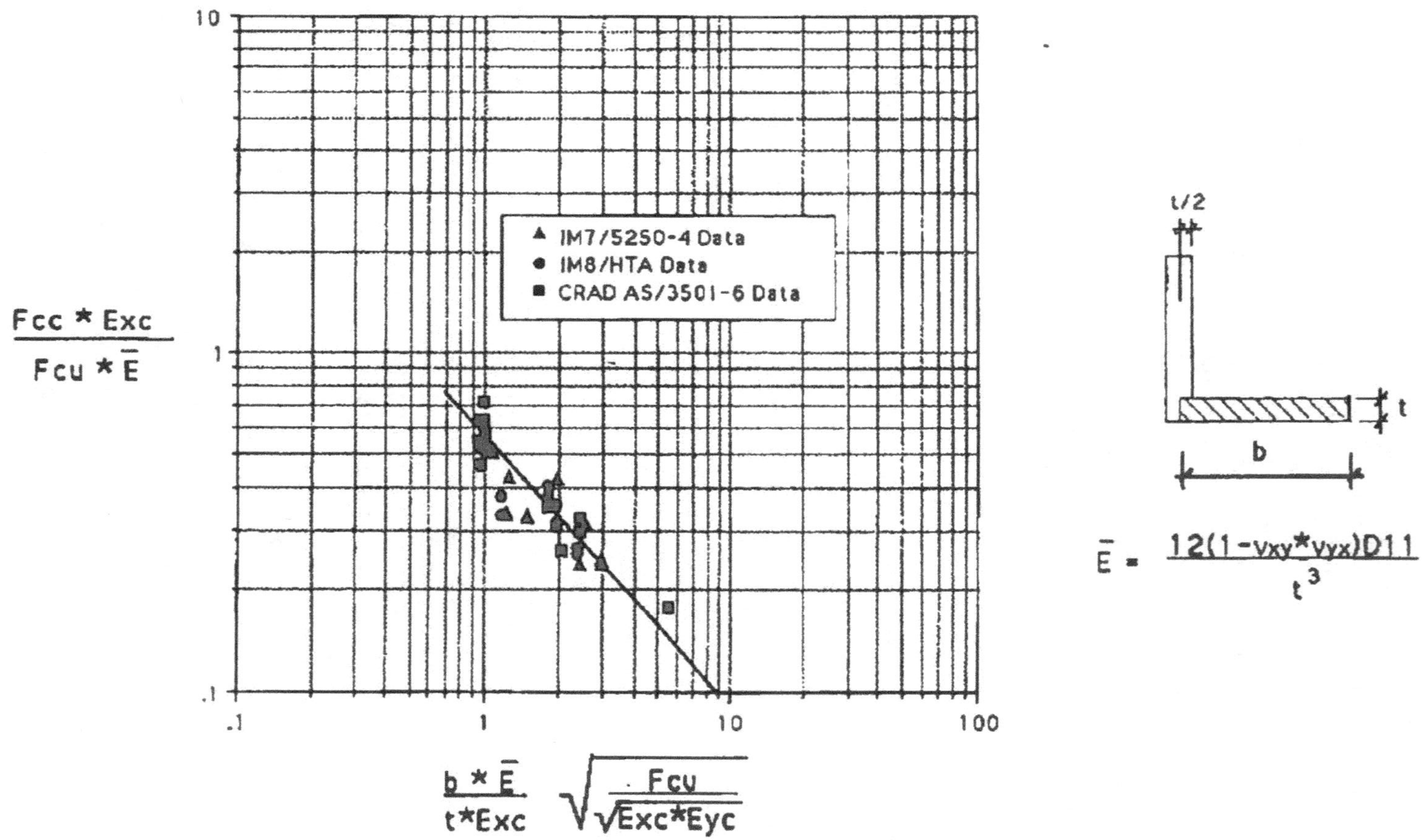

FIGURE 4.7.2.4(a) *One-edge-free crippling test results.*

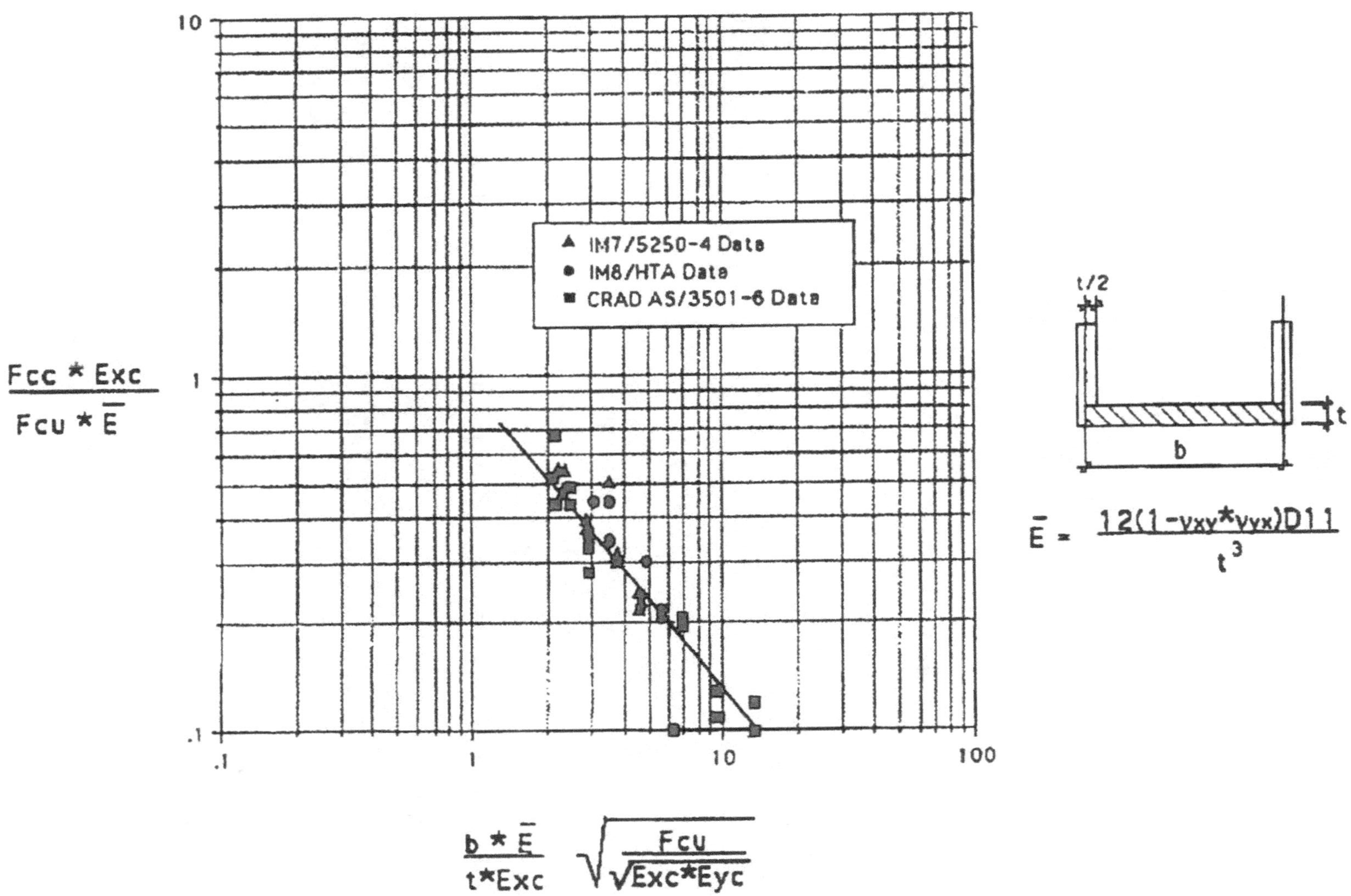

FIGURE 4.7.2.4(b) *No-edge-free crippling test results.*

4.7.2.5 Effects of corner radii and fillets

In channel, zee, or angle section stiffeners where crippling rather than delamination is the primary mode of failure, the corner radii do not appear to have an appreciable effect on the ultimate strength of the section. The opposite is true, however, for I or J stiffeners, where the corner radii do play an important role. It has been common practice to use unidirectional tape material to fill the corners of these stiffeners, as shown in Figure 4.7.2.5. The addition of this very stiff corner material increases the crippling strength of the stiffener. Since the cross-sectional area of the fillet, and thus the amount of 0° material, is proportional to the square of the radius, the increase in crippling strength may be significant for stiffeners with large corner radii. A conservative estimate for the increase in crippling strength may be obtained from the following expression:

$$\bar{F}^{cc} = F^{cc} \frac{1 + \dfrac{E_f A_f}{\Sigma E_i b_i t_i}}{1 + \dfrac{A_f}{\Sigma b_i t_i}} \qquad 4.7.2.5$$

which is based on the assumption that the critical strain in the corner region is no greater than that for a stiffener without the additional filler material.

4.7.2.6 Slenderness correction

As the unsupported length increases, the stiffener may fail in a global buckling mode rather than by local crippling. The usual procedure to account for this is to apply a correction factor to the crippling strength, F_{cc}, based on the slenderness ratio (L'/ρ) of the column. The critical stress for the stiffener now becomes

$$F^{cr} \infty F^{cc} \left[1 - \frac{F^{cc}}{4\pi^2 E_x^c}\left(\frac{L'}{\rho}\right)^2\right] \qquad 4.7.2.6(a)$$

The radius of gyration for the cross-section of a composite column is defined as

$$\rho = \sqrt{\frac{(EI)_{st}}{(EA)_{st}}} \qquad 4.7.2.6(b)$$

where $(EA)_{st}$ and $(EI)_{st}$ are the extensional and bending stiffnesses of the stiffener.

4.7.3 Summary

- The buckling strength, or stability, of flat and curved composite skin panels is strongly affected by geometry, stacking sequence, boundary conditions, and loading conditions. In many cases, it may be estimated using existing closed form solutions for orthotropic plates $(r/t > 100)$, such as equations 4.7.1.3 - 4.7.1.7.

4.8 CARPET PLOTS

4.9 CREEP AND RELAXATION

4.10 FATIGUE

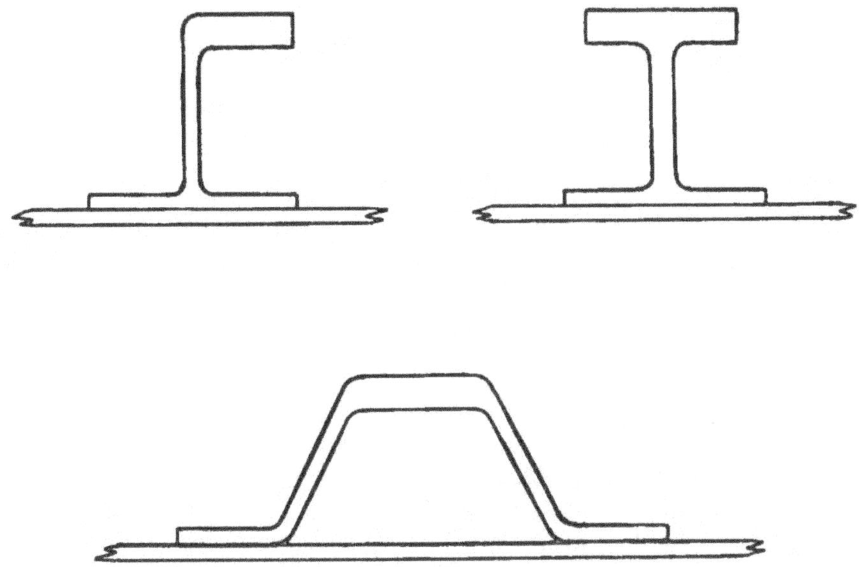

FIGURE 4.7.2.4(c) *Non-uniform thickness stiffener configurations.*

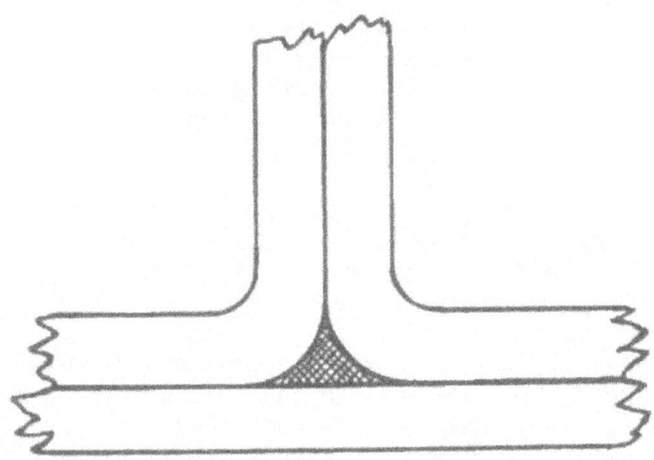

FIGURE 4.7.2.5 *Corner fillet.*

4.11 OTHER STRUCTURAL PROPERTIES

4.11.1 Damage tolerance

4.11.1.1 Background

Damage tolerance is defined as a measure of the structure's ability to sustain a level of damage or presence of a defect and yet be able to perform its operating functions. Consequently, the concern with damage tolerance is ultimately with the damaged structure having adequate residual strength and stiffness to continue in service safely until the damage can be detected by scheduled maintenance inspection and repaired, or if undetected, for the remainder of the aircraft's life. Thus, safety is the primary goal of damage tolerance.

4.11.1.2 Types of damage and concerns

There are basically two types of damage which are categorized by their occurrence during the fabrication and use of the part, i.e., damage occurring during manufacturing or damage occurring in service. The former commonly includes consideration of defects such as porosity, microcracking, and delaminations resulting from processing discrepancies and also such items as inadvertent edge cuts, surface gouges and scratches, damaged fastener holes, and impact damage. The inadvertent (non-process) damage most commonly occurs in detail parts or components during assembly or transport.

It is expected that the occurrence of the majority of manufacturing associated damage, if beyond specification limits, will be detected by routine quality inspection. Nevertheless, some "rogue" defects or damage beyond specification limits may go undetected and consequently, their occurrence must be assumed in the design procedure. An example of such a damage tolerance design criteria would be the assumption of inadvertent delaminations existing in parts. Even though the capability to detect a 0.5 in diameter delamination by in-process ultrasonic inspection is considered typical, in designing to damage tolerance criteria, we may assume the presence of a larger delamination is missed by inspection. Quantifying the size of the "rogue" or missed flaw would be part of the contractor/customer criteria development process.

Another example of a manufacturing defect is when prepreg backing paper or separation film is inadvertently left in the laminate between plies during layup. Current inspection methods may not detect this type of discrepancy which may lead to a large delamination. Consequently, until more adequate inspection techniques are developed, in-process quality controls must be sufficiently rigid to preclude this occurrence. A damage tolerance criterion might be required to design for the occurrence of the flaw. Obviously a balance is needed between designing all possible "damage" into a part (with associated weight penalties) and what realistically can and should be eliminated from the fabrication/assembly process.

Service damage concerns are similar to those for manufacturing. Types of service damage include edge and surface gouges and cuts caused by improper tool use, or foreign object collision and blunt object impact damage caused by such incidences as dropped tools or contact with service equipment. Of these, cuts and gouges, are usually the more severe for tension loading while impact damage is generally the more severe damage type for compression, especially for thick laminates.

Delaminations can also be critical defects. However, unless they are very large, i.e., more than 2.0 inches (50 mm) in diameter, the problem is mostly with thin laminates. Effects of manufacturing defects such as porosity and flawed fastener holes are usually less severe. They are generally accounted for by the use of design allowable properties that have been obtained by testing specimens with notches, i.e., stress concentrations. Most commonly, these are specimens with a centered hole. Open holes are typically used for compression specimens while either open or filled holes (holes with an installed fastener) are used for tension testing. (Open holes are more critical then filled holes for compression. Filled holes may be more critical in tension, especially for laminates with ply orientations with a predominate number of plies oriented in the load direction). Consequently, the design allowables thus derived may be used to account for a nominal design stress concentration caused by an installed or missing fastener, at least to a 0.25 inch (6.4 mm) diameter, as well as many manufacturing defects.

One of the primary concerns with damage tolerance of composites is detection of the damage. This is true both during manufacture and once in service. For the latter, the threshold of detectability depends on the type of inspection scheduled in service:

- walk around: long distance visual inspection to detect punctures, fiber breakage, i.e., readily detectable damage.

- visual detailed inspection using grazing light on a clean element, lens, etc., to detect barely visible impact damage. In this case the dent depth at the threshold of detectability (δ_d) should be defined.

- special detailed inspection for non-visible damage using ultrasonic, x-ray, shearography, etc.

Dent depths have been found to decay with time, aging, thermal cycling, and mechanical fatigue due to visco-elasticity phenomena. In the case of visual detailed inspections, the dent depth established should be that which after decay is at least equal to or above the threshold of detectability, (δ_d). In some cases, the initial impact indentation dent depth (δ_I) may be as much as 3 times that of δ_d.

Fortunately, most of the damage which is critical to tension loading such as cuts and gouges is, to some degree, visible. Tests have shown that for tension loading, the residual strength of a laminate with a cutout is primarily dependent on the width of the cutout and essentially independent of the cutout shape. Thus design values reduced to account for the presence of a 0.25 inch diameter hole also account for an equivalent length edge cut, however, and testing might be required to define appropriate design values. Cuts of this type that might be produced during manufacturing are a special problem since they may be filled with paint, and consequently, not detected. Sufficient testing should be done as part of design verification programs to ensure that cuts and gouges that are on the threshold of visibility will not degrade the structure enough to jeopardize its safety.

Low velocity impacts, e.g., impacts from dropped tools as opposed to ballistic impacts, present a special problem. Impacts on the laminate surface, especially those made by a blunt object, may cause considerable internal damage without producing visible indications on the surface. Damage to the resin may be particularly severe as evidenced by transverse shear cracks and delaminations. Consequently, the resin loses its ability to stabilize the fibers in compression and the local failure may initiate total structural collapse. Similarly, the impact may damage fibers and cause local stress concentrations which could result in significant loss of tensile strength. With conventional graphite/epoxy systems, which are quite brittle, losses in tension and compression strength for nondetectible impact may approach 50% and 60% respectively.

Low velocity impact damage, such as from dropped tools is a key problem in thin gage structures. The damage is characteristically different from damage in thick laminates. Laminate internal delamination is less of a problem because thin laminate damage is more typically in the form of fiber breakage and, possibly, penetration. For sandwich materials with thin faces, impact can result in visible core damage which has been shown to reduce the compressive and shear strengths. Impact damage which causes a break in the facesheet of the sandwich (as well as porosity, a manufacturing defect) also presents a long term durability problem in that it can allow water intrusion into the core.

It is considered possible that nonvisible damage might occur early in the aircraft's life and go undetected during subsequent service inspections. Thus, unless detection is ensured by more discriminating inspection procedures, the damage or defect must be assumed to be present for the entire life of the aircraft. This "duration of damage or defect" factor is important in determining the probability that the aircraft with a defect or damage will encounter a load which might cause failure. That is, structure having undetected damage, such as low-velocity impact damage, will have a probability of encountering a particularly high load that is directly proportional to the service time during which that load may occur. In contrast, structure that has damage that is discovered and repaired in a short time period will have less chance of encountering the same high load.

4.11.1.3 Evolving military and FAA specifications

The "duration of damage or defect" factor based on degree of detectability has been the basis for establishing minimum Air Force damage tolerance residual strengths for composite structures in requirements proposed for inclusion in AFGS-87221A, "General Specification for Aircraft Structures". These strength requirements are identical to those for metal structure having critical defects or damage with a comparable degree of detectability. Requirements for cyclic loading prior to residual strength testing of test components are also identical. The nondetectable damage to be assumed includes a surface scratch, a delamination, and impact damage. The impact damage includes both a definition of dent depth, i.e., detectability, and a maximum energy cutoff. Specifically, the impact damage to be assumed is that "caused by the impact of a 1.0 inch diameter hemispherical impactor with a 100 ft-lb of kinetic energy, or that kinetic energy required to cause a dent 0.10 inch deep, whichever is least." For relatively thin structure, the detectability, i.e., the 0.1 inch (2.5 mm) depth, requirement prevails. For thicker structure, the maximum assumed impact energy becomes the critical requirement. The associated load to be assumed is the maximum load expected to occur in an extrapolated 20 lifetimes. This is a one time static load requirement. These requirements are coupled with assumptions that the damage occurs in the most critical location and that the assumed load is coincident with the worst probably environment.

In developing the requirements, the probability of encountering undetected or undetectable impact damage above the 100 ft-lb (136 J) energy level was considered sufficiently remote that when coupled with other requirements a high level of safety was provided. For the detectability requirement, it is assumed that having greater than 0.10 inch (2.5 mm) in depth will be detected and repaired. Consequently, the load requirement is consistent with those for metal structure with damage of equivalent levels of detectability. Provisions for multiple impact damage, analogous to the continuing damage considerations for metal structure, and for the lesser susceptibility of interior structure to damage are also included.

In metal structure, a major concern about damage tolerance is not only initial damage but also growth of damage prior to the time of detection. Consequently, much development testing for metals has been focused on evaluating crack growth rates associated with defects and damage, and the time for the defect/damage size to reach residual strength criticality. Typically, the critical loading mode has been in tension. Crack growth, even at comparatively low stress amplitudes, may be significant. In general, damage growth rates for metals are consistent and, after test data has been obtained, can be predicted satisfactorily. Thus, knowing the expected stress history for the aircraft, inspection intervals have been defined that confidently ensure crack detection before failure.

By contrast, the fibers in composite laminates act to inhibit tensile crack growth. Through-thickness damage growth occurs only at relatively high stress levels. Consequently, through-the-thickness damage growth in composites has generally not been a problem. In-plane damage growth associated with delaminations or impact damage, however, must be considered. Detection again, is an important consideration. Unlike cracks in metal, growth of delaminations or impact damage in composites, with probably not be detected if it does occur. Additionally, interlaminar damage growth in composites, especially that associated with impact damage, cannot be predicted satisfactorily. This has been considered in development of the referenced Air Force specification. Hence, it is required that damage be monitored for growth during cyclic tests which are conducted during development and validation. Resulting design values must be established with sufficient margins to ensure that damage growth due to repeated loads will not occur.

The damage tolerance design procedures for civil/commercial aircraft are expressed more generally but with equal effectivity. Guidelines are addressed for the BAA in Federal Aviation Regulation (FAR) 25 and in Advisory Circular AC-107A. Relative to impact damage, the FAA guidelines states "It should be shown that impact damage that can be realistically expected from manufacturing and service, but not more than the established threshold of detectability for the selected inspection procedure, will not reduce the structural strength below ultimate load capability. This can be shown by analysis supported by test evidence, or by tests at the coupon, element, or subcomponent level." The guidance is to ensure that structure with undetected damage will still meet ultimate strength requirements. Again, it assumed that more obvious damage will be detected in a timely manner and will be repaired. The difference in the Air Force specification and the FAA guideline is primarily in the residual strength value. Also, while the Air Force specification assumes visual inspection, the FAA guideline leaves the inspection method to be selected. Consequently, since specifications and guidelines differ with the type of aircraft, the manufacturer must be aware of the differences and apply those guidelines and specifications appropriate to the situation.

4.11.1.4 Material effects on damage tolerance

The ability of composite structures to resist or tolerate damage is strongly dependent on the constituent resin and fiber material properties and the material form. The properties of the resin matrix are most significant and include its ability to elongate and to deform plastically. The area under a resin's stress-strain curve indicates the material's energy absorption capability. Damage resistance or tolerance is also related to the material's interlaminar fracture toughness, G, as indicated by energy release rate properties. Depending on the application $G_{(I)}$, $G_{(II)}$, or $G_{(III)}$ may dominate the total G calculation. These parameters represent the ability of the resin to resist delamination, and hence damage, in the three modes of fracture. The beneficial influence of resin toughness on both damage resistance and tolerance, has been demonstrated by tests on new toughened thermoset laminates and with the tougher thermoplastic material systems.

Investigations have been conducted on the effect of fiber properties on impact resistance or tolerance. In general, laminates made with fabric reinforcement have better resistance to damage than laminates with unidirectional tape construction. Differences among the carbon fiber tape laminates, however, are small. Some studies have been made of composites with hybrid fiber construction, that is, composites in which two or more types of fibers are mixed in the layup. For example, a percentage of the carbon fibers are replaced with fibers with higher elongation capability, such as fiberglass or aramid. Studies in both cases have shown improvement in residual compression strength after impact. Basic undamaged properties, however, were usually reduced.

In thin gage structures such as a two- or three-ply fabric facesheet sandwich construction, materials can have a significant effect on damage tolerance. Investigations in general have shown that compression strength (both before and after impact) increases with the fiber strain-to-failure capability within a particular class of materials. Higher strain capability aramid or glass fiber structures tend to be more impact resistant than high-strength carbon fiber structure. However, the compressive strengths of the undamaged and damaged aramid and glass structures are lower than that of carbon. Structure incorporating high-modulus, intermediate-strength carbon fibers, with higher strain-to-failures offer a significant impact resistance while retaining higher strength.

In thermoset material systems, the nominal matrix toughness variations influence the impact resistance of thin gage structures but generally to a lesser extent than in thicker structures. For thermoplastic material systems, however, the generally much larger increase in the fracture toughness (G_{I_c}, G_{II_c}, etc.) of the resins do translate into significant impact resistance and residual strength improvements. For sandwich structures, where failure is not core-dependent, core density has been shown to have little effect on impact resistance. However, core density can have a significant effect on the residual strength of the sandwich if the sandwich mode is core-dependent (i.e., face wrinkling).

Methods such as through-thickness stitching have also been used to improve damage resistance. The effect has been to reduce the size of internal delaminations due to impact and similar to the hybrid high elongation fibers, arrest damage growth. Tests involving conventional graphite/epoxies have shown increases in the residual strength of up to 15% for comparative impact energy levels. The process is quite expensive, however, and probably should be considered for application in selected critical areas only. Additionally, the stitches tend to cause stress concentrations and the tensile strength, transverse to the stitching row, is usually reduced.

There are also other properties of composites that contribute to their damage resistance/tolerance. In contrast to metals, which have a Poisson's ratio typically near 0.3, the Poisson's ratio for composite laminates may vary widely, dependent on the particular combinations of ply layup orientations. Consequently, Poisson's ratios may be unbalanced between different sets of plies within a laminate. When the laminate is stressed, these imbalances may create high internal stresses. The consequence is that these stresses may add to damage-induced stresses thereby increasing the tendency for damage growth or causing a further reduction tensile or compression strength. Hence, ply stacking sequences that produce high internal stresses when the laminate is loaded, should be avoided.

As an example consider the influence of laminate stacking sequence (LSS) on compression after impact properties. The LSS can affect compression after impact strength (CAI) in several ways. First, the bending stiffness of a laminate and failure mechanisms that occur during an impact event is strongly dependent on LSS. Load redistribution near the impact site is dependent on the distribution of damage through the laminate thickness (e.g., the LSS of sublaminates affects their stability). Finally, damage propagation leading to final failure also depends on LSS, as was the case for notched tensile strength.

When impact damage is dominated by fiber failure (e.g., Reference 4.11.1.4(a)), it is desirable to stack primary load carrying plies in locations shielded from fiber failure. Since fiber failure typically occurs first near outer surfaces, primary load carrying plies should be concentrated towards the center of the LSS.

Many of the impact damage states studied in the past have been dominated by matrix failures. The creation of matrix cracks and delaminations which combine to form sublaminates depends strongly on LSS (Reference 4.11.1.4(b)). Homogeneous sacking sequences have been found to lead to characteristic damage states which repeat through the laminate thickness. Alternatively, plies can be stacked in a sequence which concentrates damage in specific zones on the laminate.

Recent methods have been developed and verified to predict the effects of known impact damage states on CAI (References 4.11.1.4(c) - (e)). This analysis involves four steps. First, the damage state is characterized with the help of NDI and the damage is simulated as a series of sublaminates. Second, sublaminate stability is predicted with a model that includes the effects of unsymmetric LSS. Third, the in-plane load redistribution is calculated with a model that accounts for structural geometry (e.g., finite width effects). Finally, a maximum strain failure criterion is applied to calculate CAI. Figure 4.11.1.4(a) shows typical results from this analysis procedure.

Interlaminar toughness is crucial to the extent of damage created in a given impact event; however, the CAI of laminates with equivalent damage states (size and type) was found to be independent of material toughness (References 4.11.1.4(c) - (e)). The model from Reference 4.11.1.4(c), which accounts for the in-plane stress redistribution due to sublaminate buckling, has worked equally well for tough and brittle resin systems studied. Since delamination growth may be possible with some materials and LSS, a more general model would also account for out-of-plane stresses.

Experience to date suggests that a homogeneous LSS might be best for overall CAI performance dominated by matrix damage (Reference 4.11.1.4(b)). Figure 4.11.1.4(b) shows experimental data indicating that LSS has a strong effect on CAI. The combined influence of impact damage resistance and CAI performance is evident in the figure.

Basic information on laminate stacking sequence effects is found in Section 4.6.5.

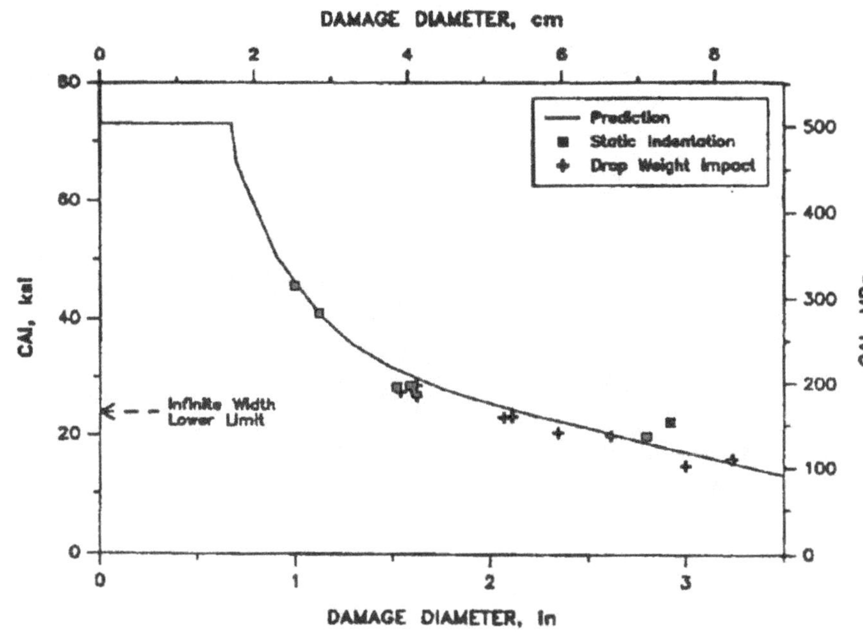

FIGURE 4.11.1.4(a) *Analysis and test results for 5 in. wide specimens, AS6/3501-6, $[45/0/-45/90]_{5s}$, and ply thickness = 0.0074 in. (from Reference 4.11.1.4(d).*

4.11.2 Durability

Durability of a structure is its ability to maintain strength and stiffness throughout the service life of the structure. A structure must have adequate durability when subjected to the expected service loads and environment spectra to prevent excessive maintenance, repair or modification costs over the service life. Thus, durability is primarily an economical consideration.

The major factors limiting the life of metallic structure are corrosion and fatigue. In composites, it has been demonstrated that one of the most common damage growth mechanisms is intercracking and fiber breakage usually occurs. Also, in three-dimensional composite structures out-of-plane loadings can either cause failure directly or induce failure indirectly. Because delamination growth is an important growth mechanism, composites are most sensitive to compression dominated fatigue loading.

A second common composite fatigue failure mode is fastener holes wear caused by high bearing stresses. In this failure mode, the hole gradually elongates may lead to a bearing failure or internal load redistribution.

Fatigue behavior of composites has been extensively investigated both experimentally and analytically. A partial list of these investigations is given as References 4.11.2(a) through 4.11.2(aa). In general, composites under in-plane loads have relatively flat fatigue stress-life (S-N) curves with high fatigue thresholds (endurance limits). This behavior is observed for various test specimen types from laminate to structural element. The fatigue threshold for a commonly used carbon/epoxy composite can be as high as 70% of the static strength under constant amplitude cyclic loading. The value of fatigue threshold is even higher when the composite is tested under applications spectra, such as that of a typical fighter aircraft spectrum (References 4.11.2(r) and (s)). Experimental data generated in Reference 19 showed that the typical fatigue failure mode was quasi-static failure when open hole specimens were tested under compression dominated

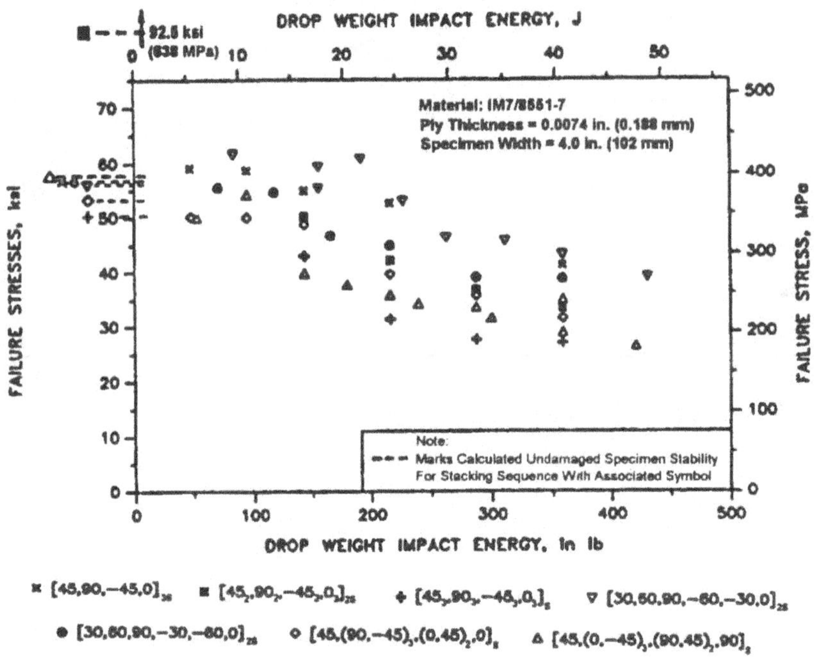

FIGURE 4.11.1.4(b) *Test data for CAI performance as a function of LSS (from Reference 4.11.1.4(e).*

fatigue spectrum loading. However, in high-cycle applications such as rotorcraft, fatigue structure endurance limits at 10^7 to 10^9 cycles must be established and these data are not readily available.

Accompanying the flat S-N curve is the significantly higher fatigue life scatter for composites compared to metals. Extensive statistical data analyses were conducted in References 4.11.2(ab) through (ae) to characterize the fatigue life data scatter of composites. The data compiled in these references included a variety of test variables such as geometry, environment and loading mode. The results of these analyses showed that the mean Weibull shape parameter for carbon/epoxy coupon data is approximately 2. This value is compared to 4.4 for commonly used aluminum alloys under constant amplitude fatigue loading and 7.7 under spectrum loading. This high variability in fatigue life is considered typical to most carbon fiber resin matrix composites and must be fully realized in order to assure adequate durability.

Traditionally, metal airframe structures are fatigue tested under spectrum loading to a minimum of two lifetimes to assure adequate durability. Based on the material variability observed in a commonly used aluminum alloys, a 2-4 lifetime fatigue test provides a very high structural reliability (approximately 0.999). However, because of the high variability of fatigue life test data observed in composites a 2-4 lifetime life factor provides a very low reliability (approximately 0.333). Therefore, the 2-4 lifetime fatigue test may not universally assure adequate durability for a composite structure. However, the use of significantly larger life factors in full-scale durability test programs may not be economically feasible. Alternate approaches may be used to reduce the test duration and at the same time demonstrate adequate reliability. Since these methods continue to evolve, early discussion with appropriate certification agency is recommended.

The load enhancement factor approach, evaluated in References 4.11.1 (ac) and (ad), is one of several alternative durability testing approaches. The objective of this approach is to increase the applied loads in the fatigue test so that the same high level of reliability can be achieved with a shorter test duration. In this approach, the fatigue life data variability, the residual strength scatter and the characteristics of composite fatigue S-N curves are integrated into a mathematical relationship. The required load enhancement factor and the test duration are then determined based on the material variability parameters and the number of tests. A simple method for determining the load enhancement factor for any composite material and failure mode is described in Reference 4.11.2.(ae).

A second approach, the ultimate strength approach, takes advantage of the excellent fatigue response of composites and assuming no damage growth during the lifetime of the structure. The objective of the ultimate strength approach is to have operating stresses below the fatigue threshold or endurance limit, depending upon application. This is possible in practice because composites have flat S-N curves where fatigue threshold is a high proportion of ultimate strength. This approach and other durability testing approaches, such as residual strength approach, are described in detail in References 4.11.2(ac) and (ad).

4.11.2.1 Design development/certification implications

The high cost of full-scale structural test prohibits generation of sufficient data for reliability analysis. Thus, for meaningful interpretation of full-scale tests, a building-block approach is desirable for certification of composite structures. This approach fully utilized coupon, element, subcomponent and component level test data so that limited full-scale structural test data can be interpreted statistically. Use of a building-block approach is discussed in References 4.11.2.1(a) through (d). The number of tests decreases from the coupon level to the component level. A relatively large number of tests is required at the coupon level to establish the data scatter and design allowables for different loading modes, failure modes and environments. A smaller number of tests is required at the element and subcomponent level to determine failure mode interaction and to demonstrate the variability in structural response. This information is then used for design and interpretation of the full-scale structural test program. The three levels of fatigue tests--allowable, design development and full-scale--are discussed in the following paragraphs.

The purpose of design allowable tests is to evaluate the material scatter and to establish fatigue life parameters for structural design. Laminate lay-up, stacking sequence effects, ply drop-offs, out-of-plane loadings, fastener holes, cutouts, etc. should be addressed at this level. Because composites are environmentally sensitive, design allowables should be obtained for structure. In planning the fatigue allowable tests, the main consideration is the test environment. The test environment depends on the relationship between the load/temperature spectrum and the material operation limit. The proper approach is to use simple, conservative constant temperature tests with a constant moisture level. A minimum of three stress levels for each test condition is required to fully characterize the fatigue behavior within the fatigue sensitive stress range. Fatigue tests should be conducted until fatigue failure occurs, except at the lowest stress level. At this stress level, because the fatigue threshold is approached, long life is expected. To reduce the test time on extremely long fatigue tests (greater than 10^6 cycles of 24000 flight hours), test may be censored at a specified lifetime. The stress level used in the fatigue tests should be selected so that the fatigue threshold can be established. For common carbon/epoxy composites under typical spectra, the threshold stress level would be approximately 60% of the mean static strength. This may not be true for rotorcraft high cycle (10^7 - 10^9) fatigue structures which require special evaluation to establish thresholds. The test data should be interpreted for B-basis life/stress relationship (see for example Reference 4.11.2(q)).

Several factors determine the test complexity of composite design development tests. These are: structural geometry complexity, hygrothermal environment simulation, fatigue load spectrum simulation including load rates, and mixed composite/metal structure interaction. The levels of complexity in the design development testing should be functions of the design feature being validated and the failure mode. Special attention should be given to correct failure mode simulation since failure modes are frequently dependent on the test environment. In particular, the influence of complex loading on the local stress at a given design feature must be evaluated. In composites, out-of-plane stresses can be detrimental to structural integrity and, therefore, require careful evaluation.

The environmental complexity necessary for fatigue design development testing will depend on the structural hygrothermal history. Three factors must be considered. These are: structural temperature for each mission profile, the load/temperature relationships for the aircraft, and the moisture content as a function of the aircraft usage and structure thickness. In order to obtain these data, it is necessary to derive the real time load-temperature profiles for each mission in the aircraft's history, unless a simple end-of-lifetime moisture content and applicable temperature can be used. These relationships will have a significant influence on the environmental fatigue test requirements.

An example of this testing philosophy is given in Reference 4.11.1.1(e). These relationships strongly depend on the aircraft type, configuration and mission requirements and must be carefully developed on a case by case basis. The structural material should be selected to meet these mission requirements without exceeding the material operating limit.

No significant load sequence effect on fatigue life has been observed in composite materials. However, studies on fixed wing aircraft structural load spectrum variations have shown that carbon/epoxy composites are extremely sensitive to variation in the number of high loads in the fatigue spectrum (Reference 4.11.2(r)). In contrast, truncation of low load does not significantly affect fatigue life. Therefore, high loads in the fatigue spectrum must be carefully simulated in developing load spectrum for fatigue testing. Low loads may be truncated to save test time. It has been observed that the temperature spectrum has not significant effects on fatigue life, however, temperature spectrum truncation should be a consideration in planning fatigue tests. Rotorcraft exception from this generalization. Fatigue testing of mixed metal/composite structure may introduce conflicting requirements and should be evaluated on an individual basis.

The number of replicates for each design development test should be sufficient to verify the critical failure mode and provide a reasonable estimate of the required fatigue reliability. The test effort should be concentrated on the most critical design features of the structure. The number of replicates should be increased for these areas of concern.

Full-scale durability test may not always be necessary for structures with non-fatigue critical metal parts, provided the design development testing and full-scale static test are successful. For mixed composite/metal structures, with fatigue critical metal parts, a two lifetime ambient test should be required to demonstrate durability validation of the metal parts.

4.11.3 Damage resistance

In its normal operation, the aircraft can be expected to be subjected to potential damage from sources such as maintenance personnel and tools, runway debris, service equipment, exposure to hail and lightning, etc. Even during initial manufacturing and assembly, parts are subject to dropped tools, bumps during transportation to assembly locations, etc. The aircraft structure must be able to endure a reasonable level of such incidents without requiring costly rework or downtime. Providing this necessary damage resistance is an important design function. Unfortunately for the designer, providing adequate damage resistance may not always be the most popular task. Resistance to damage requires robustness. This requirement commonly necessitates the addition of extra material above that necessary to carry the structural loads. As a result, there are many pressures to compromise because of competing goals for minimum weight and cost.

In order to establish minimum levels of damage resistance, various requirements for aircraft structure have been identified. The Air Force requirements are defined in their General Specification for Aircraft Structures, AFGS-87221A. In general, the Specification defines the type and level of low energy impact which must be sustained without structural impairment, moisture ingestion or a requirement for repair. It provides provision for such incidents as dropped tools, hail, and impact from runway debris. The aircraft is zoned depending on whether the region has high or low susceptibility to damage and its required damage resistance defined by other government agencies, industry companies and the Aerospace Industries Association (AIA).

In defining the requirements for damage resistance, the type of structure is pertinent. For example, the level of impact energy which typically must be sustained by honeycomb sandwich control surfaces without requiring repair or allowing moisture ingestion is quite low, e.g., 4 to 6 in-lb (0.5 to 0.7 J). This is partially because these parts must be kept very light, consequently, damage resistance has admittedly been compromised. Repair is facilitated somewhat because these parts can usually be readily replaced with spares while repairs are being accomplished in the shop. Because of their light construction, however, they must be handled carefully to prevent further damage during processing or transport. By contrast, the damage resistance requirement for primary laminate structure, which is not normally readily removable from the aircraft, is much higher, e.g., 48 in-lb (5.4 J). In addition to such impact induced loads, there also needs to be requirements of resistance to damage from normal handling and step loads that might be encountered in manufacturing and airline environments. Suggested for these have been the following:

Handling loads:

Difficult access	50 lb (23 kg) over a 4 sq-in (26 cm^2) area
Overhead easy access	150 lb (68 kg) over a 4 sq-in (26 cm^2) area

"Difficult access" is interpreted as finger tips only.
"Overhead access" is interpreted as the ability to grip and hang by one hand.

Step loads:

Difficult access	300 lb (140 kg) over a 20 sq-in (130 cm^2) area
Easy access from above	600 lb (270 kg) over a 20 sq-in (130 cm^2) area

"Difficult access" is interpreted as allowing a foothold on a structure with difficulty.
"Easy access from above" is interpreted as allowing a 2g step or"hop" onto the structure. The area of 20 square inches (130 square centimeters) represents a single foot pad.

There are certain damage susceptible regions of the airplane that require special attention. Examples of these are the lower fuselage and adjacent fairings, lower surfaces of the inboard flaps and areas around doors. These need to be reinforced with heavier structure and perhaps fiberglass reinforcement, i.e., in lieu of graphite. In addition to the above, structure in the wheel well area needs special attention because of damage susceptibility from tire disintegration. Similarly, structure in the vicinity of the thrust reversers is damage prone due to ice or other debris thrown up from the runway.

Minimum weight structure, such as that used for fairings, can cause excess maintenance problems if designed too light. For example, sandwich with honeycomb core that is too low density is unacceptable for commercial aircraft application. Also, face sheets must have a minimum thickness to prevent moisture entrance to the core. The design should not rely on the paint to provide the moisture barrier. Experience has shown that too often the paint erodes or is abraded and then moisture enters.

Honeycomb sandwich areas with thin skins adjacent to supporting fittings are particularly vulnerable to damage during component installation and removal. Consequently, use solid laminate construction within a reasonable working distance of fittings.

Trailing edges of control panels are most vulnerable to damage. The aft four inches are especially subject to ground collision and handling and also to lightning strike. Repairs in this region can be difficult because both the skins and the trailing edge reinforcement may be involved. A desirable approach for the design is to provide a load carrying member to react loads forward of the trailing edge, and material for the trailing edge, itself, that will be easily repairable and whose damage will not compromise the structural integrity of the component.

In general, damage resistance is improved by thicker laminates and, in the case of sandwich, by the use of denser core. The use of a core with some minimum thickness is also desirable. The reason is that this provides a degree of protection of the inner face, which if it does become damaged, is difficult to repair. The use of reinforcement fibers with higher strain, such as fiberglass or aramid as opposed to graphite in laminates, also provides improved damage resistance. Similarly, improvement is gained by the use of higher modulus, higher strain graphite fibers.

A most effective method for obtaining improved damage resistance is by the use of tougher resin systems. Notable demonstrations have been made comparing tough thermoplastic resin laminates with less tough conventional thermosets. The energy required to initiate damage in the thermoplastics, as measured by instrumented impactors, was much higher. Additionally, when damage did occur, the damaged region was much smaller. Likewise, in service demonstrations, the thermoplastics have shown marked damage resistance improvement. An example is the tests on comparative landing gear doors conducted on F-5 and T-38 aircraft by Northrop. Similarly, substantial improvements are expected from the use of toughened epoxy or bismaleimide systems as opposed to the untoughened resins.

Other items that either improve damage resistance or aid in the repair include having a layer of fabric as the exterior ply of tape laminates. The outer fabric ply is more resistant to scratches and abrasion and allows drilling of the laminate without the otherwise occurrence of fiber breakout.

Laminate edges should not be positioned so they are directly exposed to the air stream since they are then subject to delamination. Options include:

a) Provide non-erosive edge protection such as a cocured metal edge member.
b) Provide an easily replaceable sacrificial material to wrap the edges.
c) Locate the forward edge below the level of the aft edge of the next panel forward.

High energy lightning strikes can cause substantial damage to composite surface structure. Resistance at fasteners and connections, in particular, generates heat that causes burning and delaminations. Minor attachments, also, can cause significant damage, particularly to the tips and trailing edges. The following are guidelines to reduce the repair requirement:

a) At critical locations, provide easily replaceable conductive material with adequate conductive area.
b) Provide protection at tips and along trailing edge spans.
c) Make all conductive path attachments easily accessible.

Another item, although not directly damage resistance, involves the judicious use of blind fasteners. Such fasteners that must be drilled out for removal result in a high percentage of damaged holes. A suggestion is to use nut plates with reusable screws for closeout panel attachment, where access to the far side of a panel is restricted or otherwise where use of a normal type fastener is prevented.

4.11.4 Summary

- Damage is categorized as having occurred during manufacture or in-service.

- Routine quality inspections should uncover manufacturing-related damage. However, "rogue defects" will escape detection and should be accounted for in the design process.

- Draft manufacturing and in-service impact damage tolerance requirements typically address both minimum detection measured by dent depth, and impact energy. Given that damage can go undetected, it is important to demonstrate by both analysis and test that the maximum realistic undetectable damage will not reduce the structural strength below design ultimate load.

- A composite material's delamination damage tolerance can be determined, qualitatively, by its laminate matrix fracture toughness properties, G_I, G_{II}, and G_{III}.

- High strain-to-failure aramid and glass fiber composites typically exhibit greater damage tolerance than carbon fiber composites, but also exhibit lower strengths. Integrating glass surface plies or through thickness stitching represent two means of increasing a laminate's damage tolerance.

In general, organic matrix composites exhibit different durability characteristics compared to metals:

- Composite laminates, loaded in-plane, are less fatigue insensitive; they exhibit greater lives at stresses closer to ultimate strength.

- They are more sensitive to compression-dominated fatigue or reversed loading ($R<0$).

- In-plane fatigue damage usually begins with matrix cracking and ends with fiber breakage leading eventually to catastrophic failure.

4.12 VIBRATION

4.12.1 Introduction

4.12.2 Stacking sequence effects

Vibration characteristics of laminated plates are also sensitive to laminate stacking sequence (LSS). As was the case with bending and buckling of laminated plates, complex interactions between LSS, plate geometry and boundary conditions will not allow simple rules relating LSS to vibrations. Instead, such rules must be established for specific structure and boundary conditions. This indicates a need to use proven analysis methods as design tools for predicting dimensional stability of composite structure subjected to dynamic load conditions.

Figure 4.12.2 is one example of the complex interactions between LSS, plate geometry, and the natural frequency in the first vibrational mode.[1] A design equation from Reference 4.7.1.9(c) which was based on analysis from Reference 4.12.2 was used to make the predictions shown in the figure. Note that the relative difference in fundamental frequencies for various LSS changes with plate geometry. Higher frequencies occur for square plates with preferential stacking of ±45° plies in outer layers. The strongest effect of LSS occurs for rectangular plates in which preferential stacking of outer plies oriented perpendicular to the longest plate dimension have the highest fundamental frequencies. Basic information on laminate stacking sequence effects is found in Section 4.6.5.

4.13 COMPUTER PROGRAMS

Numerous programs for finite element analysis and prediction of composite material properties are available. Information on many of these programs can be found in Reference 4.13. In addition, there are programs available from NASA through COSMIC, Computer Software Management Information Center, 112 Barrow Hall, The University of Georgia, Athens, Georgia, 30602, (404) 542-3265. It should be noted that the use of and the results from these computer codes rely on the model developed, the material properties selected, and the experience of the user.

4.14 CERTIFICATION REQUIREMENTS

---

[1]The LSS used in Figure 4.12.2 were chosen for illustrative purposes only and do not represent optimal LSS for a given application.

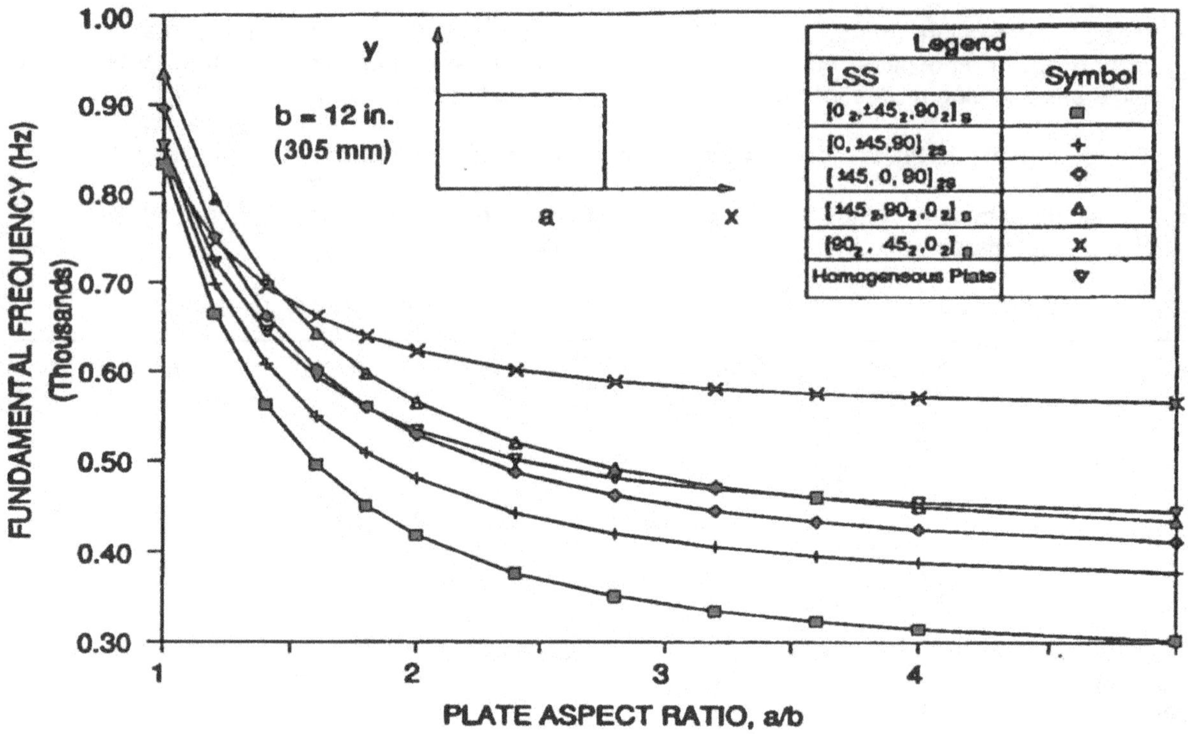

FIGURE 4.12.2 *Vibration analysis results for four sides simply-supported plates with variable aspect ratio.*

REFERENCES

4.2.2(a) Hashin, Z., "Theory of Fiber Reinforced Materials," NASA CR-1974, 1972.

4.2.2(b) Christensen, R.M., *Mechanics of Composite Materials*, Wiley-Interscience, 1979.

4.2.2.1(a) Pickett, G., AFML TR-65-220, 1965.

4.2.2.1(b) Adams, D.F., Doner, D.R., and Thomas, R.L., *J. Composite Materials*, Vol 1, 1967, pp. 4, 152.

4.2.2.1(c) Sendeckyj, G.P., "Elastic Behavior of Composites," *Mechanics of Composite Materials*, Vol. II, ed. Sendeckyj, G.P., Academic Press, 1974.

4.2.2.1(d) Hashin, Z. and Rosen, B.W., "The Elastic Moduli of Fiber-Reinforced Materials," *J. Appl. Mech.*, Vol 31, 1964, p.223.

4.2.2.1(e) Hashin, Z., "Analysis of Properties of Fiber Composites with Anisotropic Constituents," *J. Appl. Mech.*, Vol 46, 1979, p.543.

4.2.2.2(a) Christensen, R.M., *Theory of Viscoelasticity*, Academic Press, 1971.

4.2.2.2(b) Hashin, Z., "Viscoelastic Behavior of Heterogeneous Media," *J. Appl. Mech.*, Vol 32, 1965, p.630.

4.2.2.2(c) Hashin, Z., "Viscoelastic Fiber Reinforced Materials," *AIAA Journal*, Vol 4, 1966, p. 1411.

4.2.2.2(d) Schapery, R.A., "Viscoelastic Behavior of Composites," in *Mechanics of Composite Materials*, Vol. II, ed. Sendeckyj, G.P., Academic Press, 1974.

4.2.2.3(a) Levin, V.M., "On the Coefficients of Thermal Expansion of Heterogeneous Materials," *Mekh. Tverd. Tela*, Vol 1, 1967, p. 88. English translation: *Mechanics of Solids*, Vol 2, 1967, p.58.

4.2.2.3(b) Schapery, R.A., "Thermal Expansion Coefficients of Composite Materials Based on Energy Principles," *J. Composite Materials*, Vol 2, 1968, p.380ff.

4.2.2.3(c) Rosen, B.W., "Thermal Expansion Coefficients of Composite Materials," Ph.D. Dissertation, Univ. of Pennsylvania, 1968.

4.2.2.3(d) Springer, G.S., "Environmental Effects on Epoxy Matrix Composites," ASTM STP 674, ed. Tsai, S.W., 1979, p.291.

4.2.2.3(e) Tsai, S.W. and Hahn, H.T., *Introduction to Composite Materials*, Technomic, 1980.

4.2.3.1.1(a) Gucer, D.E. and Gurland, J., "Comparison of the Statistics of Two Fracture Modes," *J. Mech. Phys. Solids*, 1962, p. 363.

4.2.3.1.1(b) Zweben, C., "Tensile Failure Analysis of Composites," *AIAA Journal*, Vol 2, 1968, p. 2325.

4.2.3.1.2 Rosen, B.W., "Tensile Failure Analysis of Fibrous Composites," *AIAA Journal*, Vol 2, 1964, p. 1982.

4.2.3.1.3(a) Zweben, C., "A Bounding Approach to the Strength of Composite Materials," *Eng. Frac. Mech.*, Vol 4, 1970, p. 1.

4.2.3.1.3(b) Harlow, D.G. and Phoenix, S.L., "The Chain-of-Bundles Probability Model for the Strength of Fibrous Materials. I. Analysis and Conjectures, II. A Numerical Study of Convergence," *J. Composite Materials*, Vol 12, 1978, pp. 195, 300.

4.2.3.1.4 Rosen, B.W. and Zweben, C.H., "Tensile Failure Criteria for Fiber Composite Materials," NASA CR-2057, August 1972.

4.2.3.2(a) Dow, N.F. and Grundfest, I.J., "Determination of Most Needed Potentially Possible Improvements in Materials for Ballistic and Space Vehicles," GE-TIS 60SD389, June 1960.

4.2.3.2(b) Rosen, B.W., "Mechanics of Composite Strengthening," *Fiber Composite Materials*, Am. Soc. for Metals, Metals Park, Ohio, 1965.

4.2.3.2(c) Schuerch, H., "Prediction of Compressive Strength in Uniaxial Boron Fiber-Metal Matrix Composite Materials," *AIAA Journal*, Vol 4, 1966.

4.2.3.2(d) Rosen, B.W., "Strength of Uniaxial Fibrous Composites," in *Mechanics of Composite Materials*, Pergamon Press, 1970.

4.2.3.2(e) Chen, C.H. and Cheng, S., "Mechanical Properties of Fiber Reinforced Composites," *J. Composite Materials*, Vol 1, 1967, p. 30.

4.2.3.3(a) Drucker, D.C., Greenberg, H.J., and Prager, W., "The Safety Factor of an Elastic-Plastic Body in Plane Strain," *J. Appl. Mech.*, Vol 18, 1951, p. 371.

4.2.3.3(b) Koiter, W.T., "General Theorems for Elastic-Plastic Solids," *Progress in Solid Mechanics*, Sneddon and Hill, ed., North Holland, 1960.

4.2.3.3(c) Shu, L.S. and Rosen, B.W., "Strength of Fiber Reinforced Composites by Limit Analysis Method," *J. Composite Materials*, Vol 1, 1967, p. 366.

4.2.4(a) Tsai, S.W., "Strength Characteristics of Composite Materials," NASA CR-224, 1965.

4.2.4(b) Ashkenazi, E.K., *Zavod. Lab.*, Vol 30, 1964, p. 285; *Mezh. Polim*, Vol 1, 1965, p. 60.

4.2.4(c) Tsai, S.W. and Wu, E.M., "A General Theory of Strength for Anisotropic Materials," *J. Composite Materials*, Vol 5, 1971, p. 58.

4.2.4(d) Wu, E.M., "Phenomenological Anisotropic Failure Criterion," *Mechanics of Composite Materials*, ed. Sendeckyj, G.P., Academic Press, 1974.

4.2.4(e) Hashin, Z., "Failure Criteria for Unidirectional Fiber Composites," *J. Appl. Mech.*, Vol 47, 1980, p. 329.

4.3.2(a) Agarwal, B.D. and Broutman, L.J., *Analysis and Performance of Fiber Composites*, John Wiley & Sons, New York, NY, 1980.

4.3.2(b) Ashton, J.E. and Whitney, J.M., *Theory of Laminated Plates*, Technomic Publishing Co., Inc., Westport, CT, 1970.

4.3.2(c) Hussein, R.M, *Composite Panels/Plates: Analysis and Design*, Technomic Publishing Co., Inc., Westport, CT, 1986.

4.3.2(d) Jones, R.M., *Mechanics of Composite Materials*, Scripta Book Co., Washington, DC, 1975.

4.3.2(e) Tsai, S.W. and Hahn, H.T., *Introduction to Composite Materials*, Technomic Publishing Co., Inc., Westport, CT, 1980.

4.3.4 Shen, C. and Springer, G.S., "Moisture Absorption and Desorption of Composite Materials," *J. Composite Materials*, Vol 10, 1976, p. 1.

4.3.4.2(a) Chamis, C.C., "A Theory for Predicting Composite Laminate Warpage Resulting From Fabrication," 30th Anniversary Technical Conf., Reinforced Plastics/Composites Institute, The Society of Plastics Industry, Inc., Sec 18-C, 1975, pp. 1-9.

4.3.4.2(b) Hyer, M.W., "Some Observations on the Cured Shape of Thin Unsymmetric Laminates," *J. Composite Materials*, Vol. 15, 1981, pp. 175-194.

4.3.4.2(c) Hyer, M.W., "Calculations of Room Temperature Shapes of Unsymmetric Laminates," *J. Composite Materials*, Vol. 15, 1981, pp. 296-310.

4.3.5.5 Hashin, Z., Bagchi, D., and Rosen, B.W., "Nonlinear Behavior of Fiber Composite Laminates," NASA CR-2313, April 1974.

4.4 Flaggs, D.L. and Kural, M.H., "Experimental Determination of the In Situ Transverse Lamina Strength in Graphite/Epoxy Laminates", *Journal of Composite Materials*, Vol. 16, March, 1982, pp. 103-115.

4.4.1.1(a) Hashin, Z., "Failure Criteria for Unidirectional Fiber Composites", *Journal of Applied Mechanics*, Vol. 47, June, 1980, pp. 329-334.

4.4.1.1(b) Fiber Composite Analysis and Design, Federal Aviation Administration, DOT/FAA/CT-85/6)

4.4.4(a) Savin, G.N., "Stress Distribution Around Holes," NASA TT-F-607, November, 1970.

4.4.4(b) Whitney, J.M. and Nuismer, R.J., "Stress Fracture Criteria for Laminated Composites Containing Stress Concentrations," *J. Composite Materials*, Vol 8, 1974, p. 253.

4.4.4(c) Daniel, I.M., Rowlands, R.E., and Whiteside, J.B., "Effects of Material and Stacking Sequence on Behavior of Composite Plates With Holes," *Experimental Mechanics*, Vol. 14, 1974, pp. 1-9.

4.4.4(d) Walter, R.W. Johnson, R.W. June, R.R., and McCarty, J.E., "Designing for Integrity in Long-Life Composite Aircraft Structures," *Fatigue of Filamentary Composite Materials*, ASTM STP 636, pp. 228-247, 1977.

4.4.4(e) Aronsson, C.G., "Stacking Sequence Effects on Fracture of Notched Carbon Fibre/Epoxy Composites," *Composites Science and Technology*, Vol. 24, pp. 179-198, 1985.

4.4.4(f) Lagace, P.A., "Notch Sensitivity and Stacking Sequence of Laminated Composites," ASTM STP 893, pp. 161-176, 1986.

4.4.4(g) Harris C.E. and Morris, D.H., "A Fractographic Investigation of the Influence of Stacking Sequence on the Strength of Notched Laminated Composites," ASTM STP 948, 1987, pp. 131-153.

4.4.5(a) Murthy P.L.N, and Chamis, C.C., "Free-Edge Delamination: Laminate Width and Loading Conditions Effects," *J. of Composites Technology and Research*, JCTRER, Vol. 11, No. 1, 1989, pp. 15-22.

4.4.5(b) O'Brien, T.K., "The Effect of Delamination on the Tensile Strength of Unnotched, Quasi-isotropic, Graphite/Epoxy Laminates," *Proc. of the SESA/JSME International Conf. on Experimental Mechanics*, Honolulu, Hawaii, May, 1982.

4.4.5(c) Bjeletich, J.G., Crossman F.W., and Warren, W.J., "The Influence of Stacking Sequence on Failure Modes in Quasi-isotropic Graphite-Epoxy Laminates," *Failure Modes in Composites IV*, AIME, 1977, pp. 118.

4.4.5(d) Crossman, F.W., "Analysis of Delamination," *Proc. of a Workshop on Failure Analysis and Mechanisms of Failure of Fibrous Composite Structures*, NASA Conf. Publ. 2278, 1982, pp. 191-240.

4.4.5(e) O'Brien, T.K., "Analysis of Local Delaminations and Their Influence on Composite Laminate Behavior," *Delamination and Debonding of Materials*, ASTM STP 876, 1985, pp. 282-297.

4.4.5(f) Sun C.T. and Zhou, S.G., "Failure of Quasi-Isotropic Composite Laminates with Free Edges," *J. Reinforced Plastics and Composites*, Vol. 7, 1988, pp. 515-557.

4.4.5(g) O'Brien, T.K., "Characterization of Delamination Onset and Growth in a Composite Laminate," *Damage in Composite Materials*, ASTM STP 775, 1982, pp. 140-167.

4.4.5(h) Whitcomb J.D. and Raju, I.S., "Analysis of Free-Edge Stresses in Thick Composite Laminates," *Delamination and Debonding of Materials*, ASTM STP 876, 1985, pp. 69-94.

4.4.5(i) Lekhnitskii, S.G., *Anisotropic Plates*, Gordon and Breach Science Publ., 1968.

4.4.5(j) Garg, A.C., "Delamination-A Damage Mode In Composite Structures," *Eng. Fracture Mech.*, Vol. 29, No. 5, 1988, pp. 557-584.

4.4.5(k) Lagace, P.A., "Delamination in Composites: Is Toughness the Key?" *SAMPE J.*, Nov/Dec, 1986, pp. 53-60.

4.4.5(l) Soni S. R. and Kim, R.Y., "Delamination of Composite Laminates Stimulated by Interlaminar Shear," *Composite Materials: Testing and Design (Seventh Conf.)*, ASTM STP 893, 1986, pp. 286-307.

4.4.5(m) Chan, W.S., Rogers, C., and Aker, S., "Improvement of Edge Delamination Strength of Composite Laminates Using Adhesive Layers," *Composite Materials: Testing and Design (Seventh Conf.)*, ASTM STP 893, 1986, pp. 266-285.

4.4.5(n) Chan W.S. and Ochoa, O.O., "Suppression of Edge Delamination in Composite Laminates by Terminating a Critical Ply near the Edges," Presented at the AIAA/ASMR/ASCE/AHS 29th Structures, Structural Dynamics and Materials Conf., AIAA Paper #88-2257, 1988, pp. 359-364.

4.4.5(o) Vizzini, A.J., "Prevention of Free-Edge Delamination via Edge Alteration," Presented at the AIAA/ASME/ASCE/AHS 29th Structures, Structural Dynamics and Materials Conference, AIAA Paper #88-2258, 1988, pp. 365-370.

4.4.5(p) Sun, C.T., "Intelligent Tailoring of Composite Laminates," *Carbon*, Vol. 27, No. 5, 1989, pp. 679-687.

4.4.5(r) Lagace P.A. and Cairns, D.S., "Tensile Response of Laminates to Implanted Delaminations," *Proc. of 32nd Int. SAMPE Sym.*, April 6-9, 1987, pp. 720-729.

4.4.5.1(a) Shivakumar K.N. and Whitcomb, J.D., "Buckling of a Sublaminate in a Quasi-Isotropic Composite Laminate," NASA TM-85755, Feb, 1984.

4.4.5.1(b) Chai H. and Babcock, C.D., "Two-Dimensional Modelling of Compressive Failure in Delaminated Laminates," *J. of Composite Materials*, 19, 1985, pp. 67-98.

4.4.5.1(c) Vizzini A.J. and Lagace, P.A., "The Buckling of a Delaminated Sublaminate on an Elastic Foundation," *J. of Composite Materials*, 21, 1987, pp. 1106-1117.

4.4.5.1(d) Kassapoglou, C., "Buckling, Post-Buckling and Failure of Elliptical Delaminations in Laminates Under Compression," *Composite Structures*, 9, 1988, pp. 139-159.

4.4.5.1(e) Yin, W.L., "Cylindrical Buckling of Laminated and Delaminated Plates," Presented at the AIAA/ASMR/ASCE/AHS 27th Structures, Structural Dynamics and Materials Conference, AIAA Paper #86-0883, 1986, pp. 165-179.

4.4.5.1(f) Slomiana, M., "Fracture Mechanics of Delamination in Graphite-Epoxy Laminates Under Compression," PhD. Thesis, Drexel University, Philadelphia, PA, 1984.

4.4.5.1(g) Whitcomb, J.D., "Strain-Energy Release Rate Analysis of Cyclic Delamination Growth in Compressively Loaded Laminates," *Effects of Defects in Composite Materials*, ASTM STP 836, 1984, pp. 175-193.

4.4.5.1(h) Whitcomb, J.D., "Predicted and Observed Effects of Stacking Sequence and Delamination Size on Instability-Related Delamination Growth," *Proc. of American Society for Composites: 3rd Tech Conf*, 1988, pp. 678-687.

4.4.6.1.1(a) Jamison, R.D., "On the Interrelationship Between Fiber Fracture and Ply Cracking in Graphite/Epoxy Laminates," *Composite Materials: Fatigue and Fracture*, ASTM STP 907, 1986, pp. 252-273.

4.4.6.1.1(b) Kim, R.Y., "In-plane Tensile strength of Multidirectional Composite Laminates," UDRI-TR-81-84, University of Dayton Research Institute, Aug, 1981.

4.4.6.1.1(c) Ryder J.T. and Crossman, F.W., "A Study of Stiffness, Residual Strength and Fatigue Life Relationships for Composite Laminates," NASA CR-172211, Oct, 1983.

4.4.6.1.1(d) Sun C.T. and Jen, K.C., "On the Effect of Matrix Cracks on Laminate Strength," *J. Reinforced Plastics and Composites*, Vol. 6, 1987, pp. 208-222.

4.4.6.1.1(e) Bailey, J.E., Curtis, P.T., and Parvizi, A., "On the Transverse Cracking and Longitudinal Splitting Behavior of Glass and Carbon Fibre Reinforced Epoxy Cross Ply Laminates and the Effect of Poisson and Thermally Generated Strain," *Proc. R. Soc. London*, series A, 366, 1979, pp. 599-623.

4.4.6.1.1(f) Crossman, F.W. and Wang, A.S.D., "The Dependence of Transverse Cracking and Delamination on Ply Thickness in Graphite/Epoxy Laminates," *Damage in Composite Materials*, ASTM STP 775, American Society for Testing and Materials, 1982, pp. 118-139.

4.4.6.1.1(g) Flaggs, D.L. and Kural, M.H., "Experimental Determination of the In situ Transverse Lamina Strength in Graphite/Epoxy Laminates," *J. of Composite Materials*, Vol. 16, Mar. 1982, pp. 103-116.

4.4.6.1.1(h) Narin, J.A., "The Initiation of Microcracking in Cross-Ply Laminates: A Variational Mechanics Analysis," *Proc. Am Soc for Composites: 3rd Tech. Conf.*, 1988, pp. 472-481.

4.4.6.1.1(i) Flaggs, D.L., "Prediction of Tensile Matrix Failure in Composite Laminates," *J. Composite Materials*, 19, 1985, pp. 29-50.

4.4.6.1.1(j) Ilcewicz, L.B., Dost, E.F., McCool, J.W., and Grande, D.H., "Matrix Cracking in Composite Laminates With Resin-Rich Interlaminar Layers," Presented at 3rd Symposium on Composite Materials: Fatigue and Fracture, Nov. 6-7, Buena Vista, Fla., ASTM, 1989.

4.4.6.2(a) Shuart, M.J., "Short-Wavelength Buckling and Shear Failures for Compression-Loaded Composite Laminates," NASA TM-87640, Nov, 1985.

4.4.6.2(b) Shuart, M.J., "Failure of Compression-Loaded Multi-Directional Composite Laminates," Presented at the AIAA/ASME/ASCE/AHS 19th Structures, Structural Dynamics and Materials Conf, AIAA Paper No. 88-2293, 1988.

4.4.6.2(c) Hahn, H.T. and Williams, J.G., "Compression Failure Mechanisms in Unidirectional Composites," *Composite Materials: Testing and Design (Seventh Conf.)*, ASTM STP 893, 1986, pp. 115-139.

4.4.6.2(d) Hahn H.T. and Sohi, M.M., "Buckling of a Fiber Bundle Embedded in Epoxy," *Composites Science and Technology*, Vol 27, 1986, pp. 25-41.

4.6.2(a) Manson, J. A. and Seferis, J. C., "Internal Stress Determination by Process Simulated Laminates," *ANTEC '87 Conference Proceedings*, SPE, May 1987.

4.6.2(b) Kays, A.O., "Exploratory Development on Processing Science of Thick-Section Composites," AFWAL-TR-85-4090, October 1985.

4.6.3(a) Chan, W.S., Rogers, C. and Aker, S., "Improvement of Edge Delamination Strength of Composite Laminates Using Adhesive Layers," *Composite Materials: Testing and Design*, ASTM STP 893, J.M. Whitney, ed., American Society for Testing and Materials, 1986.

4.6.3(b) Chan, W.S. and Ochoa, O.O., "An Integrated Finite Element Model of Edge-Delamination Analysis for Laminates due to Tension, Bending, and Torsion Loads," *Proceedings of the 28th Structures, Dynamics, and Materials Conference*, AIAA-87-0704, 1987.

4.6.3(c) O'Brien, T.K. and Raju, I.S., "Strain Energy Release Rate Analysis of Delaminations Around an Open Hole in Composite Laminates," *Proceedings of the 25th Structures, Structural Dynamics, and Materials Conference*, 1984, pp. 526-536.

4.6.3(d) Pagano, N.J. and Soni, S.R., "Global - Local Laminate Variational Method," International *Journal of Solids and Structures*, Vol 19, 1983, pp. 207-228.

4.6.3(e) Pipes, R.B. and Pagano, N.J., "Interlaminar Stresses in Composites Under Uniform Axial Extension," *J. Composite Materials*, Vol 4, 1970, p. 538.

4.7.1.1(a) Almroth, B. O., Brogan, F. W., and Stanley, G. W., "User's Manual for STAGS," NASA Contractor Report 165670, Volumes 1 and 2, March 1978.

4.7.1.1(b) Ashton, J. E. and Whitney, J. M., *Theory of Laminated Plates*, Technomic Publishers, 1970, pp. 125-128.

4.7.1.2 *Advanced Composites Design Guide, Volume II - Analysis,* Air Force Materials Laboratory, Advanced Development Division, Dayton, Ohio, January 1973, Table 2.2.2-1, pp. 2.2.2-12.

4.7.1.3(a) Spier, E. E., "On Experimental Versus Theoretical Incipient Buckling of Narrow Graphite/Epoxy Plates in Compression," AIAA-80-0686-Paper, published in *Proceedings of AIAA/ASME/ASCE/AHS 21st Structures, Structural Dynamics, & Materials Conference*, May 12-14, 1980, pp.187-193.

4.7.1.3(b) Spier, E. E., "Local Buckling, Postbuckling, and Crippling Behavior of Graphite-Epoxy Short Thin Walled Compression Members," Naval Air Systems Command Report NASC-N00019-80-C-0174, June 1981, p. 22.

4.7.1.3(c) Arnold, R. R., "Buckling, Postbuckling, and Crippling of Materially Nonlinear Laminated Composite Plates," Ph.D. Dissertation, Stanford University, March 1983, p. 65.

4.7.1.3(d) Arnold, R. R. and Mayers, J., "Buckling, Postbuckling, and Crippling of Materially Nonlinear Laminated Composite Plates," *Internal Journal of Solids and Structures*, Vol 20, pp. 863-880. 82-0779-CP, AIAA/ASME/ASCE/AHS, published in the *Proceedings of the 23rd Structures, Structural Dynamics, and Materials Conference*, New Orleans, Louisiana, May 1982, pp. 511-527.

4.7.1.9(a) Chamis, C.C., "Buckling of Anisotropic Composites," *Journal of the Structural Division*, Am. Soc. of Civil Engineers, Vol. 95, No. 10, 1969, pp. 2119-2139.

4.7.1.9(b) Whitney, J.M., *Structural Analysis of Laminated Anisotropic Plates*, Technomic Publishing Co., Inc., Westport, CT, 1987.

4.7.1.9(c) "DOD/NASA Advanced Composites Design Guide, Volume II Analysis," Structures/Dynamics Division, Flight Dynamics Laboratory, Air Force Wright Aeronautical Laboratories, Wright-Patterson Air Force Base, OH, 1983.

4.7.1.9(d) Davenport, O.B., and Bert, C.W., "Buckling of Orthotropic, Curved, Sandwich Panels Subjected to Edge Shear Loads," *Journal of Aircraft*, Vol. 9, No. 7, 1972, pp. 477-480.

4.7.1.9(e) Spier, E.E., "Stability of Graphite/Epoxy Structures With Arbitrary Symmetrical Laminates," *Experimental Mechanics*, Vol. 18, No. 11, 1978, pp. 401-408.

4.7.2(a) Spier, E.E., "On Experimental Versus Theoretical Incipient Buckling of Narrow Graphite/Epoxy Plates in Compression," AIAA-80-0686-Paper, published in Proceedings of AIAA/ASME/ASCE/AHS 21st Structures, Structural Dynamics, & Materials Conference, May 12-14, 1980, pp. 187-193.

4.7.2(b) Spier, E.E., "Local Buckling, Postbuckling, and Crippling Behavior of Graphite-Epoxy Short Thin Walled Compression Members," Naval Air Systems Command Report NASC-N00019-80-C-0174, June 1981.

4.7.2(c) Spier, E.E., "On Crippling and Short Column Buckling of Graphite/Epoxy Structure with Arbitrary Symmetrical Laminates," Presented at SESA 1977 Spring Meeting, Dallas, TX, May 1977.

4.7.2(d) Spier, E.E. and Klouman, F.K., "Post Buckling Behavior of Graphite/Epoxy Laminated plates and Channels," Presented at Army Symposium on Solid Mechanics, AMMRC MS 76-2, Sept. 1976.

4.7.2(e) Renieri, M.P. and Garrett, R.A., "Investigation of the Local Buckling, Postbuckling and Crippling Behavior of Graphite/Epoxy Short Thin-Walled Compression Members," McDonnell Aircraft Report MDC A7091, NASC, July 1981.

4.7.2(f) Bonanni, D.L., Johnson, E.R., and Starnes, J.H., "Local Crippling of Thin-Walled Graphite-Epoxy Stiffeners," AIAA Paper 88-2251.

4.7.2(g) Spier, E.E., "Postbuckling Fatigue Behavior of Graphite-Epoxy Stiffenersm," AIAA Paper 82-0779-CP, AIAA/ASME/ASCE/AHS, published in the Proceedings of the 23rd Structures, Structural Dynamics, & Materials Conference, New Orleans, LA, May 1982, pp. 511-527.

4.7.2(h) Deo, R.B., et al, "Design Development and Durability Validation of Postbuckled Composite and Metal Panels," Air Force Flight Dynamics Laboratory Report, WRDC-TR-89-3030, 4 Volumes, November 1989.

4.11.1.4(a) Cairns, D.S., "Impact and Post-Impact Response of Graphite/Epoxy and Kevlar/Epoxy Structures," PhD Dissertation, Telac Report #87-15, Dept. of Aeronautics and Astronautics, Massachusetts Institute of Technology, Aug, 1987.

4.11.1.4(b) Gosse J.H. and Mori, P.B.Y., "Impact Damage Characterization of Graphite/Epoxy Laminates," *Proc. of the American Society for Composites: 3rd Tech Conf*, 1988, pp. 344-353.

4.11.1.4(c) Dost, E.F., Ilcewicz, L.B., and Gosse, J.G., "Sublaminate Stability Based Modeling of Impact-Damaged Composite Laminates," *Proc. of the American Society for Composites: 3rd Tech Conf*, 1988, pp. 354-363.

4.11.1.4(d) Ilcewicz, L.B., Dost, E.F., Coggeshall, R.L., "A Model for Compression After impact Strength Evaluation," *Proc. of 21st International SAMPE Tech. Conf.*, 1989, pp. 130-140.

4.11.1.4(e) Dost, E.F., Ilcewicz, L.B., and Avery, W.B., "The Effects of Stacking Sequence On Impact Damage Resistance and Residual Strength for Quasi-Isotropic Laminates," Presented at 3rd Symposium on Composite Materials: Fatigue and Fracture, Nov. 6-7, Buena Vista, Fla., ASTM, 1989.

4.11.2(a) Broutman, L.J., and Sahu, S., "A New Theory to Predict Cumulative Fatigue Damage in Fiberglass Reinforced Plastics," *Composite Materials: Testing and Design*, ASTM STP 497, 1972, pp. 170-188.

4.11.2(b) Halpin, J.C., Jerina, K.L., and Johnson, T.A., "Characterization of Composites for the Purpose of Reliability Prediction," *Analysis of Test Methods for High Modulus Fibers and Composites*, ASTM STP 521, 1973, pp. 5-65.

4.11.2(c) McLaughlin, P.V., Jr., Kulkarni, S.V., Huang, S.N., and Rosen, B.W., "Fatigue of Notched Fiber Composite Laminates: Part I-Analysis Model," NASA Contract Report No. CR-132747, March 1975.

4.11.2(d) Hahn, H.T., and Kim, R.Y., "Proof Testing of Composite Materials," *Journal of Composite Materials*, Vol 2, No. 3, July 1975, pp. 297-311.

4.11.2(e) Yang, J.N., and Liu, M.D., "Residual strength Degradation Model and Theory of Periodic Proof Tests for Graphite/Epoxy Laminates," AFML-TR-76-225, 1976.

4.11.2(f) Kulkarni, S.V., McLaughlin, P.V., Jr., and Pipes, R.B., "Fatigue of Notched Fiber Composite Laminates: Part II - Analytical and Experimental Evaluation," NASA Contract Report No. CR-145039, April 1976.

4.11.2(g) Ryder, J.T., and Walker, E.K., "Ascertainment of the Effect of Compression Loading on the Fatigue Lifetime of Graphite/Epoxy Laminates for Structural Applications," AFML-TR-76-241, December 1976.

4.11.2(h) Sendeckyj, G.P., and Stalnaker, H.D., "Effect of Time at Load on Fatigue Response of $[(0/\pm45/90)_s]_2$ T300/5208 Graphite-Epoxy Laminate," *Composite Materials: Testing and Design*, ASTM STP 617, 1977.

4.11.2(i) Grimes, G.C., "Structures Design Significance of Tension-Tension Fatigue Data on Composites," *Composite Materials: Testing and Design*, ASTM STP 617, 1977, pp. 106-119.

4.11.2(k) Williams, R.S., and Reifsnider, K.L., "Strain Energy Release Rate Method for Predicting Failure Modes in Composite Materials," ASTM STP 677, 1979, pp. 629-650.

4.11.2(l) Ratwani, M.M., and Kan, H.P., "Compression Fatigue Analysis of Fiber Composites," Report No. NADC-78049-60, September 1979.

4.11.2(m) Ryder, J.T., and Walker, E.K., "The Effect of Compressive Loading on the Fatigue Lifetime of Graphite/Epoxy Laminates," AFML-TR-79-4128, October 1979.

4.11.2(n) Rotem, Al, "Fatigue Mechanism of Multidirectional Laminate Under Ambient and Elevated Temperature," *Proceedings, Third International Conference on Composite Materials*, Paris, 1980, pp. 146-161.

4.11.2(o) Saff, C.R., "Compression Fatigue Life Prediction Methodology for Composite Structures-Literature Survey," Report No. NADC-78203-60, June 1980.

4.11.2(p) Lauraitis, K.N., Ryder, J.T., and Pettit, D.E., "Advanced Residual Strength Degradation Rate Modeling for Advanced Composite Structures," Vol. I, II, and III, AFWAL-TR79-3095, July 1981.

4.11.2(q) Sendeckyj, G.P., "Fitting Models to Composite Materials Fatigue Data," ASTM STP 734, 1981, pp. 245-260.

4.11.2(r) Badaliance, R., and Dill, H.D., "Effects of Fighter Attack Spectrum on Composite Fatigue Life," AFWAL-TR-81-3001, March 1981.

4.11.2(s) Ratwani, M..M. and Kan, H.P., "Development of Analytical Techniques for Predicting Compression Fatigue Life and Residual Strength of Composites," Report No. NADC-82104-60, March 1982.

4.11.2(t) Liechti, K.M., Reifsnider, K.L., Stinchcomb, W.W., and Ulman, D.A., "Cumulative Damage Model for Advanced Composite Materials: Phase I Final Report," AFWAL-TR-82-4094, July 1982.

4.11.2(u) Chou, P.C., Wang, A.S.D., and Miller, H.R., "Cumulative Damage Model for Advanced Composite Materials: Phase I Final Report," AFWAL-TR-82-4083, September 1982.

4.11.2(v) Badaliance, R., and Dill, H.D., "Compression Fatigue Life Prediction Methodology for Composite Structures, Volume I-Summary and Methodology Development," Report No. NADC-78203-60, September 1982.

4.11.2(w) Baldini S.E., and Badaliance, R., "Compression Fatigue Life Prediction Methodology for Composite Structures, Volume II-Test Data and Damage Tracking," Report No. NADC-78203-60, September 1982.

4.11.2(x) Ryder, J.T., and Crossman, F.W., "A Study of Stiffness, Residual Strength and Fatigue Life Relationships for Composite Laminates," NASA Contract No. CR-172211, October 1983.

4.11.2(y) Wang, A.S.D., Chou, P.C., and Lai, C.S., "Cumulative Damage Model for Advanced Composite Materials: Phase II Final Report," AFWAL-TR-84-4004, March 1984.

4.11.2(z) Miller, H.R., Reifsnider, K.L., Stinchcomb, W.W., Ulman, D.A., and Bruner R.D., "Cumulative Damage Model for Advanced Composite Materials: Final Report," AFWAL-TR-85-4069, June 1985.

4.11.2(aa) Wang, A.S.D., Chou, P.C., Lai, C.S., and Bucinell, R.B., "Cumulative Damage Model for Advanced Composite Materials," Final Report, AFWAL-TR-85-4104, October 1985.

4.11.2(ab) Whitehead, R.S., and Schwarz, M.G., "The Role of Fatigue Scatter in the Certification of Composite Structures," presented at ASTM Symposium on the Long Term Behavior of Composites, Williamsburg, Virginia, March 1982.

4.11.2(ac) Whitehead, R.S., Kan, H.P., Cordero, R., and Saether, E.S., "Certification Testing Methodology for Composite Structures," Report No. NADC-86132-60, January 1986.

4.11.2(ad) Sanger, K.B., "Certification Testing Methodology for Composite Structures," Report No. NADC-86132-60, January 1986.

4.11.2(ae) Wong, R. and Abbot, R., "Durability and Damage Tolerance of Graphite/Epoxy Honeycomb Structures," *SAMPE 35th International Symposium and Exhibition*, April 1990.

4.11.2(af) Rogers, Chan, and Martin, "Design Criterion for Damage Tolerance Helicopter Primary Structure of Composite Materials," Volume I-Final Design Criteria, USAAVSCOM-TR-87-D-3A, June 1987.

4.11.2.1(a) Whitehead, R.S., "A Review of the Rationale for Durability Validation of Composite Structures," Proceedings of the 5th DoD/NASA Conference on Fibrous Composites in Structural Design, New Orleans, Louisiana, January 1981.

4.11.2.1(b) Whitehead, R.S., "Durability Certification Data for Composite Structures," presented at the 6th DoD/NASA Conference on Fibrous Composites in Structural Design, New Orleans, Louisiana, January 1983.

4.11.2.1(c) Whitehead, R.S., Ritchie, G.L., and Mullineaux, J.L., "Durability of Composites," presented at 9th Mechanics of Composites Review, Dayton, Ohio, October 1983.

4.11.2.1(d) Whitehead, R.S. and Deo, R.B., "Wing/Fuselage Program Durability Methodology," presented at the 6th Mechanics of Composites Review, Dayton, Ohio, November 1980.

4.11.2.1(e) Whitehead, R.S., et al, "Composite Wing Fuselage Program," Contract No. F33615-79-C-3203, Interim Reports 1-11, October 1979 - October 1984.

4.12.2 Lackman, L.M., Lin, T.H., Konishi, D.Y., and Davidson, J.W., "Advanced Composites Data for Aircraft Structural Design - Volume III," AFML-TR-70-58, Vol. III, Los Angeles Div/Rockwell, Dec, 1970.

4.13 Brown, R. T., "Computer Programs for Structural Analysis," *Engineered Materials Handbook, Volume 1 Composites*, ASM International, Metals Park, Ohio, 1987, pp. 268 - 274.

This page intentionally left blank

CHAPTER 5 STRUCTURAL BEHAVIOR OF JOINTS

Page

5.1 INTRODUCTION

It would be difficult to conceive of a structure that did not involve some type of joint. Joints often occur in transitions between major composite parts and a metal feature or fitting. In aircraft, such a situation is represented by articulated fittings on control surfaces as well as on wing and tail components which require the ability to pivot the element during various stages of operation. Tubular elements such as power shafting often use metal end fittings for connections to power sources or for articulation where changes in direction are needed. In addition, assembly of the structure from its constituent parts will involve either bonded or mechanically fastened joints or both.

Joints represent one of the greatest challenges in the design of structures in general and in composite structures in particular. The reason for this is that joints entail interruptions of the geometry of the structure and often, material discontinuities, which almost always produce local highly stressed areas, except for certain idealized types of adhesive joint such as scarf joints between similar materials. Stress concentrations in mechanically fastened joints are particularly severe because the load transfer between elements of the joint have to take place over a fraction of the available area. For mechanically fastened joints in metal structures, local yielding, which has the effect of eliminating stress peaks as the load increases, can usually be depended on; such joints can be designed to some extent by the "P over A" approach, i.e., by assuming that the load is evenly distributed over load bearing sections so that the total load (the "P") divided by the available area (the "A") represents the stress that controls the strength of the joint. In organic matrix composites, such a stress reduction effect is realized only to a minor extent, and stress peaks predicted to occur by elastic stress analysis have to be accounted for, especially for one-time monotonic loading. In the case of composite adherends, the intensity of the stress peaks varies with the orthotropy of the adherend in addition to various other material and dimensional parameters which affect the behavior of the joint for isotropic adherends.

In principle, adhesive joints are structurally more efficient than mechanically fastened joints because they provide better opportunities for eliminating stress concentrations; for example, advantage can be taken of ductile response of the adhesive to reduce stress peaks. Mechanically fastened joints tend to use the available material inefficiently. Sizeable regions exist where the material near the fastener is nearly unloaded, which must be compensated for by regions of high stress to achieve a particular required average load. As mentioned above, certain types of adhesive joints, namely scarf joints between components of similar stiffness, can achieve a nearly uniform stress state throughout the region of the joint.

In many cases, however, mechanically fastened joints can not be avoided because of requirements for disassembly of the joint for replacement of damaged structure or to achieve access to underlying structure. In addition, adhesive joints tend to lack structural redundancy, and are highly sensitive to manufacturing deficiencies, including poor bonding technique, poor fit of mating parts and sensitivity of the adhesive to temperature and environmental effects such as moisture. Assurance of bond quality has been a continuing problem in adhesive joints; while ultrasonic and X-ray inspection may reveal gaps in the bond, there is no present technique which can guarantee that a bond which appears to be intact does, in fact, have adequate load transfer capability. Surface preparation and bonding techniques have been well developed, but the possibility that lack of attention to detail in the bonding operation may lead to such deficiencies needs constant alertness on the part of fabricators. Thus mechanical fastening tends to be preferred over bonded construction in highly critical and safety rated applications such as primary aircraft structural components, especially in large commercial transports, since assurance of the required level of structural integrity is easier to guarantee in mechanically fastened assemblies. Bonded construction tends to be more prevalent in smaller aircraft. For non-aircraft applications as well as in non-flight critical aircraft components, bonding is likewise frequently used.

This chapter describes design procedures and analytical methods for determining stresses and deformations in structural joints for composite structures. Section 5.2 which follows deals with adhesive joints. (Mechanically fastened joints will be the subject of a future revision of the Handbook.)

In the case of adhesive joints, design considerations which are discussed include: effects of adherend thickness as a means of ensuring adherend failure rather than bond failure; the use of adherend tapering to minimize peel stresses; effects of adhesive ductility; special considerations regarding composite adherends; effects of bond layer defects, including surface preparations defects, porosity and thickness variations; and, considerations relating to long term durability of adhesive joints. In addition to design considerations, aspects of joint behavior which control stresses and deformations in the bond layer are described, including both shear stresses and transverse normal stresses which are customarily referred to as "peel" stresses when they are tensile. Finally, some principles for finite element analysis of bonded joints are described.

Related information on joints in composite structures which is described elsewhere in this handbook includes Volume 1, Chapter 7, Section 7.2 (Mechanically Fastened Joints) and 7.3 (Bonded Joints) together with Volume 3, Chapter 2, Section 2.5.9 on Adhesive Bonding.

5.2 ADHESIVE JOINTS

5.2.1 Introduction

Adhesive joints are capable of high structural efficiency and constitute a resource for structural weight saving because of the potential for elimination of stress concentrations which cannot be achieved with mechanically fastened joints. Unfortunately, because of a lack of reliable inspection methods and a requirement for close dimensional tolerances in fabrication, aircraft designers have generally avoided bonded construction in primary structure. Some notable exceptions include: bonded step lap joints used in attachments for the F-14 and F-15 horizontal stabilizers as well as the F-18 wing root fitting, and a majority of the airframe components of the Lear Fan and the Beech Starship.

While a number of issues related to adhesive joint design were considered in the earlier literature cited in Reference 5.2.1(a)- 5.2.1(h), much of the methodology currently used in the design and analysis of adhesive joints in composite structures is based on the approaches evolved by L.J. Hart-Smith in a series of NASA/Langley-sponsored contracts of the early 70's (Reference 5.2.1(i) - 5.2.1(n)) as well as from the Air Force's Primary Adhesively Bonded Structures Technology (PABST) program (Reference 5.2.1(o) - 5.2.1(r)) of the mid-70's. The most recent such work developed three computer codes for bonded and bolted joints, designated A4EG, A4EI and A4EK (References 5.2.1(s) - 5.2.1(u)), under Air Force contract. The results of these efforts have also appeared in a number of open literature publications (Reference 5.2.1(v) - (z)). In addition, such approaches found application in some of the efforts taking place under the NASA Advanced Composite Energy Efficient Aircraft (ACEE) program of the early to mid 80's (Reference 5.2.1(x) and 5.2.1(y)).

Some of the key principles on which these efforts were based include: (1) the use of simple 1-dimensional stress analyses of generic composite joints wherever possible; (2) the need to select the joint design so as to ensure failure in the adherend rather than the adhesive, so that the adhesive is never the weak link; (3) recognition that the ductility of aerospace adhesives is beneficial in reducing stress peaks in the adhesive; (4) careful use of such factors as adherend tapering to reduce or eliminate peel stresses from the joint; and (5) recognition of slow cyclic loading, corresponding to such phenomena as cabin pressurization in aircraft, as a major factor controlling durability of adhesive joints, and the need to avoid the worst effects of this type of loading by providing sufficient overlap length to ensure that some of the adhesive is so lightly loaded that creep cannot occur there, under the most severe extremes of humidity and temperature for which the component is to be used.

Much of the discussion to follow will retain the analysis philosophy of Hart-Smith, since it is considered to represent a major contribution to practical bonded joint design in both composite and metallic structures. On the other hand, some modifications are introduced here. For example, the revisions of the Goland-Reissner single lap joint analysis presented in Reference 5.2.1(k) have again been revised according to the approach presented in References 5.2.1(z) and 5.2.1(aa).

Certain issues which are specific to composite adherends but were not dealt with in the Hart-Smith efforts will be addressed. The most important of these is the effect of transverse shear deformations in organic composite adherends.

Although the main emphasis of the discussion is on simplified stress analysis concepts allowed by shear lag models for shear stress prediction and beam-on-elastic foundation concepts for peel stress prediction, a brief discussion will be provided on requirements for finite element modeling of adhesive joints. Similarly, although joint failure will be considered primarily from the standpoint of stress and strain energy considerations, some discussion of fracture mechanics considerations for adhesive joints will also be included.

5.2.2 Joint design considerations

5.2.2.1 Effects of adherend thickness: adherend failures vs. bond failures

Figure 5.2.2.1(a) shows a series of typical bonded joint configurations. Adhesive joints in general are characterized by high stress concentrations in the adhesive layer. These originate, in the case of shear stresses, because of unequal axial straining of the adherends, and in the case of peel stresses, because of eccentricity in the load path. Considerable ductility is associated with shear response of typical adhesives, which is beneficial in minimizing the effect of shear stress joint strength. Response to peel stresses tends to be much more brittle than that to shear stresses, and reduction of peel stresses is desirable for achieving good joint performance.

From the standpoint of joint reliability, it is vital to avoid letting the adhesive layer be the weak link in the joint; this means that, whenever possible, the joint should be designed to ensure that the adherends fail before the bond layer. This is because failure in the adherends is fiber controlled, while failure in the adhesive is resin dominated, and thus subject to effects of voids and other defects, thickness variations, environmental effects, processing variations, deficiencies in surface preparation and other factors that are not always adequately controlled. This is a significant challenge, since adhesives are inherently much weaker than the composite or metallic elements being joined. However, the objective can be accomplished by recognizing the limitations of the joint geometry being considered and placing appropriate restrictions on the thickness dimensions of the joint for each geometry. Figure 5.2.2.1(b),which has frequently been used by Hart-Smith (References 5.2.1(n), 5.2.2.1(a)) to illustrate this point, shows a progression of joint types which represent increasing strength capability from the lowest to the highest in the figure. In each type of joint, the adherend thickness may be increased as an approach to achieving higher load capacity. When the adherends are relatively thin, results of stress analyses show that for all of the joint types in Figure 5.2.2.1(b), the stresses in the bond will be small enough to guarantee that the adherends will reach their load capacity before failure can occur in the bond. As the adherend thicknesses increase, the bond stresses become relatively larger until a point is reached at which bond failure occurs at a lower load than that for which the adherends fail. This leads to the general principle that for a given joint type, the adherend thicknesses should be restricted to an appropriate range relative to the bond layer thickness. Because of processing considerations and defect sensitivity of the bond material, bond layer thicknesses are generally limited to a range of 0.005-0.015 in. (0.125-0.39 mm). As a result, each of the joint types in Figure 5.2.2.1(a) and 5.2.2.1(b) corresponds to a specific range of adherend thicknesses and therefore of load capacity, and as the need for greater load capacity arises, it is preferable to change the joint configuration to one of higher efficiency rather than to increasing the adherend thickness indefinitely.

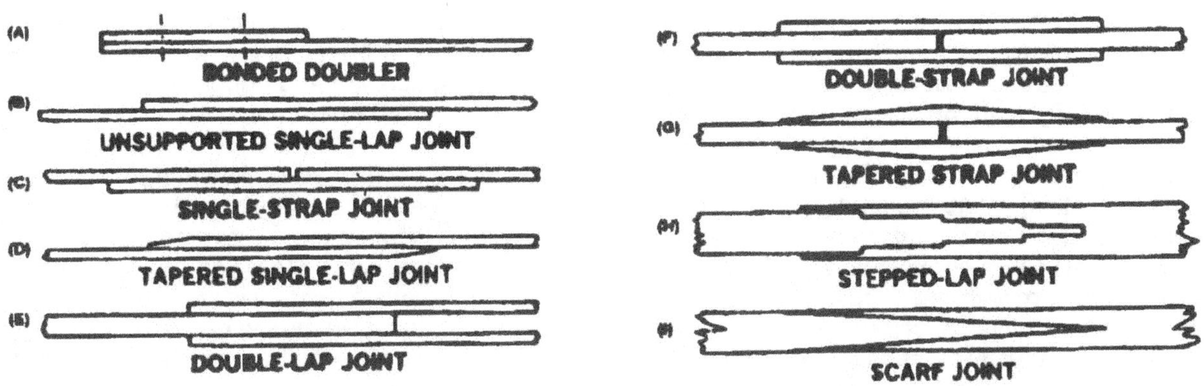

FIGURE 5.2.2.1(a) *Adhesive joint types (Reference 5.2.1(n) and 5.2.2.1(a)).*

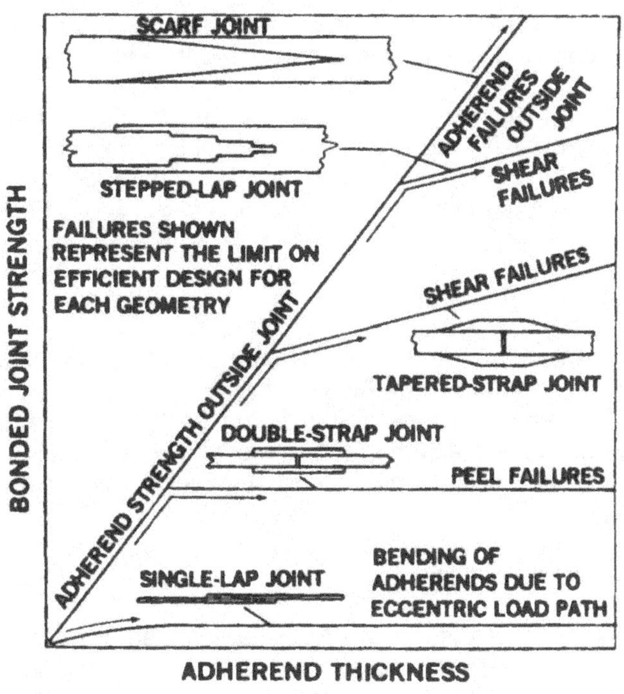

FIGURE 5.2.2.1(b) *Joint geometry effects (Reference 5.2.1(n)).*

5.2.2.2 Joint geometry effects

Single and double lap joints with uniformly thick adherends (Figure 5.2.2.1(a) - Joints (B), (E) and (F)) are the least efficient joint type and are suitable primarily for thin structures with low running loads (load per unit width, i.e., stress times element thickness). Of these, single lap joints are the least capable because the eccentricity of this type of geometry generates significant bending of the adherends that magnifies the peel stresses. Peel stresses are also present in the case of symmetric double lap and double strap joints, and become a limiting factor on joint performance when the adherends are relatively thick.

Tapering of the adherends (Figure 5.2.2.1(a) - Joints (D) and (G)) can be used to eliminate peel stresses in areas of the joint where the peel stresses are tensile, which is the case of primary concern. No tapering is needed at ends of the overlap where the adherends butt together because the transverse normal stress at that location is compressive and rather small. Likewise, for double strap joints under compressive loading, there is no concern with peel stresses at either location since the transverse extensional stresses that do develop in the adhesive are compressive in nature rather than tensile; indeed, where the gap occurs, the inner adherends bear directly on each other and no stress concentrations are present there for the compression loading case.

For joints between adherends of identical stiffness, scarf joints (Figure 5.2.2.1(a) - Joint (I)) are theoretically the most efficient, having the potential for complete elimination of stress concentrations. (In practice, some minimum thickness corresponding to one or two ply thicknesses must be incorporated at the thin end of the scarfed adherend leading to the occurrence of stress concentrations in these areas.) In theory, any desirable load capability can be achieved in the scarf joint by making the joint long enough and thick enough. However, practical scarf joints may be less durable because of a tendency toward creep failure associated with a uniform distribution of shear stress along the length of the joint unless care is taken to avoid letting the adhesive be stressed into the nonlinear range. As a result, scarf joints tend to be used only for repairs of very thin structures. Scarf joints with unbalanced stiffnesses between the adherends do not achieve the uniform shear stress condition of those with balanced adherends, and are somewhat less structurally efficient because of rapid buildup of load near the thin end of the thicker adherend.

Step lap joints (Figure 5.2.2.1(a) - Joint (H)) represent a practical solution to the challenge of bonding thick members. These types of joint provide manufacturing convenience by taking advantage of the layered structure of composite laminates. In addition, high loads can be transferred if sufficiently many short steps of sufficiently small "rise" (i.e., thickness increment) in each step are used, while maintaining sufficient overall length of the joint.

5.2.2.3 Effects of adherend stiffness unbalance

All types of joint geometry are adversely affected by unequal adherend stiffnesses, where stiffness is defined as axial or in-plane shear modulus times adherend thickness. Where possible, the stiffnesses should be kept approximately equal. For example, for step lap and scarf joints between quasi-isotropic carbon/epoxy (Young's modulus = 8 Msi (55 GPa)) and titanium (Young's modulus = 16 Msi (110 GPa)) ideally, the ratio of the maximum thickness (the thickness just beyond the end of the joint) of the composite adherend to that of the titanium should be 16/8=2.0.

5.2.2.4 Effects of ductile adhesive response

Adhesive ductility is an important factor in minimizing the adverse effects of shear and peel stress peaks in the bond layer. If peel stresses can be eliminated from consideration by such approaches as adherend tapering, strain energy to failure of the adhesive in shear has been shown by Hart-Smith to be the key parameter controlling joint strength (Reference 5.2.1(j)); thus the square root of the adhesive strain energy density to failure determines the maximum static load that can be applied to the joint. The work of Hart-Smith has also shown that for predicting mechanical response of the joint, the detailed stress-strain curve of the adhesive can be replaced by an equivalent curve consisting of a linear rise followed by a constant stress plateau (i.e., elastic-perfectly plastic response) if the latter is adjusted to provide the same strain energy density to failure as the actual stress-strain curve gives. Test methods for adhesives should be aimed at providing data on this parameter (see Volume 1, Section 7.3). Once the equivalent elastic-perfectly plastic stress-strain curve has been identified for the selected adhesive in the range of the most severe environmental conditions (temperature and humidity) of interest, the joint design can proceed through the use of relatively simple one-dimensional stress analysis, thus avoiding the need for elaborate finite element calculations. Even the most complicated of joints, the step lap joints designed for root-end wing and tail connections for the F-18 and other aircraft, have been successfully designed (References 5.2.1(t) and 5.2.1(u), Reference 5.2.2.1(b)) and experimentally demonstrated using such approaches. Design procedures for such analyses which were developed on Government contract have been incorporated into public domain in the form of the A4EG, A4EI and A4EK computer codes (Reference 5.2.1(s) - 5.2.1(u)). Note that the A4EK code permits analysis of bonded joints in which local disbonds are repaired by mechanical fasteners.

5.2.2.5 Behavior of composite adherends

Polymer matrix composite adherends are considerably more affected by interlaminar shear and tensile stresses than metals, so that there is a significant need to account for such effects in stress analyses of joints. Transverse shear and thickness-normal deformations of the adherends have an effect analogous to thickening of the bond layer, corresponding to a lowering of both shear and peel stress peaks. In addition, the adherend matrix is often weaker than the adhesive in shear and transverse tension, as a result of which the limiting element in the joint may be the interlaminar shear and transverse tensile strengths of the adherend rather than the bond strength. Ductile behavior of the adherend matrix can be expected to have an effect similar to that of ductility in the adhesive in terms of response of the adherends to transverse shear stresses, although the presence of the fibers probably limits this effect to some extent, particularly in regard to peel stresses.

The effect of the stacking sequence of the laminates making up the adherends in composite joints is significant. For example, 90-degree layers placed adjacent to the bond layer theoretically act largely as additional thicknesses of bond material, leading to lower peak stresses, while 0-degree layers next to the bond layer give stiffer adherend response with higher stress peaks. In practice it has been observed that 90-degree layers next to the bond layer tend to seriously weaken the joint because of transverse cracking which develops in those layers, and advantage cannot be taken of the reduced stresses. Large disparity of thermal expansion characteristics between metal and composite adherends can pose severe problems. Adhesives with high curing temperatures may be unsuitable for some uses below room temperature because of large thermal stresses which develop as the joint cools below the fabrication temperature.

Composite adherends are relatively pervious to moisture, which is not true of metal adherends. As a result, moisture is more likely to be found over wide regions of the adhesive layer, as opposed to confinement near the exposed edges of the joint in the case of metal adherends, and response of the adhesive to moisture may be an even more significant issue for composite joints than for joints between metallic adherends.

5.2.2.6 Effects of bond defects

Defects in adhesive joints which are of concern include surface preparation deficiencies, voids and porosity, and thickness variations in the bond layer.

Of the various defects which are of interest, surface preparation deficiencies are probably the greatest concern. These are particularly troublesome because there are no current nondestructive evaluation techniques which can detect low interfacial strength between the bond and the adherends. Most joint design principles are academic if good adhesion between the adherends and bond layer is poor. The principles for achieving this (Reference 5.2.2.6(a) - 5.2.2.6(c)) are well established for adherend and adhesive combinations of interest. Hart-Smith, Brown and Wong (Reference 5.2.2.6.(a)) give an account of the most crucial features of the surface preparation process. Results shown in Reference 5.2.2.6.(a) suggest that surface preparation which is limited to removal of the peel ply from the adherends may be suspect, since some peel plies leave a residue on the bonding surfaces that makes adhesion poor. (However, some manufacturers have obtained satisfactory results from surface preparation consisting only of peel ply removal.) Low pressure grit blasting (Reference 5.2.2.6(b)) is preferable over hand sanding as a means of eliminating such residues and mechanically conditioning the bonding surfaces.

For joints which are designed to ensure that the adherends rather than the bond layer are the critical elements, tolerance to the presence of porosity and other types of defect is considerable (Reference 5.2.1(t)). Porosity (Reference 5.2.1(z)) is usually associated with over-thickened areas of the bond, which tend to occur away from the edges of the joint where most of the load transfer takes place, and thus is a relatively benign effect, especially if peel stresses are minimized by adherend tapering. Reference 5.2.1(z) indicates that in such cases, porosity can be represented by a modification of the assumed stress-strain properties of the adhesive as determined from thick-adherend tests, allowing a straightforward analysis of the effect of such porosity on joint strength as in the A4EI computer code. If peel stresses are significant, as in the case of over-thick adherends, porosity may grow catastrophically and lead to non-damage-tolerant joint performance.

In the case of bond thickness variations (Reference 5.2.1(aa)), these usually take place in the form of thinning due to excess resin bleed at the joint edges, leading to overstressing of the adhesive in the vicinity of the edges. Inside tapering of the adherends at the joint edges can be used to compensate for this condition; other compensating techniques are also discussed in Reference 5.2.1(aa). Bond thicknesses per se should be limited to ranges of 0.005-0.01 in. (0.12-0.24 mm) to prevent significant porosity from developing, although greater thicknesses may be acceptable if full periphery damming or high minimum viscosity paste adhesives are used. Common practice involves the use of film adhesives containing scrim cloth, some forms of which help to maintain bond thicknesses. It is also common practice to use mat carriers of chopped fibers to prevent a direct path for access by moisture to the interior of the bond.

5.2.2.7 Durability of adhesive joints

In Reference 5.2.1(t), Hart-Smith discusses differences in durability assessment of adhesive joints between concepts related to creep failure under cyclic loading and those related to crack initiation and propagation which require fracture mechanics approaches for their interpretation. In summary, Hart-Smith suggests that if peel stresses are eliminated by adherend tapering or other means, and if the principle discussed in Section 5.2.2.1 of limiting the adherend thickness to ensure failure of the adherends rather than the adhesive is followed, crack-type failures will not be observed under time-varying loading, failures being related primarily to creep fatigue at hot-wet conditions, in joints with short overlaps which are subject to relatively uniform distributions of shear stress along the joint length. Additional discussion of viscoelastic response of bonded joints is given in References 5.2.2.7(a) - 5.2.2.7(c).

On the other hand, there is an extensive body of literature (References 5.2.2.7(d) - 5.2.2.7(j) for example) on fracture mechanics approaches to joint durability, based on measurement of energy release rates for various adhesives together with analytical efforts aimed at applying them to joint configurations of interest. In particular, Johnson and Mall (Reference 5.2.2.7(j)) report fatigue crack initiation in bonded specimen configurations with adherend tapering aimed at reduction of peel stresses in varying degrees, in some cases practically eliminating them; data in Reference 5.2.2.7(j) indicates that crack initiation will occur even with the adhesive in pure shear, for cycling to 10^6 cycles above loading levels which are established in the course of the study described and which are probably considerably below static failure loads. The results given in Reference 5.2.2.7(j) suggest that for combinations of peel and shear stressing, total (Mode I + Mode II) cyclic energy release rate can be used to determine whether or not cracking will occur. However, Hart-Smith reported in Reference 5.2.1(t) that in "thick adherend" test specimens that provide a relatively uniform shear stress distribution in the adhesive (see Volume 1, Section 7.3) which were subjected to fatigue tests in the PABST program (Reference 5.2.1(o)), cycling to more than 10^7 cycles applied at high cycling rates (30 Hz) was achieved without failure of the adhesive, although in certain cases, namely those involving 0.25 in. (6.27 mm) adherend thicknesses, fatigue failures of the metal adherends did result. More study is needed to resolve some of the apparently contradictory results which have come out of various studies.

5.2.3 Stress analysis in adhesive joints

5.2.3.1 General

Stress analyses of adhesive joints have ranged from very simplistic "P over A" formulations in which only average shear stresses in the bond layer are considered, to extremely elegant elasticity approaches that consider fine details, e.g., the calculation of stress singularities for application of fracture mechanics concepts. A compromise between these two extremes is desirable, since the adequacy of structural joints does not usually depend on a knowledge of details at the micromechanics level, but rather only at the scale of the bond thickness. Since practical considerations force bonded joints to incorporate adherends which are thin relative to their dimensions in the load direction, stress variations through the thickness of the adherend and the adhesive layer tend to be moderate. Such variations do tend to be more significant for polymer matrix composite adherends because of their relative softness with respect to transverse shear and thickness normal stresses. However, a considerable body of design procedure has been developed based on ignoring thickness-wise adherend stress variations. Such approaches involve using one-dimensional models in which only variations in the axial direction are accounted for. Accordingly, the bulk of the material to be covered in this chapter is based on simplified one-dimensional approaches characterized by the work of Hart-Smith, and emphasizes the principles which have been obtained from that type of effort, since it represents most of what has been successfully applied to actual joint design, especially in aircraft components. The Hart-Smith approach makes extensive use of closed form and classical series solutions since these are ideally suited for making parametric studies of joint designs. The most prominent of these have involved modification of Volkersen (Reference 5.2.1(a)) and Goland-Reissner (Reference 5.2.1(b)) solutions to deal with ductile response of adhesives in joints with uniform adherend thicknesses along their lengths, together with classical series expressions to deal with variable adherend thicknesses encountered with tapered adherends, and scarf joints. Simple lap joint solutions described below calculate shear stresses in the adhesive for various adherend stiffnesses and applied loadings. For the more practical step lap joints, the described expressions can be adapted to treat the joint as a series of separate joints each having uniform adherend thickness.

5.2.3.2 Adhesive shear stresses

Figure 5.2.3.2(a) shows a joint with ideally rigid adherends in which neighboring points on the upper and lower adherends that lie along a vertical line before deformation slide horizontally with respect to each other when the joint is loaded to cause a displacement difference $\delta = u_U - u_L$ related to the bond layer shear strain by $\gamma_b = \delta / t_b$. The corresponding shear stress, τ_b, is given by $\tau_b = G_b \gamma_b$. The rigid adherend assumption implies that δ, γ_b and τ_b are uniform along the joint. Furthermore, the equilibrium relationship indicated in Figure 5.2.3.2(a) (C), which requires that the shear stress be related to the resultant distribution in the upper adherend by

$$dT_U / dx = \tau_b \qquad \text{5.2.3.2(a)}$$

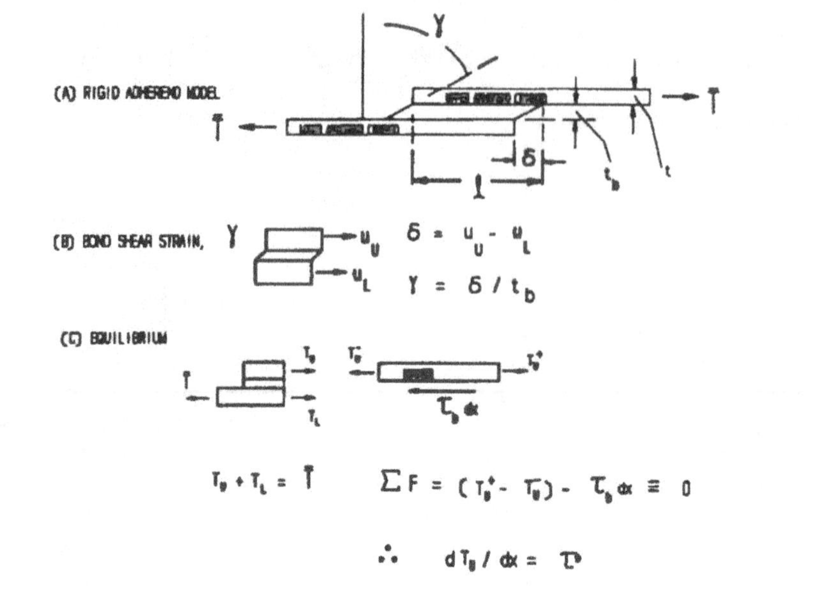

FIGURE 5.2.3.2(a) *Elementary joint analysis (rigid adherend model).*

leads to a linear distribution of T_U and T_L (upper and lower adherend resultants) as well as the adherend axial stresses σ_{xU} and σ_{xL}, as indicated in Figure 5.2.3.2(b). These distributions are described by the following expressions:

$$T_U = \bar{T}\frac{x}{\ell} \quad ; \quad T_L = \bar{T}\left(1 - \frac{x}{\ell}\right) \quad \text{i.e.} \quad \sigma_{xU} = \bar{\sigma}_x\frac{x}{\ell} \quad ; \quad \sigma_{xL} = \bar{\sigma}_x\left(1 - \frac{x}{\ell}\right)$$

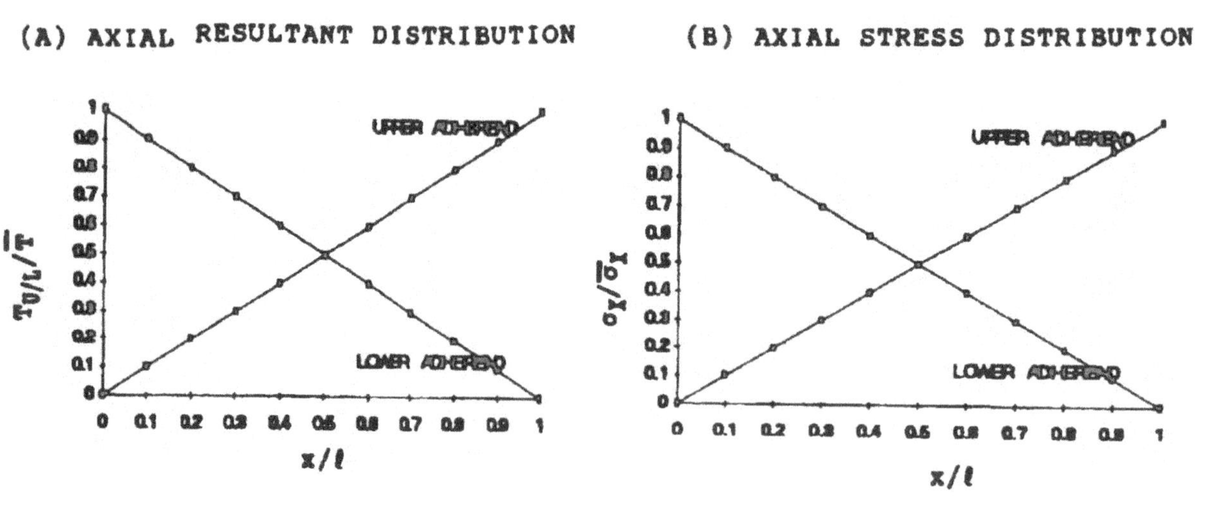

FIGURE 5.2.3.2(b) *Axial stresses in joint with rigid adherends.*

where $\bar{\tau}_x = \bar{T}/t$. In actual joints, adherend deformations will cause shear strain variations in the bond layer which are illustrated in Figure 5.2.3.2(c). For the case of a deformable upper adherend in combination with a rigid lower adherend shown in Figure 5.2.3.2(c) (A) (in practice, one for which $E_L t_L >> E_U t_U$), stretching elongations in the upper adherend lead to a shear strain increase at the right end of the bond layer. The case in which both adherends are equally deformable, shown in Figure 5.2.3.2(c) (B), indicates a bond shear strain increase at both ends due to the increased axial strain in whichever adherend is stressed at the end under consideration. For both cases, the variation of shear strain along the bond results in an accompanying increase in shear stress which, when inserted into the equilibrium equation (5.2.3.2(a)) leads to a nonlinear variation of stresses. The Volkersen shear lag analysis (Reference 5.2.1(a)) can be used to provide for calculations of adhesive shear stresses for the case of deformable adherends. This involves the solution of the following differential equation:

$$\frac{d^2T_U}{dx^2} = \frac{G_b}{t_b}\left[\left(\frac{1}{B_U}+\frac{1}{B_L}\right)T_U - \frac{1}{B_L}\bar{T}\right] \quad ; \quad B_U = E_U t_U \quad ; \quad B_L = E_L t_L \qquad 5.2.3.2(b)$$

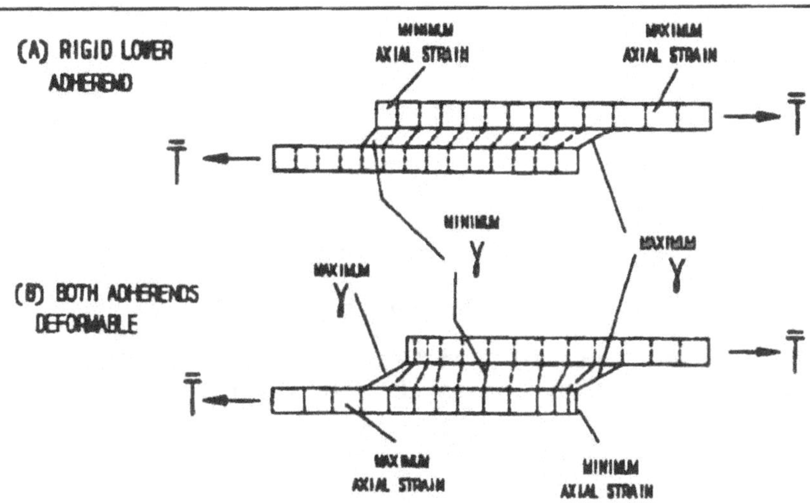

FIGURE 5.2.3.2(c) *Adherend deformations in idealized joints.*

which applies to the geometry of Figure 5.2.3.2(d) below. The solution for this equation which provides zero traction conditions at the left end of the upper adherend and the right end of the lower adherend, together with the applied load $\bar{T}$ at the loaded ends gives the resultants as

$$T_U = \bar{T}\left\{\frac{1}{1+\rho_B}\left[1+\frac{\sinh\beta(x-\ell)/\bar{t}}{\sinh\beta\ell/\bar{t}}\right] + \frac{\rho_B}{1+\rho_B}\,\frac{\sinh\beta x/\bar{t}}{\sinh\beta\ell/\bar{t}}\right\} \quad ; \quad T_L = \bar{T} - T_U \qquad 5.2.3.2(c)$$

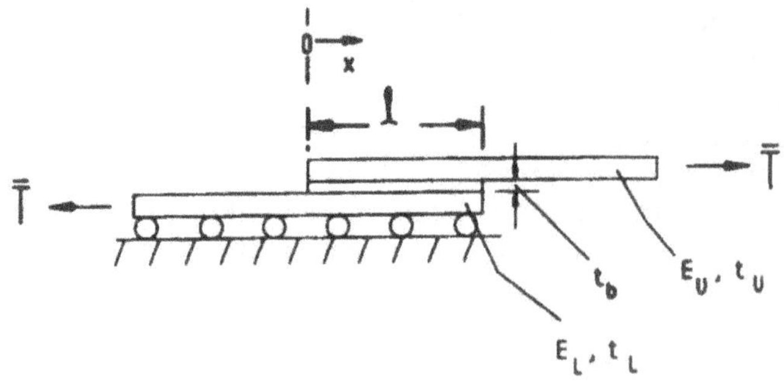

FIGURE 5.2.3.2(d) *Geometry for Volkersen solution.*

where

$$\beta = \left[G_b \frac{\bar{t}^2}{t_b} \left(\frac{1}{B_U} + \frac{1}{B_L} \right) \right]^{1/2} \quad ; \quad \bar{t} = \frac{t_U + t_L}{2} \quad ; \quad \rho_B = B_L / B_U$$

Using Equation 5.2.3.2(a) to obtain an expression for the shear stress distribution leads to

$$\tau_b = \beta \bar{\sigma}_x \left[\frac{1}{1+\rho_B} \frac{\cosh \beta (x-\ell)/\bar{t}}{\sinh \beta \ell/\bar{t}} + \frac{\rho_B}{1+\rho_B} \frac{\cosh \beta x/\bar{t}}{\sinh \beta \ell/\bar{t}} \right] \qquad 5.2.3.2(d)$$

For the case of dissimilar adherends, assuming that the lower adherend is as stiff as or stiffer than the upper, the maximum shear stress obtained from Equation 5.2.3.2(d) is given by

$$B_L \geq B_U$$

$$\tau_b \big|_{max} = \beta \bar{\sigma}_x \left(\frac{1}{1+\rho_B} \frac{1}{\sinh \beta \ell/\bar{t}} + \frac{\rho_B}{1+\rho_B} \frac{1}{\tanh \beta \ell/\bar{t}} \right) \qquad 5.2.3.2(e)$$

where

$$\bar{\sigma}_x = T/\bar{t}$$

Also of interest in the discussion which follows is the minimum shear stress in the joint. To a good approximation (which is exact for identical adherends) this occurs at $x=\ell/2$, leading to

$$\tau_b \big|_{min} = \frac{1}{2} \beta \bar{\sigma}_x \frac{1}{\sinh \beta \ell / 2t} \qquad 5.2.3.2(f)$$

In the case of ductile adhesive response, the ratio of minimum shear stress in the elastically responding part of the bond line to the yield stress in the ductilly responding part is of interest from the standpoint of joint durability, to be discussed subsequently. Figure 5.2.3.2(e) shows the distribution of axial adherend stresses and bond layer shear stress for two cases corresponding to $E_U=E_L$ vs. $E_L=10\,E_U$ with $t_U=t_L$, $\beta=0.387$ and $\ell/t=20$ for both cases (giving $\beta\ell/t=7.74$) and a nominal adherend stress $\bar{\sigma}_x=10$. As in the approximate analysis given earlier, the shear stresses given by Equation 5.2.3.2(e) are maximum at both ends for equally deformable adherends ($B_U=B_L$); for dissimilar adherends with the lower adherend more rigid ($B_L>B_U$), the maximum shear stress obtained from Equation 5.2.3.2(d) occurs at the right end of the joint where $x=\ell$, again as it did for the approximate analysis.

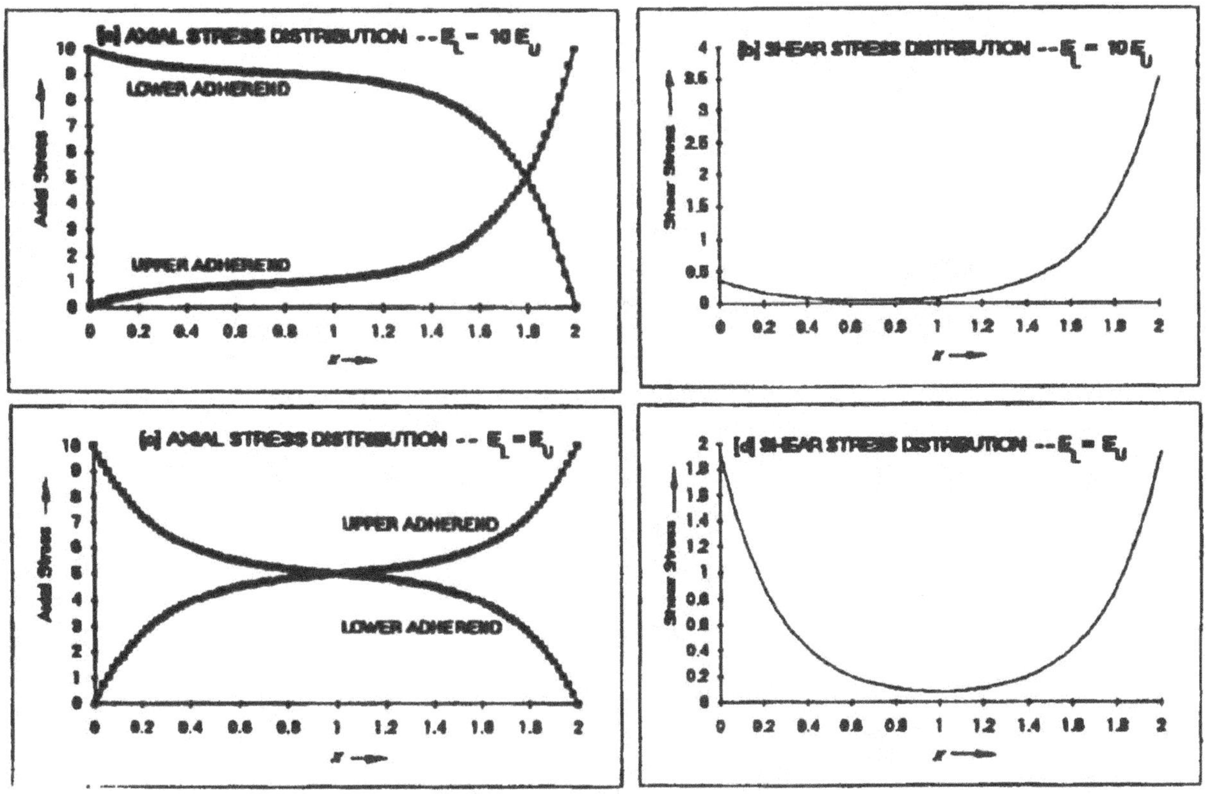

FIGURE 5.2.3.2(e) *Comparison of adherend stresses and bond shear stresses for $E_l=E_u$ vs. $E_l=10e_u$, β and adherend thicknesses equal for both cases.*

Figure 5.2.3.2(f) compares the behavior of the maximum shear stress with the average shear stress as a function of the dimensionless joint length, ℓ/t, for equal adherend stiffnesses. The point illustrated here is the fact that although the *average* shear stress continuously decreases as the joint length increases, for the *maximum* shear stress which controls the load that can be applied without failure of the adhesive, there is a diminishing effect of increased joint length when $\eta \equiv \beta\ell/t$ gets much greater than about 2.

An additional point of interest is a typical feature of bonded joints illustrated in Figure 5.2.3.2(e) Part (c) which gives the shear stress distribution for equal adherend stiffness; namely, the fact that high adhesive shear stresses are concentrated near the ends of the joint. Much of the joint length is subjected to relatively low levels of shear stress, which implies in a sense that region of the joint is structurally inefficient since it doesn't provide much load transfer; however, the region of low stress helps to improve damage tolerance of the joint since defects such as voids, and weak bond strength may be tolerated in regions where the shear stresses are low, and in joints with long overlaps this may include most of the joint. In addition, Hart-Smith has suggested (Reference 5.2.1(z)) that when ductility and creep are taken into account, it is a good idea to have a minimum shear stress level no more than 10% of the yield strength of the adhesive, which requires some minimum value of overlap length. Equation 5.2.3.2(f) given earlier can be used to satisfy this requirement for the case of equal stiffness adherends. The two special cases of interest again are for equal adherend stiffness vs. a rigid lower adherend, since these bound the range of behavior of the shear stresses. As a practical consideration, we will be interested primarily in long joints for which $\beta\ell/t >> 1$. For these cases Equation 5.2.3.2(e) reduces to

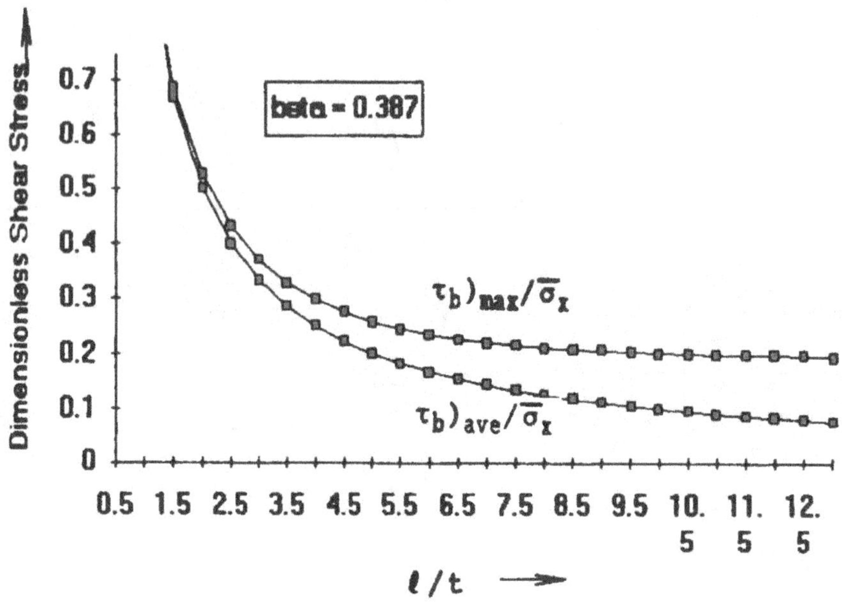

FIGURE 5.2.3.2(f) *Comparison of average and maximum shear stress vs. ℓ/t.*

$\beta\ell/t \gg 1$

$$B_L \gg B_U \ ; \ \tau_b|\max \approx \beta\bar{\sigma}_x \quad : \quad B_L = B_U \ ; \ \tau_b|\max \approx \frac{1}{2}\beta\bar{\sigma}_x \qquad \text{5.2.3.2(g)}$$

i.e., for long overlaps, the maximum shear stress for the rigid adherend case tends to be twice as great as that for the case of equally deformable adherends, again illustrating the adverse effect of adherend unbalance on shear stress peaks.

5.2.3.3 Peel stresses

Peel stresses, i.e., through the thickness extensional stresses in the bond, are present because the load path in most adhesive joint geometries is eccentric. It is useful to compare the effect of peel stresses in single and double lap joints with uniform adherend thickness, since peel stresses are most severe for joints with uniform adherend thickness. The load path eccentricity in the single lap joint (Figure 5.2.3.3(a)) is relatively obvious due to the offset of the two adherends which leads to bending deflection as in Figure 5.2.3.3(a) (B). In the case of double lap joints, as exemplified by the configuration shown in Figure 5.2.3.3(b), the load path eccentricity is not as obvious, and there may be a tendency to assume that peel stresses are not present for this type of joint because, as a result of the lateral symmetry of such configurations, there is no overall bending deflection. However, a little reflection brings to mind the fact that while the load in the symmetric lap joint flows axially through the central adherend prior to reaching the overlap region, there it splits in two directions, flowing laterally through the action of bond shear stresses to the two outer adherends. Thus eccentricity of the load path is also present in this type of joint. As seen in Figure 5.2.3.3(b) (C), the shear force, designated as F_{SH}, which represents the accumulated effect of τ_b for one end of the joint, produces a component of the total moment about the neutral axis of the upper adherend equal to $F_{SH}t/2$. (Note that F_{SH} is equivalent to T/2, since the shear stresses react this amount of load at each end.) The peel stresses, which are equivalent to the forces in the restraining springs shown in Figure 5.2.3.3(b) (B) and (C) have to be present to react the moment produced by the offset of F_{SH} about the neutral axis of the outer adherend. Peel stresses are highly objectionable. Later discussion will indicate that effects of ductility significantly reduce the tendency for failure associated with shear stresses in the adhesive. On the other hand, the adherends tend to prevent lateral contraction in the in-plane direction when the bond is strained in the thickness direction, which minimizes the availability of ductility effects that could provide the same reduction of adverse effects for the peel stresses. This is illustrated by what happens in the butt-tensile test shown in Figure 5.2.3.3(c) in which the two adherend surfaces adjacent to the bond are pulled away from each other uniformly. Here the shear stresses associated with yielding are restricted to a small region whose width is about equal to the thickness of the bond layer, near the outer edges of the system; in most of the bond, relatively little yielding can take place. For polymer matrix composite adherends, the adherends may fail at a lower peel stress level than that at which the bond fails, which makes the peel stresses even more undesirable.

It is important to understand that peel stresses are unavoidable in most bonded joint configurations. However, it will be seen that they can often be reduced to acceptable levels by selecting the adherend geometry appropriately.

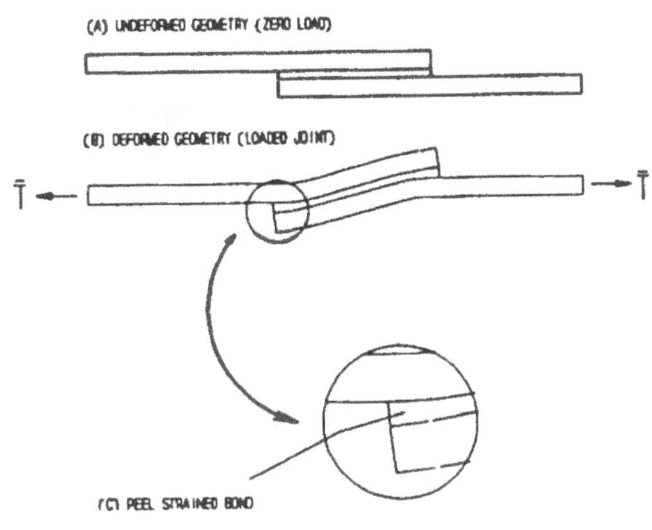

FIGURE 5.2.3.3(a) *Peel stress development in single lap joints.*

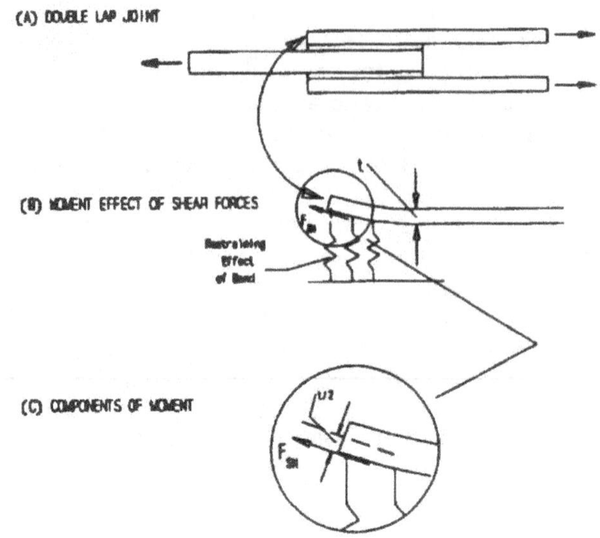

FIGURE 5.2.3.3(b) *Peel stress development in double lap joints.*

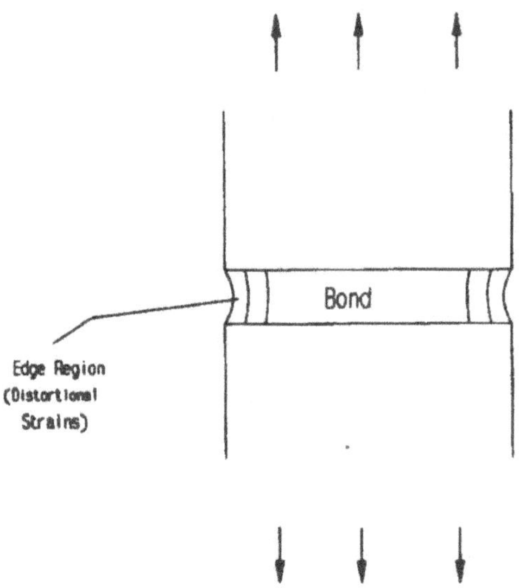

FIGURE 5.2.3.3(c) *Shear stresses near outer edges of butt tensile test..*

5.2.3.4 Finite element modeling

Finite element methods have often been used for investigating various features of bonded joint behavior, but there are serious pitfalls which the analyst must be aware of to avoid problems in such analyses, mainly because of the tendency of the bond layer thinness to unbalance the finite element model. To achieve adequate accuracy, it is especially important to provide a high degree of mesh refinement around the ends of the overlap (see Figure 5.2.3.4(a)) and yet transition the mesh to a coarser representation away from the ends of the overlap to avoid unneeded computational costs. Without such approaches, the need for limiting the aspect ratios of elements will force either a crude representation of the bond layer or an excessively over-refined mesh for the adherends. The mesh shown in Figure 5.2.3.4(a) was generated with a custom designed automated mesh generator developed by C. E. Freese of the Army Research Laboratory Materials Directorate, Watertown, MA (Reference 5.2.3.4). The elements shown consist of 8-point isoparametric quadrilaterals and 6-point isoparametric triangles, providing a quadratic distribution of displacements within each element. A number of commercially available finite element codes are presently available for developing such refined meshes. The commonly used displacement-based finite element methods are not capable of satisfying exact boundary conditions such as the traction free condition shown at the left end of the upper adherend in Figure 5.2.3.4(a) (C). In addition, a mathematical stress infinity occurs at the corner formed by the left end of the bond layer and the lower adherend.

These characteristics cannot be represented exactly, but a measure of the adequacy of the mesh refinement is provided by the degree to which the solution achieves the traction free condition shown in Figure 5.2.3.4(a) (C). Pertinent results are shown in Figure 5.2.3.4(b) which gives a solution for a double lap joint with unidirectional carbon/epoxy adherends. The finite element results represented by the "x" and "Δ" symbols are relevant to the issue under consideration. These represent the distribution of shear stresses along the interface between the upper adherend and the bond layer as indicated in the insert at the top of Figure 5.2.3.4(b). Since this line intersects the left end at a point fairly near the corner where the singularity occurs, it is reasonable to expect some difficulty in satisfying traction free conditions at the left end. The computer results did not go to zero at the end (where x=1.1953) but did show signs of heading in that direction since the end stress is slightly below the peaks for the two curves. Note that the Δ symbols represent a condition in which the bond is replaced by a continuation of the upper adherend, a considerably more difficult situation to deal with than that of the x's which allow for an actual bond layer. The third curve shown in Figure 5.2.3.4(b) indicated by open circles represents a modification of Volkersen's one-dimensional shear lag analysis which allows for transverse shear deformations in the adherends; the latter agrees surprisingly well with the prediction for the finite element analysis with the bond layer present (x's) for most of the joint length, although the peak stress predicted by the approximation is somewhat less than that of the FE analysis.

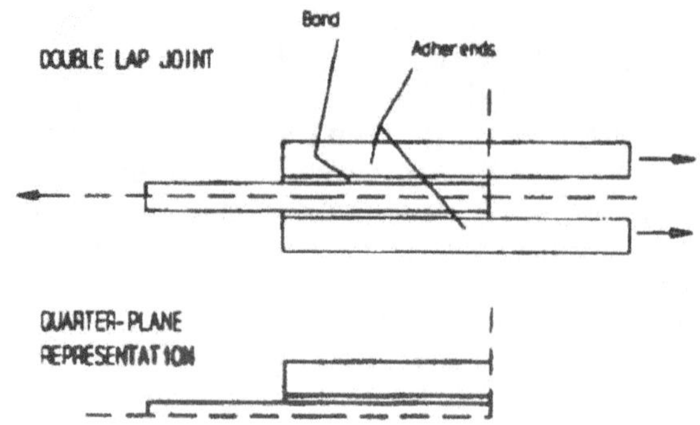

(A) Joint configuration

(B) Overall mesh

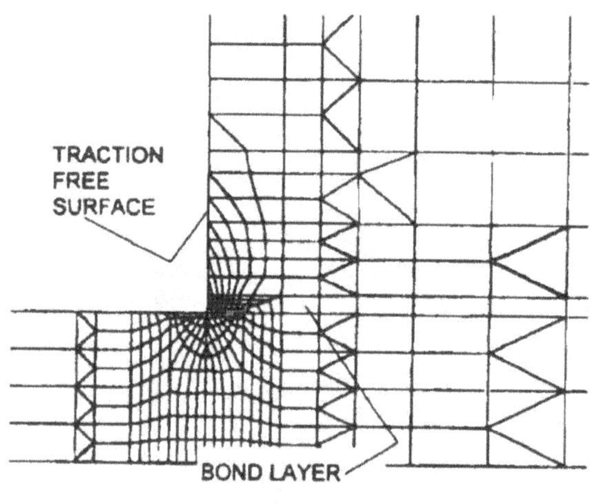

(C) Detailed Mesh

FIGURE 5.2.3.4(a) *Mesh details for finite element analysis of double lap joint.*

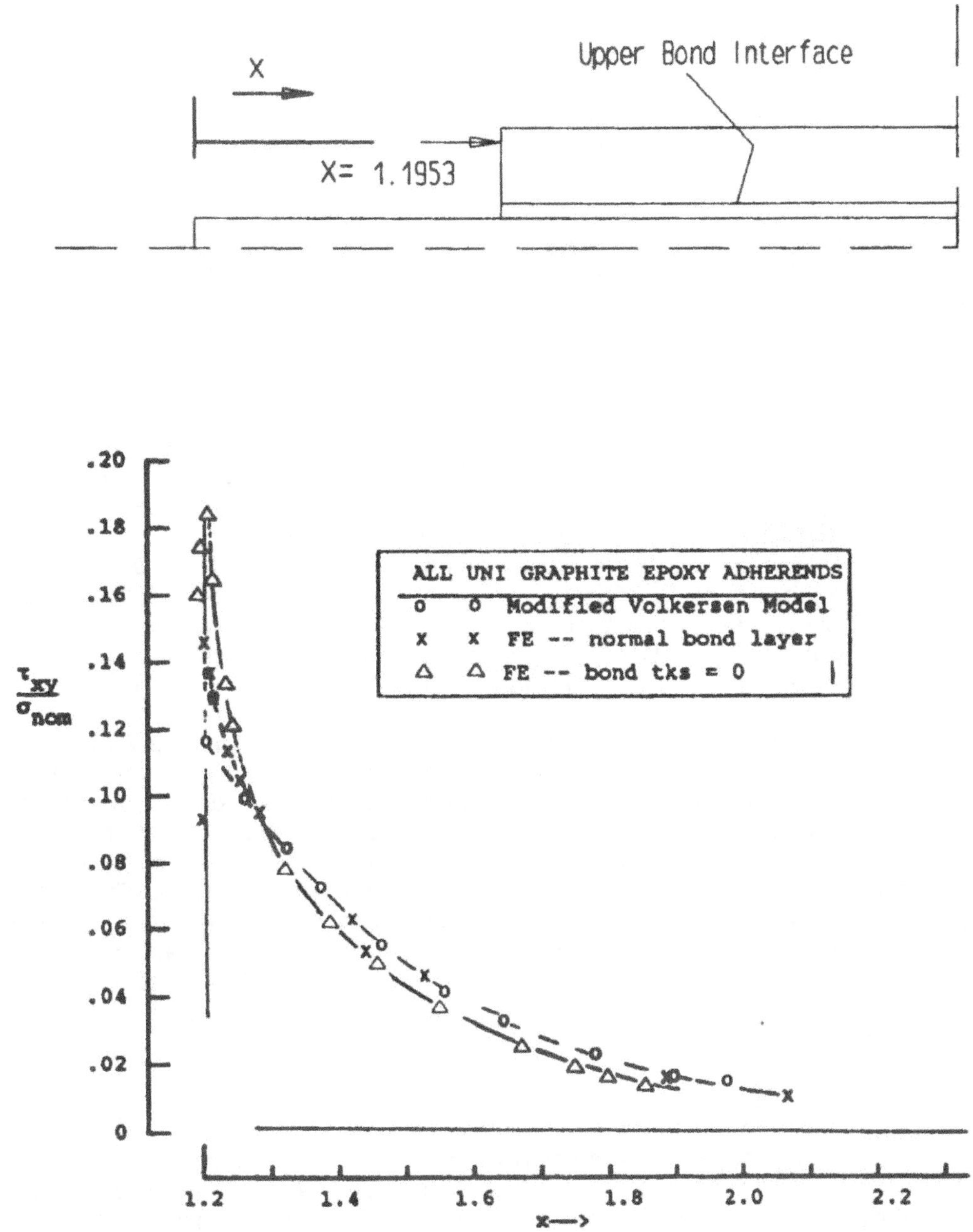

FIGURE 5.2.3.4(b) *Finite element predictions of shear stress distribution along bond-upper adherend interface, double lap joint shown in Figure 5.2.3.4(a).*

5.3 MECHANICALLY FASTENED JOINTS

5.3.1 Introduction

Mechanically fastened joints for composite structures have been studied since the mid-1960's when high modulus, high strength composites first came into use. It was found early in this period that the behavior of composites in bolted joints differs considerably from what occurs with metals. The brittle nature of composites necessitates more detailed analysis to quantify the level of various stress peaks as stress concentrations control static strength to a larger extent than in metals (no local yielding). This affects joint design as the edge distances and hole spacing have to be increased over those that are common in metal designs. Low through the thickness strength of composite laminates has led to specialized fasteners for composite and eliminated the use of rivets. The special fasteners feature larger tail footprint areas which have improved efficiency of composite joints. Galvanic corrosion susceptibility between carbon and aluminum has all but eliminated the use of aluminum fasteners.

Mechanically fastened joints can be classified into two general types by the amount of load being transferred. Examples of lightly loaded joints are the connection between substructure and skin or access panels in airframe construction. These are characterized by a single row of fasteners where each fastener carries only a very small portion of the total load being transferred. Root joint of a wing or a control surface is an example of a highly loaded joint. All the load accumulated on the aerodynamic surface is off-loaded into a fitting using a complex bolt pattern consisting of several rows and columns. Initially, aircraft designers avoided the latter type of mechanical joints preferring to use bonded joints, see Section 5.2. However, once confidence in analysis and design of bolted joints was established, highly loaded bolted joints were being implemented, the first being the B-1 horizontal stabilizer, Reference 5.3.1.

The confidence to design mechanically fastened joints in composite structures evolved mainly out of a number of DOD, NASA, and associated university programs aimed at providing a methodology which could be applied routinely to aircraft, although more generic applications have also been examined. The main focus to understand the behavior of a bolted joint has been concentrated on analysis to predict failure of a single bolt joint and its correlation with test results. This is understandable because the problem of load sharing between bolts in a multi-fastener joint is not much different from that of metal joints. The material presented here reflects the state of the art as practiced primarily in the aircraft industry. The objective is to give the reader some insight into the key factors that control the behavior of mechanically fastened joints in composite structures. The discussion which follows is arranged primarily to achieve that objective.

5.3.2 Structural analysis

5.3.2.1 Load sharing in a joint

Most of the mechanical joints encountered in aircraft structures have multiple fasteners. The number and type of fasteners needed to transfer the given loads are usually established by airframe designers by considerations of available space, producibility, and assembly. Although the resulting joint design is usually sufficient for finite element (FE) modeling purposes, further structural analyses are required before joint design drawings are released for fabrication. These analyses should consist of two distinct calculations: (1) computation of individual loads and orientation at each fastener with possible optimization to obtain near equal loading of each equal diameter fastener, and (2) stress analysis of load transfer for each critical fastener using fastener loads from previous analysis.

An example of a joint is shown in Figure 5.3.2.1(a). In order to obtain individual fastener loads for this or any other joint configuration (including single in-line row of fasteners), overall loading, geometry, plate stiffnesses, and individual fastener flexibilities must be known. Two structural analysis approaches have evolved in the aircraft industry. One performs the analysis in two steps, the first step being a calculation of individual bolt flexibilities followed by FE analysis with the fastener flexibilities as input. The second type includes the computation of the joint flexibility as a special FE in the overall FE analysis. An example of the latter is the SAMCJ code developed for the Air Force, Reference 5.3.2.1(a). Both approaches approximate a nonlinear joint load-displacement response, Figure 5.3.2.1(b), by a bilinear representation. This simplification permits the overall finite element problem to be linear. Recently a closed form analytical model has been developed and programmed for the personal computer to deal with the multiple hole joint strength problem (Reference 5.3.2.1(b)).

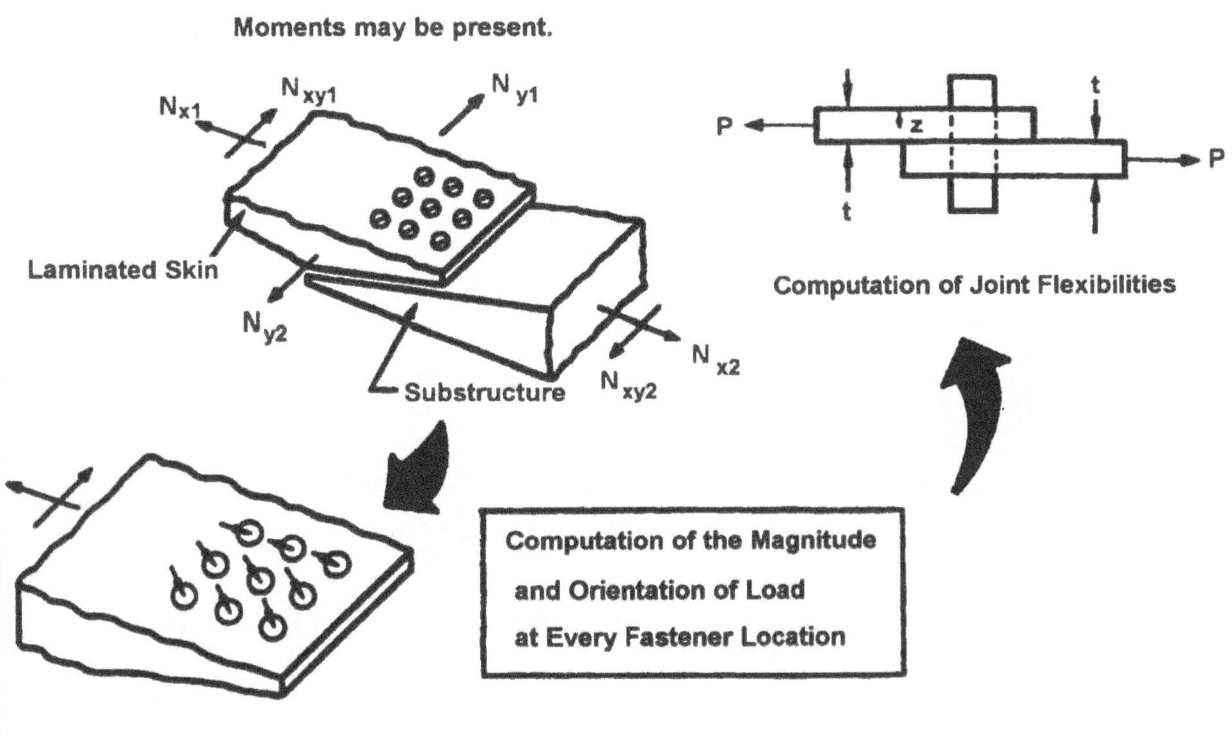

FIGURE 5.2.3.1(1) *Overview of the strength analysis of bolted structures.*

Fastener flexibility is based on joint displacement not only due to the axial extension of the joining plates but to other effects not easily modeled. These are fastener deflection in shear and bending, joint motion attributable to localized bearing distortions, and fastener rigid body rotation in single shear joints. Additionally, for composite laminates the value of joint flexibility should reflect the material orientation, ply fractions, and the stacking sequence of the laminates being joined. Other variables to be considered are the fit of the pin in the hole, presence of a free edge close to the hole, and head/tail restraint. Because of the many variables, test data for joint flexibility is the best type of input for the overall FE model of the multi-fastener joint. However, the data is not always available for all the different design situations. Hence, various modeling schemes have evolved to obtain flexibility values. Calculation of joint flexibility can be quite complex if the joint contains multiple stack-ups of plates with gaps. Analytical models to solve for the joint flexibility range from representing plates as springs to those where the fastener is idealized as a flexible beam on an elastic foundation provided by the plate or laminate.

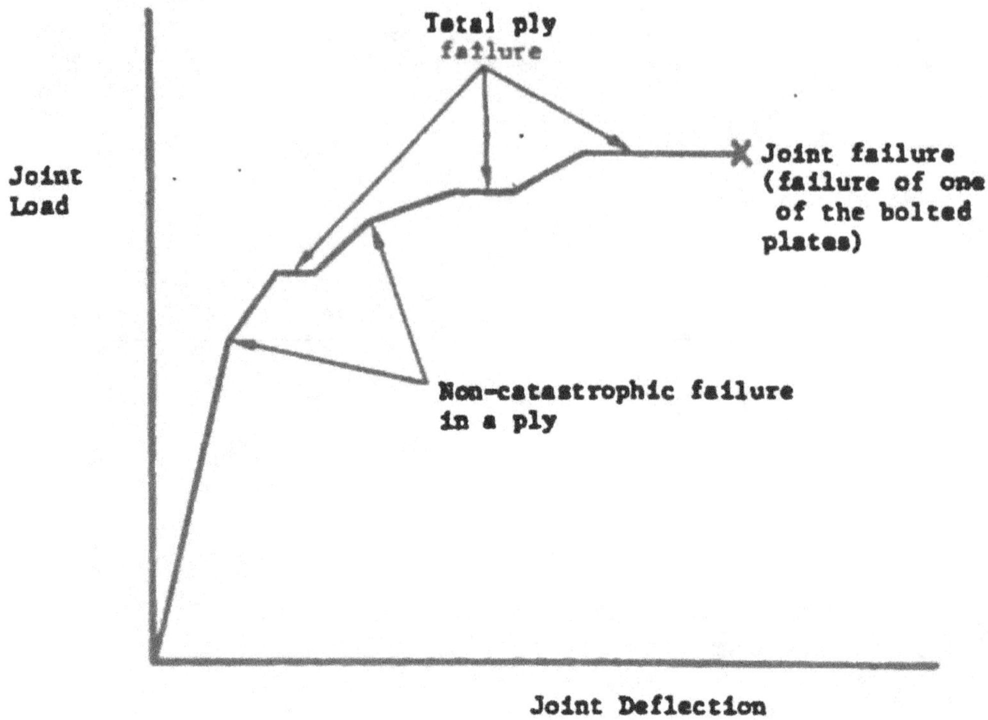

FIGURE 5.3.2.1(b) *A schematic representation of the overall load versus deflection response of the joint.*

For thick plates fasteners, flexibility may not be as important a parameter as for thin plates. Reference 5.3.2.1(c) has shown that good correlation between test and analysis for bolt load distribution using rigid inclusions to represent bolts. Reference 5.3.2.1(c) also included effects of the contact problem with and without gaps to calculate bearing stress distributions.

Load sharing in mechanically fastened joints is strongly dependent on the number and the diameter and material of the bolts, and the stiffness of joining members. For a single in-line row of bolts the first and the last bolt will be more highly loaded, if the plates are of uniform stiffness. This is illustrated in Figure 5.3.2.1(c) in which, in addition to the equal stiffness members (configuration 2), other combinations of fastener diameters/plate configurations are shown, which can alter the bolt distributions appreciably.

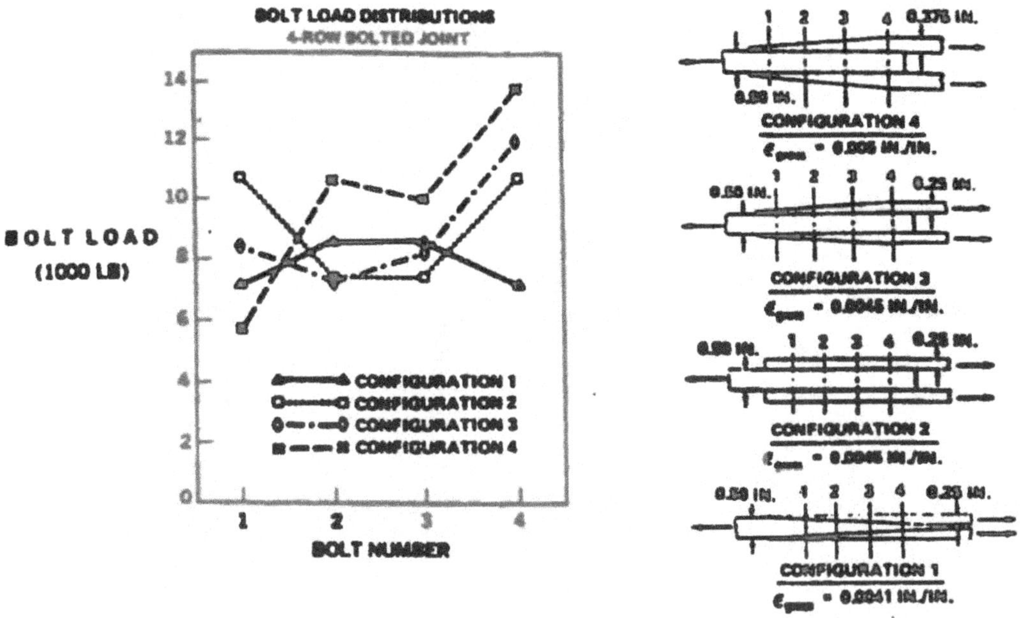

FIGURE 5.3.2.1(c) *Effect of joint configuration on fastener load distribution (Reference 5.3.2.1(d)).*

5.3.2.2 Analysis of local failure in bolted joints

Once the load sharing analysis has been performed, bolted joint analysis reduces to modeling a single bolt in a composite plate as shown in a free body diagram in Figure 5.3.2.2.(a). A number of analysis codes have been developed that perform the stress analysis and provide useful failure predictions for problem of Figure 5.3.2.2.(a). One cannot depend on analysis alone, and the design of a bolted composite joint will entail an extensive test program involving various joint configurations, laminates, and bearing/bypass ratios. However, because of the variety of laminates and load conditions present in a complex structure, testing frequently cannot cover all conditions of interest. Therefore, analytical methods are needed to extend the applicability of the test data to a wider range of cases.

There are multiple failure modes that must be considered. The first is net section failure of the composite. Alternatively, the laminate may fail immediately ahead of the bolt due to bearing pressure or the specimen will fail by pull-through. Depending on hole spacing, edge distances, or lay-up, shear-out may occur before bearing failure is reached. Delaminations may also be present but these are not the primary cause of failure. Finally, failure of the fastener must be considered. A more comprehensive description of possible failure modes is discussed in the next section.

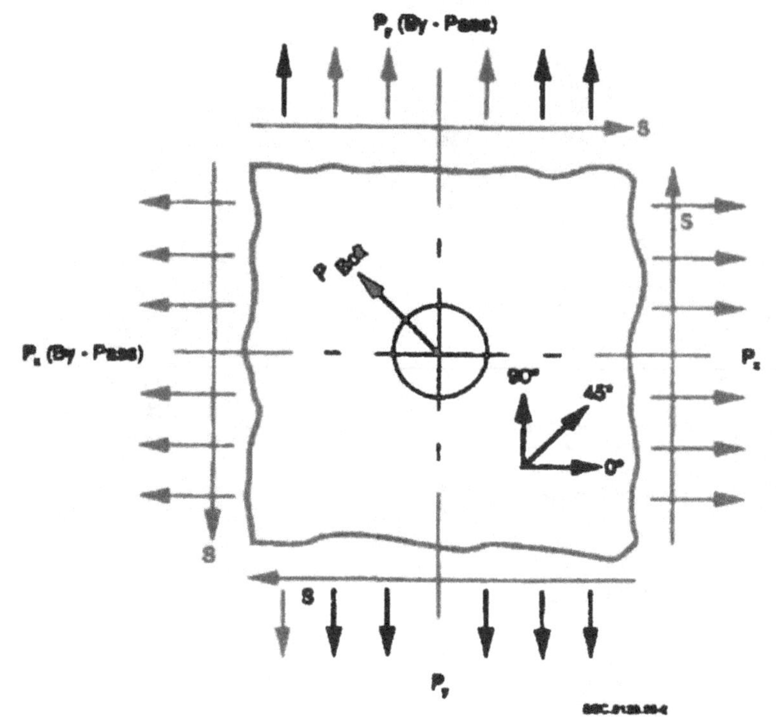

FIGURE 5.2.2.2(a) *Bolted joint under generalized loading.*

The analysis of fiber dominated in-plane failure modes, such as net-section failure, has typically been accomplished using variations of the approach by Whitney and Nuismer (Reference 5.3.2.2(a)), or the semi-empirical model of Hart-Smith (Reference 5.3.2.2(b)). The basis of the approach is to evaluate a ply-level failure criterion at a characteristic distance, d_0 away from the edge of the hole. The characteristic distance accounts for two experimentally observed effects. First, the strength of laminates containing a hole is greater than would be implied by dividing the unnotched strength by the theoretical stress concentration for the open hole. Second, the strength is observed to be a function of hole diameters, with strength decreasing as hole diameter increases. The use of a fixed d_0 simulates these effects, Figure 5.3.2.2(b).

The characteristic distance is treated as it was a laminate material property, and is determined by correlating the analysis to the ratio between the unnotched and open-hole strengths of laminates. More extensive correlations may reveal that d_0 is a function of the laminate ply fractions. The value of d_0 will also depend on the ply-level failure criterion used.

The establishment of laminate material allowables for the failure prediction must include a consideration of the material variability, and the inherent inability of current failure theories to completely account for changes in laminate stacking sequence, joint geometry, and hole size. One approach is to establish B-basis allowables for the ply-level failure criterion based on unnotched ply data. The d_0 is then selected such that the predicted values of failure are equivalent to the B-basis value of the notched laminate tests. The B-basis d_0 can also be obtained directly from notched laminate tests if sufficient number of different laminates with various hole sizes are tested.

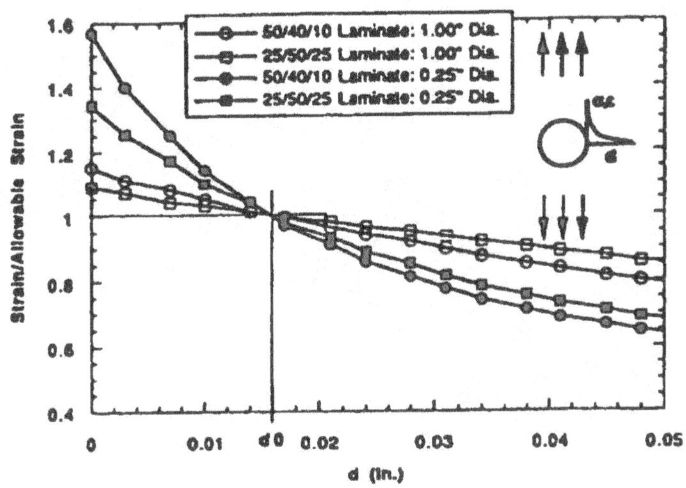

FIGURE 5.3.2.2(b) *Strain distributions near an open hole for different hole diameters and tape laminates. Applied far field load is equivalent to the expected failure load. Laminates are given as percentages of 0°/±45/90°. Crossing point of curves defines characteristic distance, d_o.*

Although the Whitney-Nuismer method was originally conceived for failure under uniaxial tension, the method has been applied to compression, and biaxial loading. The compression d_0 will be different than the tension value and the edgewise shear d_0 different from either. Reference 5.3.2.2(c) suggests a smooth characteristic curve for connecting the tension and compression values. When biaxial loads are introduced, one must search for the most critical location around the hole. A search algorithm is needed even for the case of uniaxial loading as it can be shown that the maximum circumferential stress may not occur at a point tangential to the load direction when the percentage of ±45° plies is large, or when an off-axis laminate is considered.

Use of this failure criterion for predicting failure implies that an accurate stress solution for the vicinity of the hole is available. A solution for a hole in an infinite, anisotropic sheet was given by Lekhnitskii (Reference 5.3.2.2(d)). This solution can be extended to the case of an assumed pressure distribution for a loaded bolt, and can be combined with boundary integral techniques to include the effects of nearby boundaries and multiple holes. General boundary element methods and finite element methods have also been applied. Care should be exercised in the use of finite-element techniques due to the high stress gradients present at the hole. The finite element model should be compared against the theoretical stress concentration at the edge of the hole to ensure sufficient mesh refinement.

The behavior of joints with bearing-loaded bolts has often been simulated by assuming a pressure distribution around the perimeter of the hole, although the actual behavior is governed by the displacement condition corresponding to the circular cross section of the bolt bearing into the surrounding plate. A typical assumption in the modeling of the joint is that the radial pressure due to the bolt follows a cosine function distribution over a 180° contact zone (Figure 5.3.2.21c) part (A)) and zero pressure elsewhere (with zero tangential stresses around the whole circumference). In many cases this gives satisfactory results for predicting the critical stress peaks, e.g., the peak net-section stress at the 90 degree points around the fastener. Figure 5.3.2.2(c) in Reference 5.3.2.2(e) shows a comparison of the predicted stress concentration factors for an assumed "half-cosine" radial *pressure* distribution vs. the more accurate solution which assumes a radial *displacement* condition along the edge of the hole. The "K" values tabulated at the left side of the figure represent peak stresses normalized with respect to the gross stress, P/Wt (thus the subscript "G"), including the peak net section stress (K_G^{nt}) at =90°, peak bearing stress (K_G^b) at =0° and peak shear stress (K_G^s) at = 45°. These results were predicted for W/D=2,e/W=1 and a neat fitting fastener. For these conditions, the stress concentration factors obtained from the two approaches are not substantially different, suggesting that the "half cosine" radial pressure distribution is an adequate approximation for the more accurate analysis which solves for the radial displacement distribution.

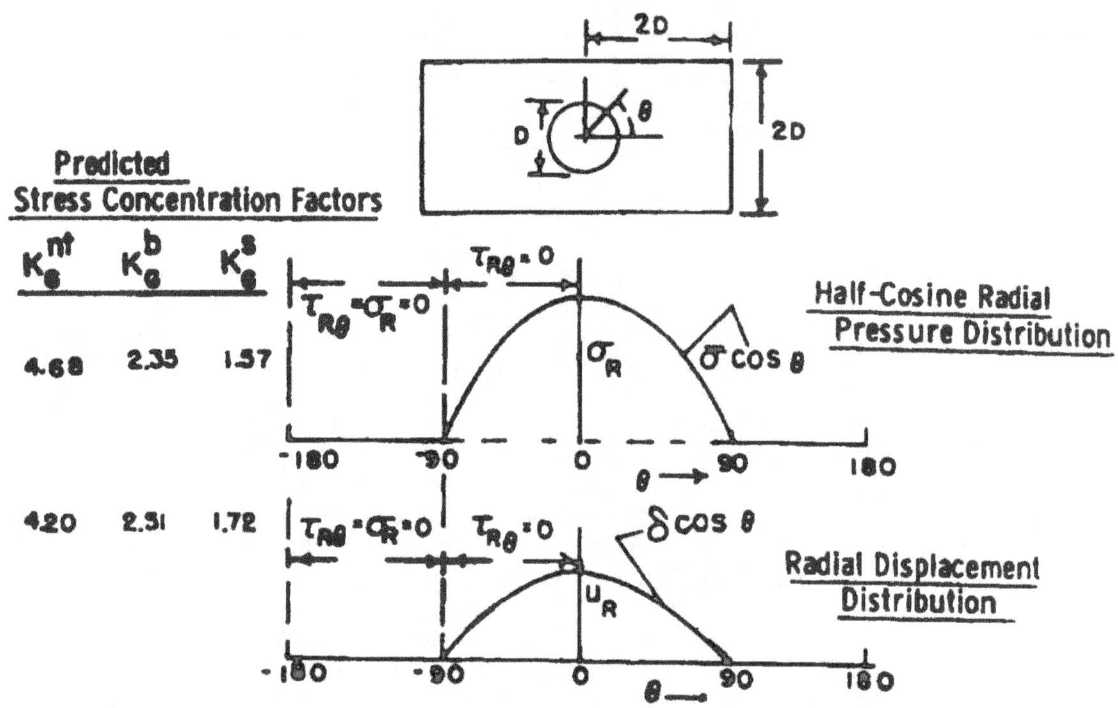

FIGURE 5.3.2.2(c) *Comparison of predicted stress concentrations for assumed radial pressure distribution vs. radial displacement distribution (Reference 5.3.2.2(e)).*

There are some important situations for which the "half cosine" pressure distribution will give poor results, however. Figure 5.3.2.2(d), which compares a variety of situations, includes one case in which the edge distance is relatively small (square symbols, e/W=0.375, W/D=2); the radial pressure distribution is characterized by a dip in the pressure near θ=0. This corresponds to the tendency for the part of the plate in front of the fastener to deform as if in beam bending (Figure 5.3.2.2(e)) in the case of short edge distances, relieving the pressure in front of the fastener so as to account for the drop in radial pressure near θ=0° which is seen in Figure 5.3.2.2(d).

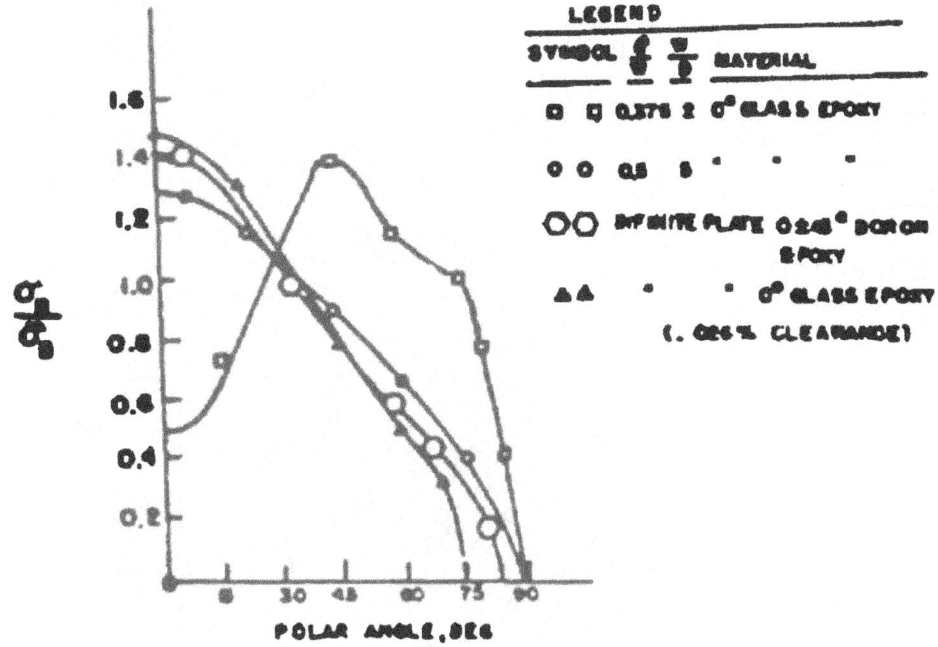

FIGURE 5.3.2.2(d) *Radial pressure distributions for various joint configurations (Reference 5.3.2.2(e)).*

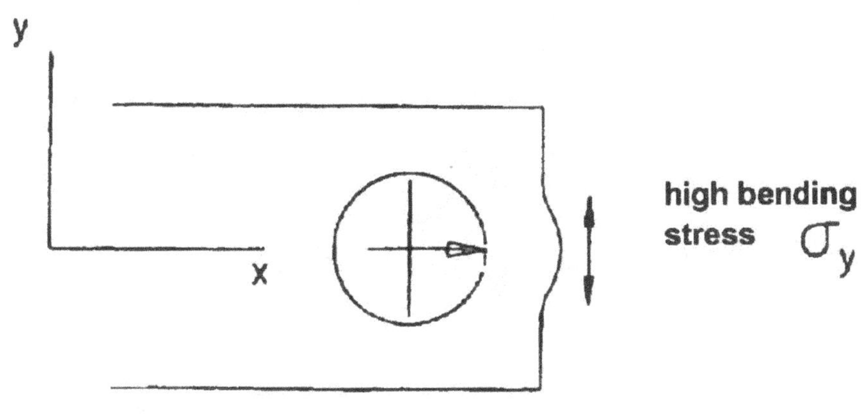

FIGURE 5.3.2.2(e) *Development of bending deflection in front of fastener for small e/D.*

In addition to the case of small edge distances, combined bearing and bypass loads can result in radial pressure distributions which deviate excessively from the "half=cosine" distribution. This can be understood in terms of the displacement behavior illustrated in Figure 5.3.2.2(f), for pure bypass loading in which there are two gaps between the plate and fastener centered about 0° and 180°, vs, the case of pure bearing load in which a single gap located between q = 90° and 270° occurs. For low bypass loads one would, therefore, expect a single region of contact centered about q = 0°, while for large bypass loads a split contact region would be expected. In terms of the notation defined in Figure 5.3.2.2(g), this type of behavior is predicted by stress analyses which correctly model the contact situation between the fastener and plate as illustrated in Figure 5.3.2.2(h). Note in Figure 5.3.2.2(g) that P_{TOT} is the total load at the left of end of the joint, which is the sum of P_F, the fastener load, and P_{BP}, the bypass load.

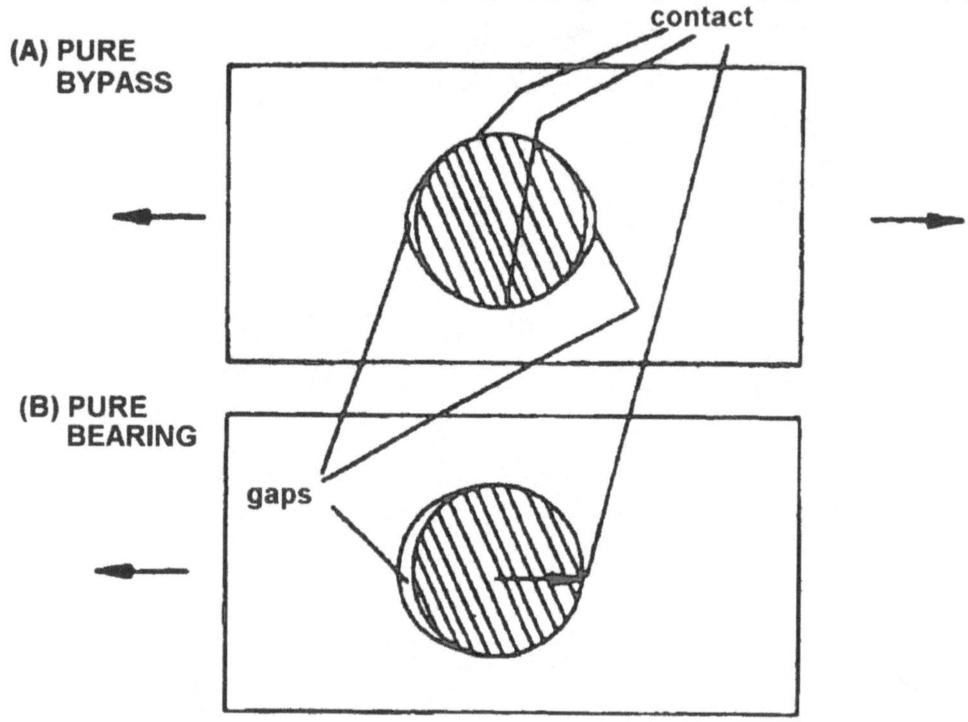

FIGURE 5.3.2.2(f) *Bolt shank/hole contact regions for pure bypass vs. pure bearing loads.*

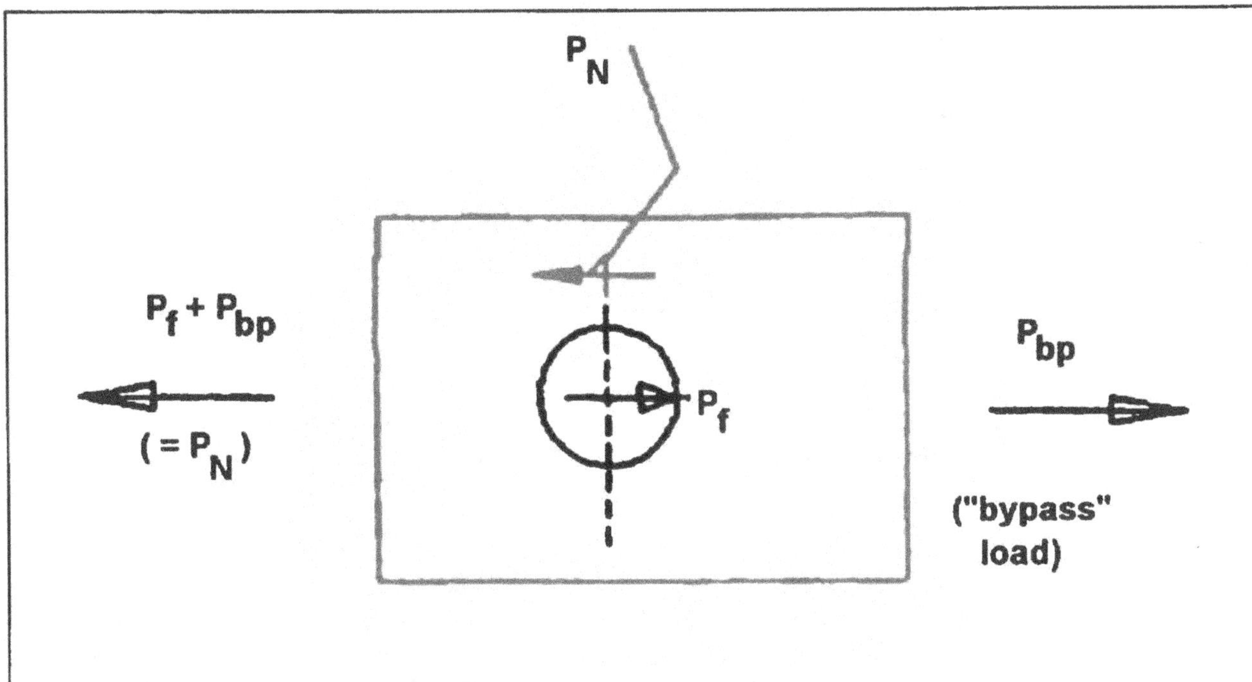

FIGURE 5.3.2.2(g) *Load definitions for combined bearing and bypass loads.*

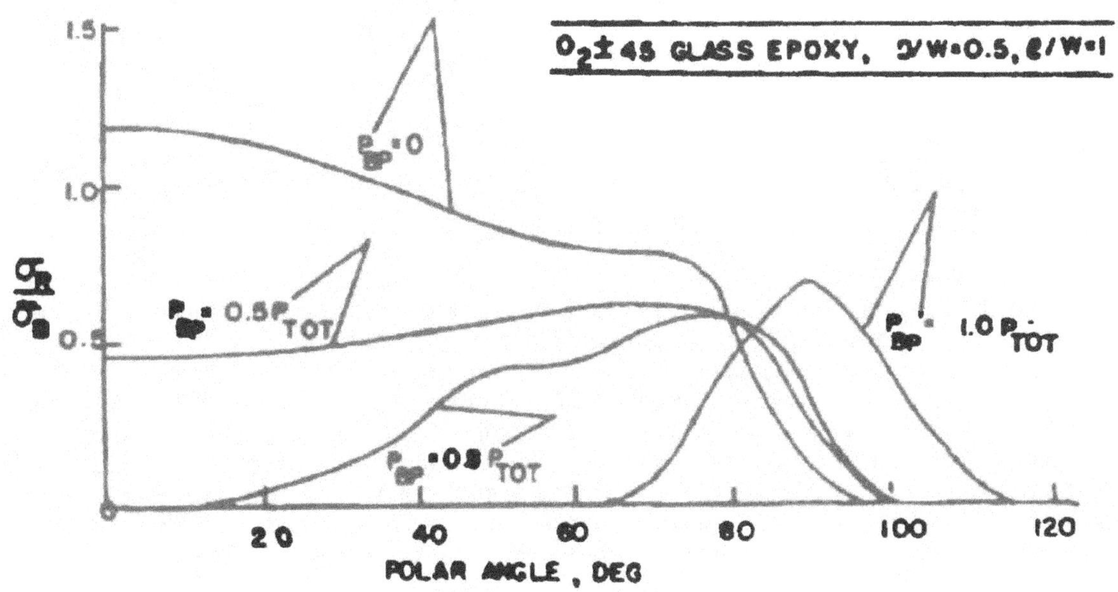

FIGURE 5.3.2.2(h) *Effect of bearing/bypass load ratio on radial pressure distribution (Reference 5.3.2.2(e)).*

Figure 5.3.2.2(i) illustrates how taking into account the effect of the radial displacements at the edge of the hole can influence predictions of the net section stress peaks. In this figure, predictions of K_g^{nl} (peak net section stress divided by gross stress) for the conventional superposition approach obtained by a linear combination of K_g^{nl} values for pure bearing load and pure bypass load (denoted "linear approximations" in Figure 5.3.2.2(i)), are compared with the corresponding results obtained when the contact problem is taken into account (open circles and squares). For the latter case, the curves are fairly flat over most of the range of load ratios, dropping rapidly near the high bypass end to a little above 3, the classical open hole value for isotropic plates having boundaries at infinity. Strength values for joints under combined bearing and bypass loading should follow similar trends with respect to the load ratio.

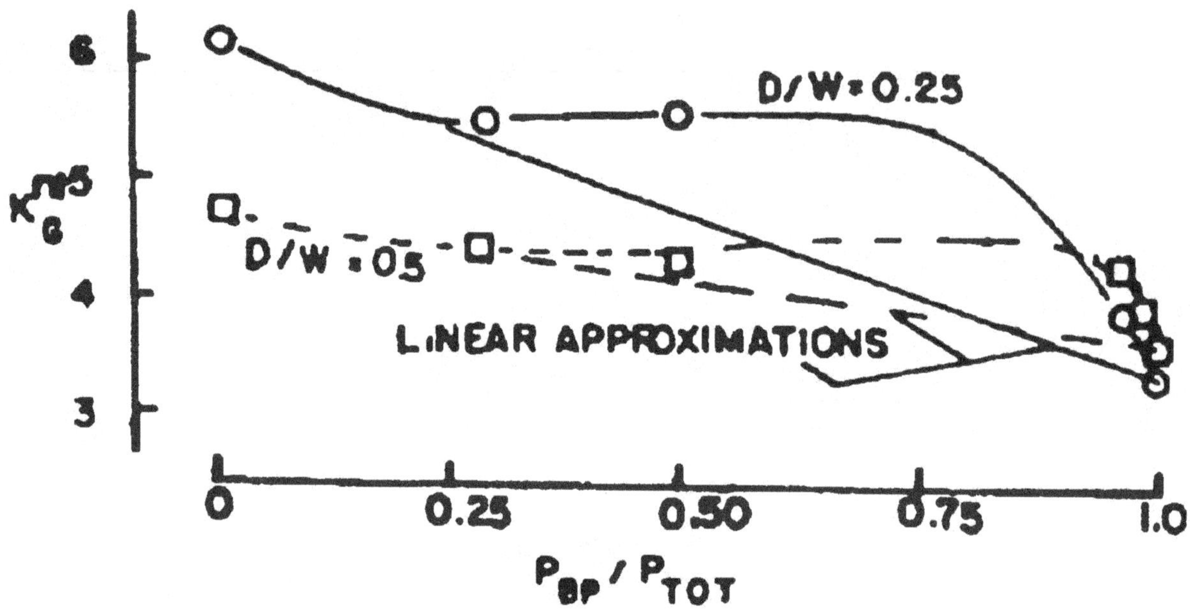

FIGURE 5.3.2.2(i) *Effect of bearing/bypass load ratio on peak net section stresses-- neat fit fastener (Reference 5.3.2.2(e)).*

The above results apply to cases of exact fastener fits. Additional complications occur with clearance fits corresponding to tolerances which are representative of available machining practice. Clearance fit cases have been analyzed extensively by Crews and Naik (Reference 5.3.2.2(f)) for clearances on the order of 0.0025 in. (0.04 mm) with fastener diameters of 0.25 in. (6.3 mm), i.e., clearances about 1% of the fastener diameter[1]. Significant changes in the radial pressure distribution occur with respect to the exact fit case. The angle subtended by the contact region becomes a function of load for this case, starting at zero for incipient loads and growing to only about 60° on either side of the axial direction for typical peak loads. The reduction of the angle of contact by the effects of clearance results in significant increases in the peak bearing stress. Again, the "half-cosine" load distribution can not be used to predict this type of behavior.

Crews and Naik also addressed the applicability of the superposition method for predicting failure under combined bearing and bypass loading, on the basis of their analytical results with the Nuismer Whitney

[1]Note that the SI equivalent dimensions provided throughout Section 5.3 are "soft" conversions, that is SI dimensions for fastener sizes are provided but sizes are not converted to SI standard sizes.

correction taken into account. They observed that the superposition approach gives adequate accuracy for predictions of the net-section tensile failures, although the predictions of radial pressure distributions are quite bad so that bearing failures cannot be treated by superposition.

The basic analytical steps described above have been implemented in several computer codes. Codes developed under government sponsorship include BJSFM (Reference 5.3.2.2(9)), SAMCJ (Reference 5.3.2.2(h)), BOLT (Reference 5.3.2.2(c) and 5.3.2.2(i)), SCAN (Reference 5.3.2.2(j)), and BREPAIR (Reference 5.3.2.2(k)). BREPAIR has been specialized for the case of bolted repairs for composites, and also computes the bolt loads from the fastener and plate flexibilities.

In principle, the analysis methods described should be able to account for the shear-out failure mode if the stress analysis method used includes the effects of multiple holes and plate edges. However, because of the variety of ply-level failure criteria used, and the details of the analysis implementation, it is recommended that additional test correlation be performed before applying these methods to cases involving small edge distances, or close hole spacing.

Furthermore, current analysis methods should not be relied upon to predict matrix dominated modes such as bearing failure. Generally, the analysis codes can be used to predict net-section failures, while bearing failure is checked by direct comparison of the average bearing stress (P/dt) to test data.

The actual bearing pressure due to a bolt varies considerably through the thickness of the laminate. For this reason, the test configuration must closely match the actual joint geometry in terms of laminate thickness, gaps and shims, and configuration (double versus single shear) and type of fastener. The bearing strength will depend on factors such as the countersink depth and angle, joint rotation under load, and the type of fastener head. The though-thickness distribution of bearing stresses can be estimated by treating the bolt as a beam, and the laminate as an elastic foundation (Reference 5.3.2.2(1)). These methods are suitable for estimating the changes in the bearing stress due to changes in gap distances or laminate thickness. They may also be useful for determining the moment and shear distribution in the bolt to predict fastener failure.

Clamp-up forces have been shown to have a significant effect on laminate failure, particularly under fatigue loading. Clamp-up can suppress delamination failure modes, and changes the fastener head restraint. This effect cannot be included in the two-dimensional analysis methods described above. Before taking advantage of the beneficial effects of clamp-up, long-term relaxation of the laminate stresses should be considered. Because of this effect, minimum clamp-up (if possible) should be used when conducting bolt bearing tests, i.e., finger tight or 10-20 in.-lb (1-2 N-m) torque up on a 1/4 in. (6.4 mm) diameter bolt. This may not be the normal torque installation of the fastener.

5.3.2.3 Failure criteria

The design of a mechanically fastened joint must assure against all possible failures of the joint. These are illustrated in Figure 5.3.2.3. Accepted design practice is to select edge distances, plate thicknesses, and fastener diameters so that of all the possible failure modes probable failures would be net section and bearing. There is no consensus whether the joint should fail in net section tension/compression or bearing. Reference 5.3.2.3(a) recommends that highly loaded structural joints be designed to fail in a bearing mode to avoid the catastrophic failures associated with net section failures. Although this is a commendable goal, particularly for single bolt joints, it is impractical in most cases as the increase in edge distances adds weight to the structure. For usual width to bolt diameter ratios of 6 both, net and bearing failures are possible, and the stress engineer is satisfied if he can show a positive margin against both failure modes. He does not try to get a higher margin for net failure than for bearing failure. Steering the joint design to have bearing failures by having large bearing allowables may result in in-service problems of bolt hole wear, fuel leakage, and fastener fatigue failures. Furthermore, net tension failure is unavoidable for multi-row joints.

In contrast to metals, load redistribution in a multi-fastened joint cannot be counted on and hence a single fastener failure in bearing constitutes failure of the joint. Failure criteria in bearing should be either bearing yield, defined either as the 0.02D or 0.04D based on actual bearing load displacement curves, Figure 7.2.2.4 of Volume 1, or B-basis ultimate load, whichever is lower-. The beneficial effects of clamp-up on bearing failure has to be evaluated in light of relaxation during service.

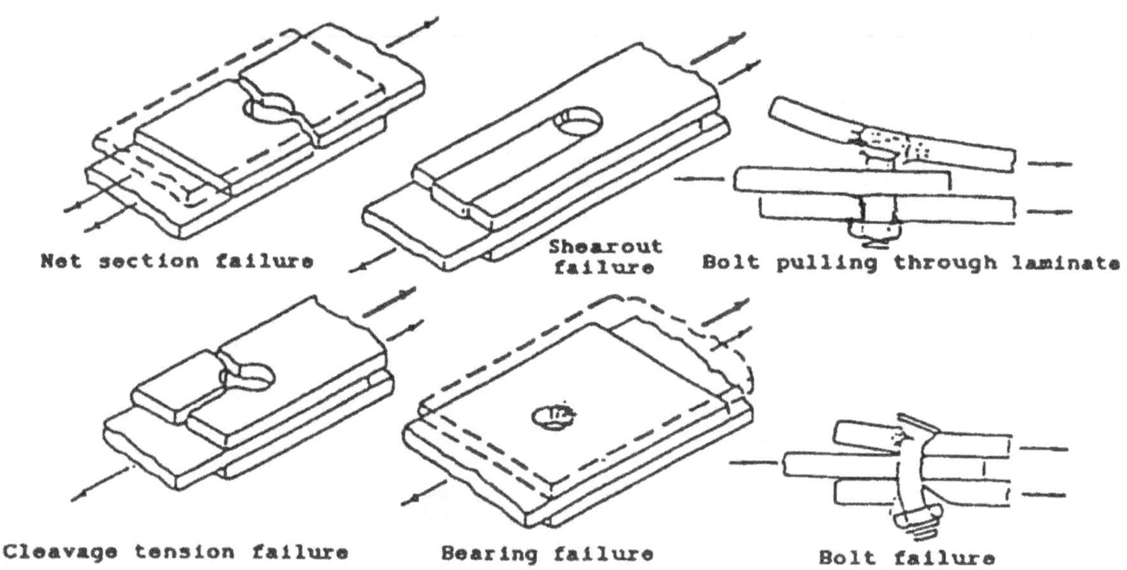

FIGURE 5.3.2.3(a) *Failure modes for mechanically fastened joints.*

Failure criteria for single fastener joint were discussed in Section 5.3.2.2. For complex loading or proximity to other fasteners, the failure location or mode identification may not be as shown in Figure 5.3.2.3 for unidirectional loading. For thick composites, recent work (Reference 5.3.2.3(b)) has shown that net section failures do not necessarily occur at 90° to the load direction but at some other locations around the hole.

5.3.3 Design considerations

5.3.3.1 Geometry

5.3.3.2 Lay-up and stacking sequence

5.3.3.3 fastener selection

The use of mechanical fasteners to join non-metallic composite structures is bound by certain constraints which do not exist in the design of metallic joints. In other words, special care must be taken to select fasteners that are appropriate with polymer composite structures. Because of these special requirements fastener manufacturers have developed fasteners especially for use with composites. These fasteners develop the full bearing capability of the composite (which, at least for carbon/epoxy, is equal or better than aluminum) without encountering local failure modes and are not susceptible to corrosion. Therefore, these fasteners or those having such properties, should be used. Nondiscriminant use of off-theshelf fasteners will lead to premature joint failures.

Design of mechanically fastened joints has always been guided by the principle that the material being joined should fail before the fastener, and this is the practice with composites. Although composites have high strength/stiffness to weight ratios with good fatigue resistance, it is a fact that today's composites must be treated very carefully when designing joints. The major structural limitation in this area is the insufficient through-thickness strength of the laminates. This has given rise to the term "pull-thru strength". It has become necessary to increase the bearing area of fastener heads (or tails) in order to reduce the axial stresses against the laminate when the fastener is loaded in tension.

Another area of concern is the bearing stress which a fastener applies to the edge of the hole in a composite laminate as its axis rotates due to secondary bending of the joint. This condition can impose a severe limitation on a joint with limited stiffness. Another problem is the composite's inability to support installation stresses of formed fasteners, such as solid rivets or blind fasteners with bulbed tails. In addition to surface damage, such as digging-in into composite, subsurface damage to the laminate may occur. For this reason, these fastener types are avoided in favor of two piece fasteners and blind fasteners which do not generate this type of loading during installation.

For the above reasons, tension head 100° countersunk fasteners rather than shear heads should be selected as the projected area of the tension head fastener is larger than that of a shear head fastener. The larger area improves pull-through and delamination resistance in composites, while reducing overturning forces from bolt bending. These fasteners are also recommended for double shear joints. Caution should be observed in the use of 130° countersunk head fasteners. Although this type of fastener increases the bearing area of the fastener and permits it to be used in thin laminates, pull-through strength and resistance to prying moments can be adversely affected.

The full bearing capability of composites can only be attained using fasteners with high fixity (good clamp-up). Fixity is a function of fastener stiffness, fastener fit, installation forces, torque and rotational resistance of the fastener head and collar or formed backside. However, because of relaxation with service usage, normal design/analysis practice uses data based on tests where the fasteners were installed finger tight or with light torque. As part of the allowables program, testing should also be done with fasteners installed per fastener supplier's recommended procedures.

Although close tolerance fit fasteners are desirable for use with composites, interference fit fasteners cannot be used due to potential delamination of plies at the fastener hole. There are exceptions to this rule. Some automatic high impact driving equipment which was used in production has been shown not to cause composite damage.

Presence of galvanic corrosion between metallic fasteners and non-metallic composite laminates has eliminated several commonly used alloys from consideration. Conventional plating materials are also not being used because of compatibility problems. The choice of fastener materials for composite joints has been limited to those alloys which do not produce galvanic reactions. The materials currently used in design include unplated alloys of titanium and certain corrosion resistant stainless steels (cres) with aluminum being eliminated. The choice is obviously governed by the makeup of the composite materials being joined, weight, cost, and operational environment. Aircraft practice has been to coat fasteners with anti-corrosion agent to further alleviate galvanic corrosion.

5.3.4 Fatigue

Fatigue performance of bolted composite joints is generally very good as compared to metal joints. Under maximum cyclic load level as high as 70% of the static strength, composite bolted joints have been observed to endure extremely long fatigue life and with minimal reduction in residual strength. The predominant damage mechanism under cyclic loads is usually bearing failure in form of hole elongation with net section failure for static residual test.

Even though the general trends of fatigue behavior of bolted composites has been well established, the influence of individual parameters on the fatigue performance needs to be investigated. For bolted composite joints, the parameters include material system, geometry, attachment details, loading mode and environment. Several government funded programs have been conducted to evaluate the influence of specific design on composite bolted joints. Typical examples are given in References 5.3.4(a) - 5.3.4(e). However, the large number of design variables makes it very difficult to develop an overall understanding of the specific influence of each of the primary design parameters. Based on the results of References 5.3.4(a) - 5.3.4(e), the following paragraphs summarize the significant effects of key design parameters on the fatigue performance of bolted composite joints. Because the parameters used in each reference are significantly different, direct comparison of the results is difficult. Only the trends of the data, based on coupon tests, are discussed.

5.3.4.1 Influence of loading mode

Under a constant amplitude fatigue situation, the most severe loading condition is fully reversed loading (R=-1). The results in Reference 5.3.4(a) indicate that fatigue failures will occur within 10^6 cycles if the maximum cyclic bearing stress is above 35% of the static bearing strength. However, the results of Reference 5.3.4(d) show that a 10^6 cycles fatigue threshold exceeds 67% of the static strength. Failure observed in the specimens exposed to fully reversed fatigue loads were induced by local bearing and excessive hole elongation. The hole elongation increases slowly for the major portion of the specimen's fatigue life, but increases rapidly near the end of the fatigue life. That is, once the bearing mode of failure is precipitated, hole elongation increases from a low value (1 to 2% of the original hole diameter) to a prohibitive value (>10%) within a few cycles. The fatigue threshold increases with decreasing R-ratio for tension-compression loading, and tension-tension loading is the least severe constant amplitude fatigue load.

Typical aircraft spectra loading were used in References 5.3.4(a), 5.3.4(c) and 5.3.4(d) to investigate the effects of variable amplitude cyclic loading on the fatigue performance of composite bolted joints. The results in Reference 5.3.4(a) show that the specimens survived two lifetimes of a typical vertical stabilizer spectrum loading without fatigue failure. The maximum spectrum load used in these tests ranges from 0.66 to 1.25 times of the static strength. Four loading spectra were tested in Reference 5.3.4(d) to investigate the influence of spectrum profile and load truncation levels. The results of these tests showed no fatigue failure and no distinguishable difference in the fatigue life for the spectrum loading investigated. The maximum spectrum stress was 78% of the static strength and the minimum stress at -49% of the static strength.

An extensive spectrum sensitivity database for bolted composite joints was generated in Reference 5.3.4(c). In this reference, the spectrum parameters investigated included load frequency, spectrum truncation, stress level, extended life, temperature and moisture, and specimen size. With approximately 600 specimens tested in the reference, there were no fatigue failures observed within the composite portion of the bolted joint specimens. This absence of composite fatigue failures confirmed that composite bolted joints are fairly insensitive to fatigue in tension loading at normal operating loads. These results also showed that composite bolted joints are insensitive to fatigue even in severe environments, such as real flight time loads and temperature, and 15 lifetimes of accelerated fatigue at 70% of the static strength in a 250°F (120°C) hot-wet condition. This does not mean that fastener failures have not occurred, sometimes precipitated by composite stiffness or fitup.

5.3.4.2 Influence of joint geometry

The influence of fastener diameter and fastener spacing on the fatigue performance of bolted composite joints is investigated in Reference 5.3.4(d). Three fastener diameters (0.25, 0.375 and 0.5 in. (6.4, 9.5, and 13 mm)) and three fastener spacing-to-diameter ratios (3.0, 4.0 and 6.0) are considered in the investigation. The results indicate that larger spacing to diameter ratio specimens have lower fatigue performance than specimens with lower ratios. The limited amount of data in the reference is not sufficient to draw a general conclusion. However, the results in Reference 5.3.4(d) are presented in terms of gross area stress, the lower fatigue performance of the wider specimens may be caused by the higher loads in the fastener and result in fastener or joint failure.

The fatigue performances of single lap joint and double lap joint are compared in Reference 5.3.4(a). Test results in the reference indicate that the threshold bearing stress value is relatively unaffected by the differences in the two joint configurations.

The effects of bolt bearing/by-pass stress interaction on the fatigue performance is also investigated in Reference 5.3.4(a). Joints with bolt-to-total load ratios of 0.0, 0.2, 0.33 and 1.0 are considered in the reference. The results of these tests show change in failure mode with bolt bearing/by-pass stress ratio. Net section failures were observed for specimens tested with a bolt bearing/by-pass ratio of 0.0 (or open hole). When 20% of the total load was introduced directly as a bearing load, half the specimens suffered a net section failure, and the other half suffered local bearing failures. For the test case where 33% of the total load was presented as the fastener bearing load, the observed failures were local bearing induced excessive hole elongation, similar to the results of full-bearing.

5.3.4.3 Influence of attachment details

The effects of attachment details on the fatigue performance of bolted composite joints are investigated in References 5.3.4(a) and 5.3.4(d). The influence of fastener fit is studied in Reference 5.3.4(d) by considering four levels of hole diameter for controlled over and under size, including slight interference. At applied cyclic load levels greater than 50% of static strength, no significant difference in fatigue performance for the different fastener fits was observed. The specimens were tested at a stress ratio of $R = -1.0$.

The effects of fastener torque on fatigue performance is studied in Reference 5.3.4(a). The results of these tests showed that there was no change in the failure mode and the fatigue performance improved with increased torque. The results also indicated that at low torque levels, hole elongation increased gradually with fatigue cycling and at high torque levels, the cyclic hole elongation rate was very abrupt.

The effect of countersink on joint performance was investigated in Reference 5.3.4(a). When countersunk (100° tension head) steel fasteners were used, approximately half of the tests resulted in fastener failure. The fasteners failed in a tensile mode near the head/shank boundary. Comparing these results with those with protruding head steel fasteners, the effect of the countersink is seen to be earlier elongation at a constant cyclic bearing stress amplitude. It is also seen that the fatigue threshold is lower when countersunk fasteners are used. When countersunk titanium fasteners were used instead of the steel fasteners, fastener failures occurred in every specimen.

5.3.4.4 Influence of laminate lay-up

The effect of laminate lay-up on the joint performance was investigated in Reference 5.3.4(a) by considering three laminate lay-ups--(50/40/10), (70/20/10) and (30/60/10). The results of this investigation indicated that despite the difference in static bearing strength of these laminates, the 10**6 cycle fatigue threshold is approximately equal.

5.3.4.5 Influence of environment

The effects of temperature and moisture are experimentally evaluated in References 5.3.4(a) and 5.3.4(c). The results of these studies indicate that the fatigue threshold may be lower under the hot/wet (218°F/wet (103°C/wet)) condition.

5.3.4.6 Influence of specimen thickness

The effect of laminate thickness on fatigue performance is examined in Reference 5.3.4(d) and the effect of specimen size is evaluated in Reference 5.3.4(c). The results of Reference 4 show that within the thickness of 0.25 to 0.50 inch (6.4 to 13 mm) the fatigue threshold is not significantly affected. In comparing the fatigue performance of small and large scale joints, Reference 5.3.4(c) showed that there is no significant scale up effect.

5.3.4.7 Residual strength

The extensive amount of residual strength data generated in Reference 5.3.4(c) suggested that bolted composite joints have an excellent capability of retaining static strength. This trend is also supported by the results of other investigations. The largest percentage of fatigue strength reduction observed in Reference 5.3.4(c), when compared with static strength, was 8%. There were no real time or environmental effects on residual strength reduction that were greater than this. Therefore, a design static tension strength reduction factor is appropriate to account for tension fatigue effects on bolted composite joints under practical service environments.

5.3.5 Test verification

In addition to joint coupon testing which is performed to obtain baseline data, element testing should be performed to verify joint analysis, failure mode, and location. This is particularly important for primary connections and where the load transfer is complex. The purpose of testing is to obtain assurance that the joint behaves in the predicted manner or where analysis is inadequate.

The structural joints to be tested are usually identified early in the design process and are part of the certification process, if the building block approach is used, see Section 2.1.1, Volume 1. The test specimens are classified by levels of complexity as elements, subcomponents, or components. Some examples of types of joints that are tested are shown for a fighter wing structure in Figures 5.3.5(a) and 5.2.5(b).

The bolted joint element or subcomponent tests are usually performed at ambient conditions with sufficient instrumentation to fully characterize load transfer details: direction and amplitude of bolt and by-pass loads. Tests at other than ambient conditions are necessary in cases when the low or elevated temperatures with associated moisture contents substantially change the load distributions.

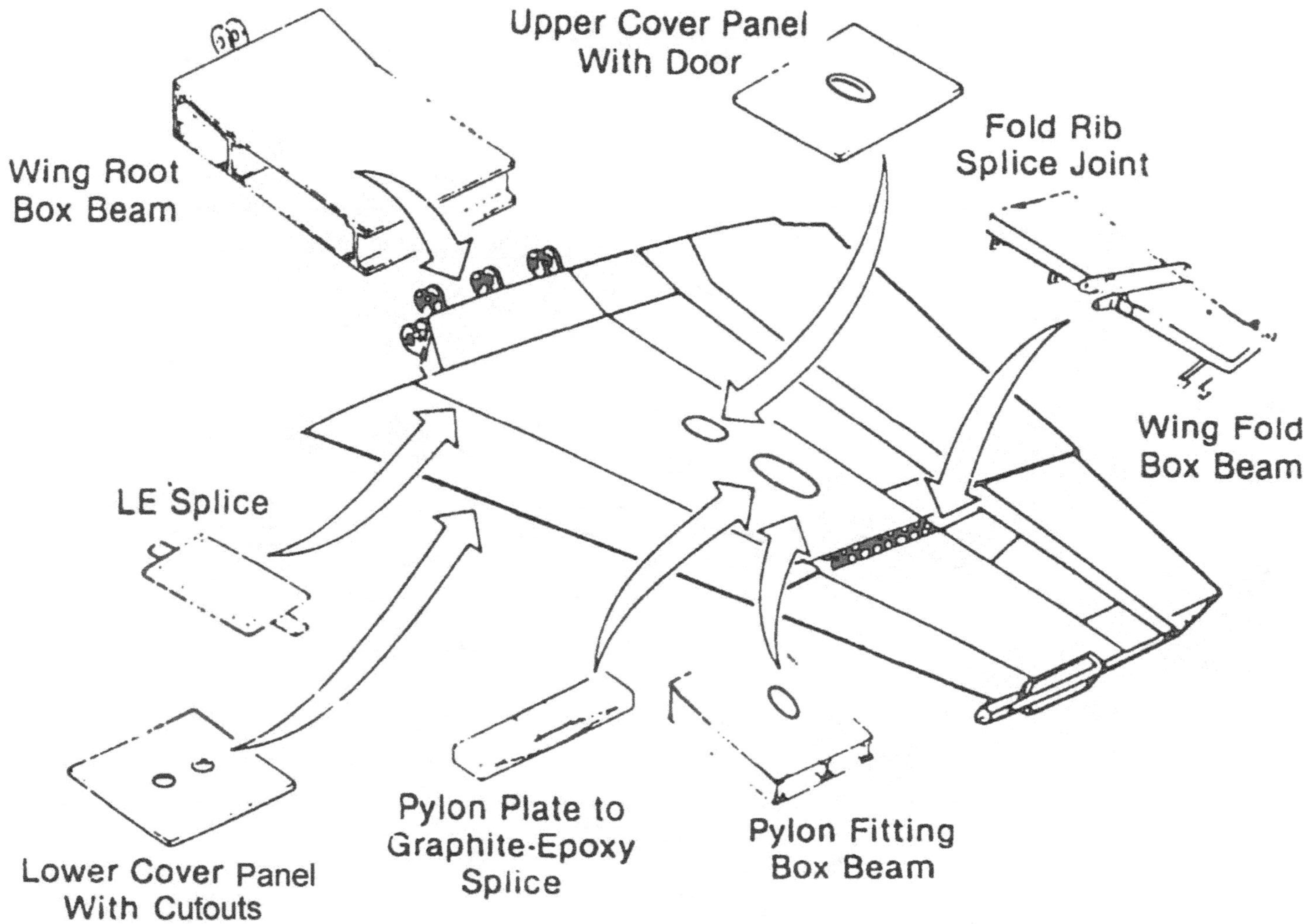

FIGURE 5.3.5(a) *Wing subcomponent tests.*

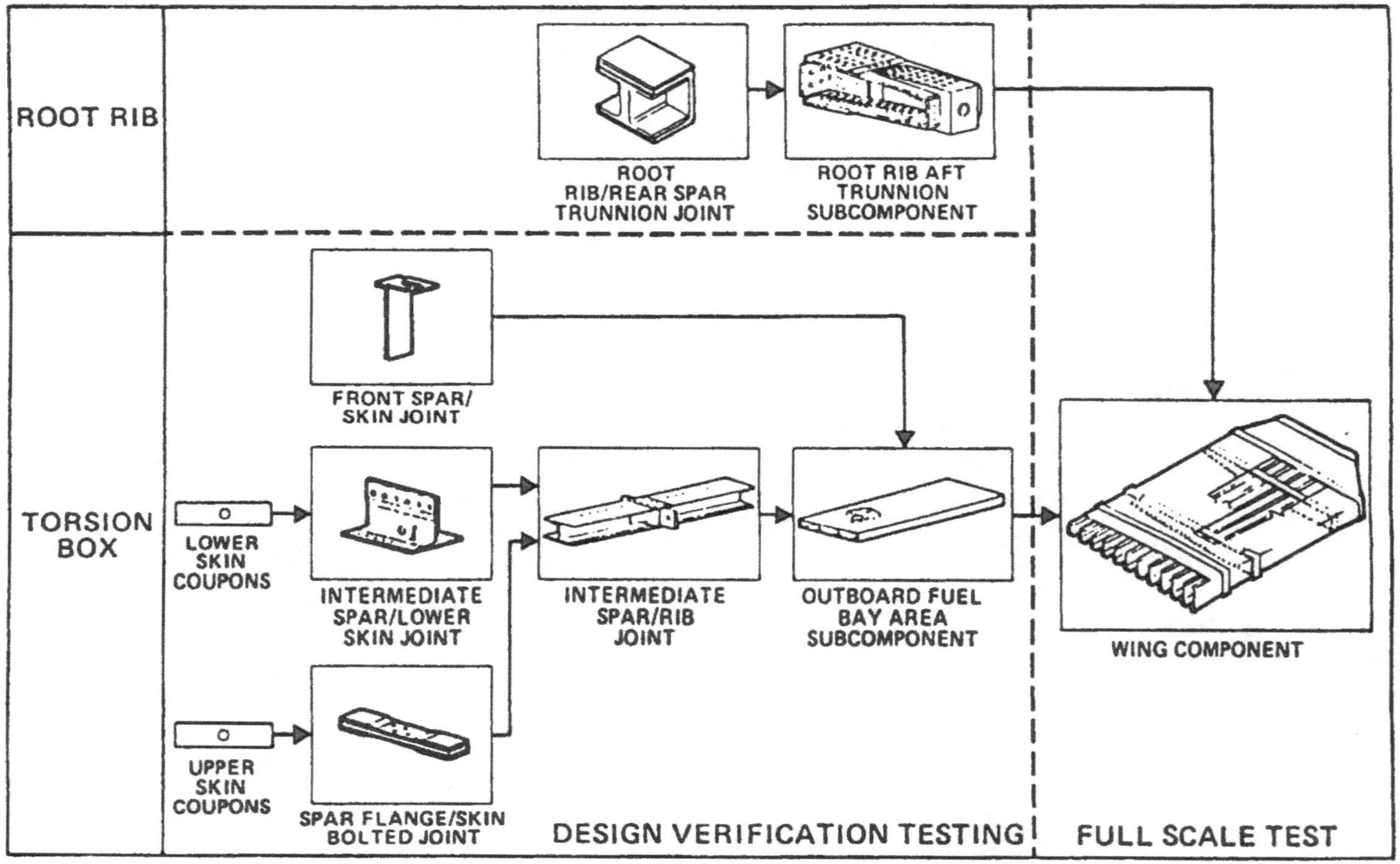

FIGURE 5.3.5(b) *Building block approach for the wing structure in the composite wing/fuselage program (Reference 5.3.5).*

REFERENCES

5.2.1 (a) Volkersen, O., "Die Nietkraftverteilung in Zugbeanspruchten Nietverbindungen mit Konstanten Laschenquerschn;tten," *Lufttahrtforschung*, Vol 15, 1938, pp. 4-47.

5.2.1(b) Goland, M. and Reissner, E., "Stresses In Cemented Joints," *Journal of Applied Mechanics*, Vol 11, 1944, pp. A 17-A27.

5.2.1(c) Kutscha, D. and Hofer, K., "Feasibility of Joining Advanced Composite Flight Vehicle Structures," Air Force Materials Laboratory Report AFM L-TR-68-391, 1968.

5.2.1 (d) Dickson, J.N., Hsu, T.N., and McSkinney, J.N., "Development of an Understanding of the Fatigue Phenomena of Bonded and Bolted Joints In Advanced Filamentary Composite Materials, Volume 1, Analysis Methods," Lockheed Georgia Aircraft Company, USAF Contract Report AFFDL-TR 72-64, Volume 1, June 1972.

5.2.1 (e) Grimes, G.C., Wah, T., et al, "The Development of Non-Linear Analysis Methods for Bonded Joints in Advanced Filamentary Composite Structures," Southwest Research Institute, USAF Contract Report AFML-TR-72-97, September 1972.

5.2.1 (f) Renton, W.J., "The Analysis and Design of Composite Materials Bonded Joints Under Static and Fatigue Loadings," PhD Thesis, University of Delaware, 1973.

5.2.1 (g) Renton, W.J. and Vinson, J.R., "The Analysis and Design of Composite Materials Bonded Joints under Static and Fatigue Loadings," Air Force Office of Scientific Research Report TR-73-1627, 1973

5.2.1 (h) Oplinger, D.W., "Stress Analysis of Composite Joints," *Proceedings of 4th Army Materials Technology Conference*, Brook Hill Publishing Co., Newton MA, 1975, pp 405-51.

5.2.1 (i) Hart-Smith, L.J., AFFDL-TR-72-130, pp 813-856.

5.2.1 (j) Hart-Smith, L.J., "Adhesive Bonded Double Lap Joints," NASA Langley Contractor Report NASA CR-112235 1973.

5.2.1 (k) Hart-Smith, L.J., "Adhesive Bonded Single Lap Joints," NASA Langley Contractor Report NASA CR-112236, 1973.

5.2.1 (l) Hart-Smith, L.J., "Adhesive Bonded Scarf and Stepped-Lap Joints," NASA Langley Contractor Report NASA CR-112237, 1973.

5.2.1(m) Hart-Smith, L.J., "Analysis and Design of Advanced Composite Bonded Joints," NASA Langley Contractor Report NASA CR-2218, 1973.

5.2.1(n) Hart-Smith, L.J., "Advances in the Analysis and Design of Adhesive-Bonded Joints in Composite Aerospace Structures," *SAMPE Process Engineering Series*, Vol 19, SAMPE, Azusa, 1974, pp 722-737.

5.2.1 (o) *Primary Adhesively Bonded Structure (PABST) Technology*, Air Force Contract F33615-75-C -3016, 1975.

5.2.1 (p) Thrall, E.W., "Primary Adhesively Bonded Structure Technology (PABST) Phase 1b: Preliminary Design," Air Force Flight Dynamics Laboratory Report AFFDL-TR-76-141, 1976.

5.2.1 (q) Shannon, R.W., et al, "Primary Adhesively Bonded Structure Technology (PABST) General Material Property Data," Air Force Flight Dynamics Laboratory Report AFFDL-TR-77-101 1977.

5.2.1 (r) Land, K.L., Lennert, F.B., et al, "Primary Adhesively Bonded Structure Technology (PABST): Tooling, Fabrication and Quality Assurance Report," USAF Technical Report AFFDL-TR-79 -3154, October 1979.

5.2.1(s) Hart-Smith, L.J., "Adhesive Bond Stresses and Strains at discontinuities and Cracks in Bonded Structures," *Transactions of the ASME, Journal of Engineering Materials and Technology*, Vol 100, January 1978, pp. 15-24.

5.2.1 (t) Hart-Smith, L.J., "Differences Between Adhesive Behavior in Test coupons and Structural Joints," Douglas Aircraft Company paper 7066, Presented to ASTM Adhesives Committee D-14 Meeting, Phoenix, Arizona, 1981.

5.2.1(u) Hart-Smith, L.J., "Design Methodology for Bonded-Bolted Composite Joints," Douglas Aircraft Company, USAF Contract Report AFWAL-TR-81-3154, Vol I and II, February 1982.

5.2.1 (v) Thrall, E.W., Jr., "Failures in Adhesively Bonded Structures," AGARD-NATO Lecture Series No. 102, "Bonded Joints and Preparation for Bonding," Oslo, Norway, and The Hague, Netherlands, April 1979 and Dayton, Ohio, October 1979.

5.2.1(w) Hart-Smith, L.J., "Further Developments in the Design and Analysis of Adhesive-Bonded Structural Joints," Douglas Aircraft Co. Paper No. 6922, presented at the ASTM Symposium on Joining of Composite Materials, Minneapolis MN April 1980

5.2.1 (x) Hart-Smith, L.J., "Adhesive Bonding of Aircraft Primary Structures," Douglas Aircraft Company Paper 6979, Presented to SAE Aerospace Congress and Exposition, Los Angeles, California, October 1980.

5.2.1(y) Hart-Smith, L.J., "Stress Analysis: A Continuum Analysis Approach" in *Developments in Adhesives - 2*, ed. A. J. Kinloch, Applied Science Publishers, England, 1981, pp. 1-44.

5.2.1 (z) Hart-Smith, L.J., "Effects of Adhesive Layer Edge Thickness on Strength of Adhesive-Bonded Joints" Quarterly Progress Report No. 3, Air Force Contract F33615-80-C-5092, 1981.

5.2.1(aa) Hart-Smith, L.J., "Effects of Flaws and Porosity on Strength of Adhesive-Bonded Joints" Quarterly Progress Report No. 5, Air Force Contract F33615-80-C-5092, 1981.

5.2.1(ab) Hart-Smith, L.J. and Bunin, B.L., "Selection of Taper Angles for Doublers, Splices and Thickness Transition in Fibrous Composite Structures," *Proceedings of 6th Conference On Fibrous Composites in Structural Design*, Army Materials and Mechanics Research Center Manuscript Report AMMRC MS 83-8, 1983.

5.2.1 (ac) Nelson, W.D., Bunin, B.L., and Hart-Smith, L.J., "Critical Joints in Large Composite Aircraft Structure," *Proceedings of 6th Conference On Fibrous Composites in Structural Design*, Army Materials and Mechanics Research Center Manuscript Report AMMRC MS 83-8, 1983.

5.2.1(ad) Oplinger, D.W., "A Layered Beam Theory for Single Lap Joints," U.S. Army Materials Technology Laboratory Report MTL TR 91 -23, 1991.

5.2.1(ae) Oplinger, D.W., "Effects of Adherend Deflections on Single Lap Joints," *Int. J. Solids Structures*, Vol 31, No. 18, 1994, pp. 2565-2587.

5.2.2.1(a) Hart-Smith, L.J., "Adhesively Bonded Joints in Fibrous Composite Structures," Douglas Aircraft Paper 7740; presented to the International Symposium on Joining and Repair of Fibre-Reinforced Plastics, Imperial College, London, 1986.

5.2.2.1(b) Hart-Smith, L.J., "Induced Peel Stresses in Adhesive-Bonded Joints," Douglas Aircraft Company, Technical Report MDC-J9422A, August 1982.

5.2.2.6(a) Hart-Smith, L.J., Brown, D. and Wong, S., "Surface Preparations for Ensuring that the Glue will Stick in Bonded Composite Structures," *10th DoD/NASA/FAA Conference on Fibrous Composites in Structural Design*, Hilton Head Is, SC, 1993.

5.2.2.6(b) Hart-Smith, L.J., Ochsner, W., and Radecky, R.L., "Surface Preparation of Fibrous Composites for Adhesive Bonding or Painting," *Douglas Service Magazine*, First quarter 1984, pp 1 2-22.

5.2.2.6(c) Hart-Smith, L.J., Ochsner, W., and Radecky, R.L., "Surface Preparation of Fibrous for Adhesive Bonding or Painting," *Canadair Service News*, Summer 1985, 1985, pp. 2-8.

5.2.2.7(a) Frazier, T.B. and Lajoie, A.D., "Durability of Adhesive Joints," Air Force Materials Laboratory Report AFML TR-74-26, Bell Helicopter Company, 1974.

5.2.2.7(b) Becker, E.B., et al, "Viscoelastic Stress Analysis Including Moisture Diffusion for Adhesively Bonded Joints," Air Force Materials Laboratory Report AFWAL-TR-84-4057, 1984.

5.2.2.7(c) Jurf, R. and Vinson, J., "Effects of Moisture on the Static and Viscoelastic Shear Properties of Adhesive Joints," Dept. of Mechanical and Aerospace Engineering Report MAE TR 257, University of Delaware, 1984.

5.2.2.7(d) Mostovoy, S., Ripling, E.J., and Bersch, C.F., "Fracture Toughness of Adhesive Joints," *J. Adhesion*, Vol 3, 1971, pp. 125-44.

5.2.2.7(e) DeVries, K.L., Williams, M.L., and Chang, M.D., "Adhesive Fracture of a Lap Shear Joint," Experimental Mechanics, Vol 14, 1966, pp 89-97.

5.2.2.7(f) Trantina, G.G. "Fracture Mechanics Approach to Adhesive Joints," University of Illinois Dept. of Theoretical and Applied Mechanics Report T&AM 350, Contract N00019-71-0323, 1971.

5.2.2.7(g) Trantina, G.G., "Combined Mode Crack Extension in Adhesive Joints," University of Illinois Dept. of Theoretical and Applied Mechanics Report T&AM 350, Contract N00019-71-C0323, 1971.

5.2.2.7(h) Keer, L.M., "Stress Analysis of Bond Layers," *Trans. ASME J. Appl. Mech.*, Vol 41, 1974, pp 679-83.

5.2.2.7(i) Knauss, J.F., "Fatigue Life Prediction of Bonded Primary Joints," NASA Contractor Report NASA-CR-159049, 1979.

5.2.2.7(j) Wang, S.S. and Yau, J.F., "Analysis of Interface Cracks in Adhesively Bonded Lap Shear Joints," NASA Contractor Report NASA-CR-165438, 1981.

5.2.2.7(k) Johnson, W.S. and Mall, S., "A Fracture Mechanics Approach for Designing Adhesively Bonded Joints."

5.2.3.4 Oplinger, D.W. "Effects of Mesh Refinement on Finite Element Analysis of Bonded Joints," U.S. Army Research Laboratory Study (Unpublished) 1983.

5.3.1 Whitman, B., Shyprykevich, P., and Whiteside, J.B., "Design of the B-1 Composite Horizontal Stabilizer Root Joint," *Third NASA/USAF Conference on Fibrous Composites in Flight Vehicles Design*, Williamsburg, VA, November 4-6 1976.

5.3.2.1(a) Ramkumar, R.L., Saether, E.S., Appa, K., "Strength Analysis of Laminated and Metallic Plates Bolted Together by Many Fasteners," AFWAL-TR-86-3034, July, 1986.

5.3.2.1(b) Xiong, Y. and Poon, C., "A Design Model for Composite Joints with Multiple Fasteners," National Research Council, Canada, IAR-AN-80, August 1994.

5.3.2.1(c) Griffin, O.H., et. al., "Analysis of Multifastener Composite Joints," *Journal of Spacecraft and Rockets*, Vol 31, No. 2, March-April 1994.

5.3.2.1(dl ACEE Composite Structures Technology, *Papers by Douglas Aircraft Company*, ed. M. Klotzsche, NASA-CR-172359, August 1984.

5.3.2.2(a) Whitney, J.M. and Nuismer, R.J., "Stress Fracture Criteria for Laminated Composites Containing Stress Concentrations," *J. Composite Materials*, Vol 8, July, 1974, pp. 235-265.

5.3.2.2(b) Hart-Smith, J., "Mechanically-Fastened Joints for Advanced Composites Phenomenological Considerations and Simple Analysis," *Fibrous Composites in Structural Design*, ed. Edward M. Lenoe, Donald W. Oplinger, John J. Burke, Plenum Press, 1980.

5.3.2.2(c) Chang, F., Scott, R.A., Springer, G.S., "Strength of Mechanically Fastened Composite Joints," Air Force Wright Aeronautical Laboratories Technical Report AFWAL-TR-82-4095.

5.3.2.2(d) Lekhnitskii, S.G., *Anisotropic Plates*, Gordon and Breach Science Publishers, New York, 1968.

5.3.2.2(e) Oplinger, D.W., "On the Structural Behavior of Mechanically Fastened Joints" in *Fibrous Composites in Structural Design*, ed. Edward M. Lenoe, Donald W. Oplinger, John J. Burke, Plenum Press, 1980.

5.3.2.2(f) Crews, J.H., and Naik, R.A., "Combined Bearing and Bypass Loading on a Graphite/Epoxy Laminate," *Composite Structures*, Vol 6, 1968, pp. 21-40.

5.3.2.2(g) Garbo, S.P. and Ogonowski, J.M., "Effect of Variances and Manufacturing Tolerances on the Design Strength and Life of Mechanically Fastened Composite Joints, Volume 3 - Bolted Joint Stress Field Model (BJSFM) Computer Program User's Manual," Air Force Wright Aeronautical Laboratories Technical Report AFWAL-TR-81-3041, April 1981.

5.3.2.2(h) Ramkumar, R.L., Saether, E.S., and Appa, K., "Strength Analysis of Laminated and Metallic Plates Bolted Together by Many Fasteners," Air Force Wright Aeronautical Laboratories Technical Report AFWAL-TR-86-3034, July 1986.

5.3.2.2(i) Chang, F., Scott, R.A., and Springer, G.S., "Strength of Bolted Joints in Laminated Composites," Air Force Wright Aeronautical Laboratories Technical Report AFWAL-TR-84- 4029.

5.3.2.2(j) Hoehn, G., "Enhanced Analysis/Design Methodology Development for High Load Joints and Attachments for Composite Structures," Naval Air Development Center Technical Report.

5.3.2.2(k) Bohlmann, R.E., Renieri, G.D., Horton, D.K., "Bolted Repair Analysis Methodology," Naval Air Development Center Technical Report NADC-81063-60, Dec. 1982.

5.3.2.2(l) Harris, H.G., Ojalvo, I. U., and Hooson, R.E., "Stress and Deflection Analysis of Mechanically Fastened Joints," Air Force Flight Dynamics Laboratory Technical Report AFFDL-TR-70-49, May 1970

5.3.2.3(a) Ramkumar, R.L., Saether, E.S., Cheng, D., "Design Guide for Bolted Joints in Composite Structures," Air Force Wright Aeronautical Report AFWAL-TR-86-3035, March 1986.

5.3.2.3(b) Cohen, D., Hyer, M. W., Shuart, M. J., Griffin, O. H., Prasad, C., Yalamanchili, S. R., "Failure Criterion for Thick Multifastener Graphite-Epoxy Composite Joints," Journal of Composites Technology & Research, JCTRER, Vol . 17, No. 3, July 1995, pp. 237-248.

5.3.4(a) Ramkumar, R.L., and Tossavainen, E.W., "Bolted Joints in Composite Structures: Design, Analysis and Verification, Task I Test Results--Single Fastener Joints," AFWAL-TR-84-3047, August 1984.

5.3.4(b) Garbo, S.P., and Ogonowski, J.M., "Effects of Variances and Manufacturing Tolerances on the Design Strength and Life of Mechanically Fastened Composite Joints," Vol 1, 2 and 3, AFWAL-TR-81-3041, April 1981.

5.3.4(c) Jeans, L.L., Grimes, G.C., and Kan, H.P., "Fatigue Spectrum Sensitivity Study for Advanced Composite Materials, Volume I - Technical Summary," AFWAL-TR-80-3130, Vol 1, December 1980.

5.3.4(d) Walter, R.W., and Tuttle, M.M., "Investigation of Static and Cyclic Failure Mechanisms for GR/EP Laminates," *Proceedings of the Ninth DoD/NASA/FAA Conference on Fibrous Composites in Structural Design*, DOT/FAA/CT-92-25, September 1992, p. I-167.

5.3.4(e) Walter, R.W., and Porter, T.R., "Impact of Design Parameters on Static, Fatigue and Residual Strength of GR/EP Bolted Joints," *Proceedings of the Tenth DoD/NASA/FAA Conference on fibrous Composites in Structural Design*, NAWCADWAR-94096-60, p.111-75 April 1994.

5.3.5 Whitehead, R.S., et al., "Composite Wing/Fuselage Program," AFWAL-TR-883098, Vol1-4., February, 1989.

CHAPTER 6 STRUCTURAL RELIABILITY

6.1 INTRODUCTION

Reliability is commonly defined (References 6.1(a) and (b)) as "the probability of a device performing its purpose adequately for the period of time intended under the operating conditions encountered". There are four elements to the definition that must be considered. First, *probability* refers to the *likelihood* that a device or structural component will work properly. These terms imply acceptance of some degree of uncertainty. The second element refers to adequate performance. In order to determine whether a component has performed adequately, a standard is needed to define what is meant by adequate performance. The third element is the intended period of time. This is the mission endurance or lifetime of the structure under consideration. The final element of the definition is the operating conditions. Environmental conditions play a large role in reliability of composite materials, particularly polymer matrix composites. Simply stated, structural reliability is a yardstick of the capability of a structure to operate without failure when put into service. In the broadest sense, structural reliability includes events that are safety and non-safety related.

Until recently, structural reliability was not routinely analyzed or quantified in the design process. Reliability was accounted for tacitly by the factor-of-safety approach to design. Also guidelines and lessons learned helped to improve reliability. The structural designer/analyst does not perform a formal risk analysis on newly designed structure. This task is performed by reliability specialists who employ methodologies that are empirically based. The reliability assessment is usually conducted after a drawing or concept is produced and bears little relationship to the structural margin-of-safety.

As implied in the definition, structural failure and, hence, reliability, is influenced by many factors. In its simplest form, the measure of reliability is made by comparing a component's stress to its strength. Failure occurs when the stress exceeds the strength. The larger this gap, the greater the reliability and the heavier the structure. Conversely, the smaller the gap, the lower the r, but the lighter the structure. The gap between stress and strength, enforced by the factor-of-safety, generally produces adequate although unmeasured reliability.

The complications that mask the ability to quantify reliability reside in the stochastic nature of design inputs. The calculations are relatively easy; statistical characterizations of the strength and stress distributions are compared mathematically and a probability of failure calculated. Definition of these distributions however, can be an imposing, if not impossible, task. Each is influenced by many considerations with relatively unknown effects.

The primary purpose for establishing a factor-of-safety for design is to ensure safety. Until recently, no objective analysis has gone into the choice for factor-of-safety. Consequently, no evaluations are performed on the factor-of-safety as new materials or technologies are developed. As suggested by methodologies developed in Reference 6.1(c), these evaluations can now be performed. This fact suggests that future design and design processes might benefit greatly by focusing on reliability targets rather than factors-of-safety. This may be particularly true for composite materials.

The following sections discuss some of the important factors that affect composite structure reliability.

6.2 FACTORS AFFECTING STRUCTURAL RELIABILITY

6.2.1 Static strength

An aircraft structure's capability to sustain operational flight loads is commonly assessed by comparing material performance parameters to limit or ultimate loads. Limit loads are generally defined as the maximum load expected during the life of the aircraft. Ultimate loads are obtained by multiplying limit loads by the factor-of-safety. Limit loads are derived by considering the extremes of flight envelopes, gross weight, load factors, environments, and pilot inputs. In some cases, the likelihood of encountering limit load is very remote. The 1.5 factor-of-safety used to obtain ultimate design loads from applied loads has been widely accepted by generations of engineers, mostly without questioning the origin of the factor. Reference 6.2.1 provides an excellent historical review of the evolution of the 1.5 factor-of-safety in the United States.

From the beginning of flight, occupant safety has been a primary concern in designing manned vehicles and the "factor-of-safety" has been a prominent design criteria. Like many design requirements, the implementation of the 1.5 factor-of-safety evolved over a period of time and was influenced by many concerns.

Design criteria require structures to withstand ultimate loads without failure and limit loads with no permanent deformation. This has led to the impression that the 1.5 factor-of-safety was due to the performance of metals, 2024 in particular. At the time the 1.5 factor-of-safety was established, 2024 aluminum had a ratio of ultimate to yield stress of approximately 1.5. However, in the early 1930'2 when the 1.5 factor-of-safety was formally established by the Air Corps, material properties were not considered. Mr. A. Epstein, who worked for the United States Army Air Corps Material Center from 1929 to 1940, prepared the original Air Corps Structures Specification X-1803 in 1936. Mr. Epstein noted (Reference 6.2.1) that "the factor-of-safety of 1.5 has withstood many moves to alter it, but there was a period in 1939 when the Chief of the Structures Branch of Engineering Division at Wright Field thought seriously of reducing the value of the factor. Newer aluminum alloys were becoming available with higher ratios of yield to ultimate strength and he interpreted the factor as the ratio of ultimate to yield. However, no action was taken when the following explanation was offered: 'The factor-of-safety is not a ratio of ultimate to yield strength, but is tied in with the many uncertainties in airplane design, such as fatigue, inaccuracies in stress analysis, and variations of material gages from nominal values. It might also be considered to provide an additional margin of safety for an airplane subjected to shellfire.'" Thus, while the factor-of-safety does much to promote reliability, it was defined independently of any specific reliability goal.

Generally speaking, composite structures are sized by comparing ultimate internal stresses to statistically reduced material parameters (e.g., B-basis strengths). The internal stresses are a result of applied design ultimate loads (1.5 x DLL). In general the deterministic approach produces adequate reliability, but not necessarily the same as metallic structure. This is because composite materials exhibit different statistical distribution and variation from metals (see Figure 6.2.1). The result is that even though materials may have equivalent B-basis strengths, their reliabilities may be quite different. Reliability-based design procedures may be necessary to account for this difference (see Reference 6.1(c)).

6.2.2 Environmental effects

Composite material components are subjected to a wide range of environments. The operating conditions in which the aircraft must perform are not well characterized. Environmental factors of major importance include a combination of humidity and temperature. Many studies have been conducted to investigate moisture absorption as well as the reduction of mechanical properties due to temperature and moisture exposure. The current approach used to account for environmental factors is to define exposures that are extremes and selectively evaluate by test the effects on material properties. These extremes are then considered to be invariant during the lifetime of the structure. Strength values are reduced to coincide with the environmental extremes.

6.2.3 Fatigue

Composite materials exhibit higher fatigue threshold stresses than metals. Once this threshold is exceeded, composites show more scatter in fatigue than metals and might tend toward lower reliability performance if the composite structures were stressed that highly. Because of this high threshold stress, fatigue is not the limiting factor in the design of composite structures. Design criteria such as damage tolerance limit the stress levels in composite structures to such low values that fatigue does not generally represent a design constraint. However, this is not necessarily true for high-cycle fatigue (e.g., $n > 10^7$) dynamic system components in rotorcraft. (For more information on fatigue or durability of composite structures, see Volume 3, Sections 4.10 and 4.11.2).

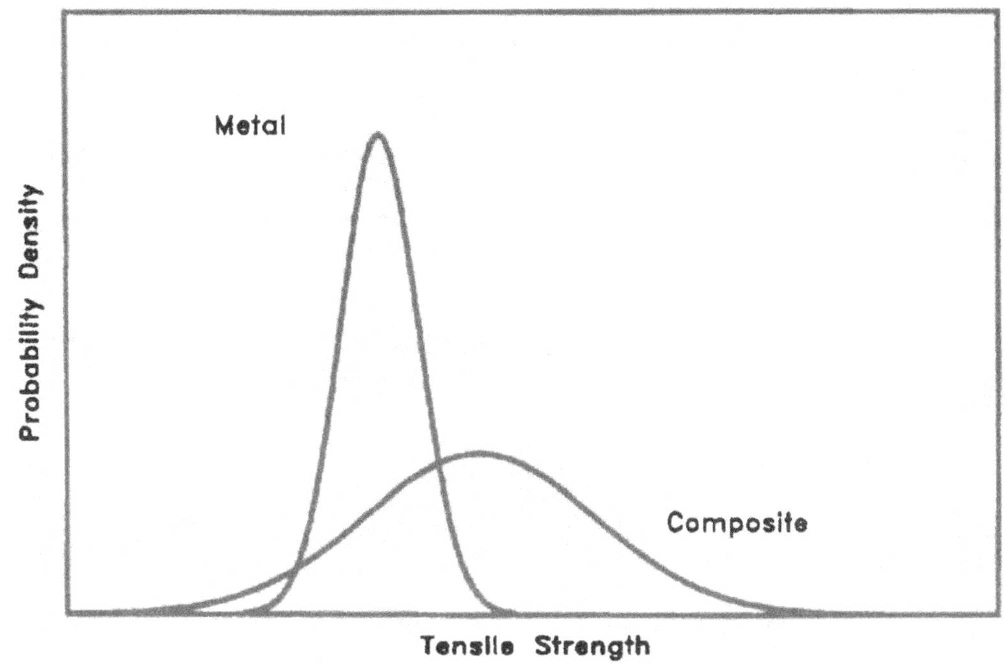

FIGURE 6.2.1 *Composites generally exhibit variation in material performance different from their metallic counterparts.*

6.2.4 Damage tolerance

As stated in Volume 3, Section 4.11.1, "damage tolerance is defined as a measure of a structure's ability to sustain a level of damage or presence if a defect and yet be able to safely perform its operating functions." Damage to composite structures can occur during manufacturing or operational usage. In order to design the structure to operate safely after sustaining such damage, a common practice is to limit the stress allowed in the composite structure. Typically, composite structures are designed to withstand the most severe of either of the following tow conditions: a 0.25-inch open hole in any location at ultimate load or damage sustained when objects of specified size strike the surface (representative of barely visible impact damage threats). Both criteria assume the defect exists for the life of the part. These criteria reduce the allowable strength.

6.3 RELIABILITY ENGINEERING

Reliability is of such concern to the military that specific requirements are defined in contractual documents. Industry satisfies these requirements by employing engineers who specialize in reliability to guide the design. Currently, structural reliability is being increasingly emphasized. The reasons are twofold. First, advances in materials technology have resulted in higher performance materials that often possess detrimental side effects (e.g., high strength steels that exhibit low fracture toughness). Second, the need for higher vehicle performance has pushed operating stresses to higher levels in order to reduce structural weight.

Structural designs are documented via engineering drawings. Drawings are not "released" until they undergo scrutiny by several technical disciplines. Reliability is one of the concerns that is dealt with by the technical disciplines. Reliability specialists ensure that these concerns are incorporated into the design.

Customer reliability criteria typically specify three goals. These are: Mean Time Between Failure (MTBF), mission reliability, and Failure Modes Effects Criticality Analysis (FMECA). MTBF is measured in unscheduled maintenance events per million flight hours. Mission reliability is an indication of the probability of having to abort a flight. FMECA determines the impact of specific failures on mission performance, safety, and utilization.

In addition to supplying input to design, reliability engineering output is supplied to maintainability groups for maintenance man-hour predictions. Their results are used by logistics persons to establish provisioning requirements for spare parts.

6.4 RELIABILITY DESIGN CONSIDERATIONS

The following is a list of general composite structures considerations which provide insight on improved reliability and causes of poor reliability:

- Eliminate/minimize potential galvanic corrosion and/or thermal expansion problems by selecting compatible materials.
- Allow for the difference in thermal expansion when mating composites to metals. The coefficient of thermal expansion for composites is low.
- Assess carefully the use of honeycomb sandwich panels which utilize thin facesheets in areas where Foreign Object Damage (FOD) and bird strikes are likely to occur. Thin facesheets are susceptible to impact damage.
- Protect the structure for possible lightning strikes. Good electrical contact between all metallic and carbon/epoxy structural components must provide for the dissipation of static and lightning-induced electrical currents.
- Fasteners: Use titanium alloy or other materials that are compatible with carbon/epoxy to prevent galvanic corrosion.
- The current ability to detect flaws in composite structures, especially honeycomb, is evolving. Designs that enhance access for inspection tend to promote reliability.
- Improved reliability can be obtained by avoiding anomalies such as wrinkling and porosity in integral stiffeners. The ability to detect such flaws is limited.
- Extreme care should be taken during the repair of composite structures. Avoid damaging additional plies during patch or repair operations as it may result in a decline in reliability.
- Variations in manufacturing processes such as curing and machining can be responsible for a range of part strengths thus influencing reliability.

- The supplier's prepreg material should be closely monitored (i.e., acceptance testing) to assure incoming material consistency and conformance to design values.

6.5 RELIABILITY ASSESSMENT AND DESIGN

6.5.1 Background

Advanced composite materials offer sizable improvements in weight savings, maintainability, durability, and reliability. There are a number of performance factors that have limited their success. Thus far, composite design and treatment of unique performance factors have been handled in a traditional metals approach in the aircraft industry. This approach is characteristically deterministic in nature. Probabilistic methods offer a different technology that can be used as a design tool, or, in a more conservative manner, as a risk analysis. The application of probabilistic methods opens up technical information not available in traditional approaches.

Probabilistic methods represent a technology that cannot be implemented without careful development. It is, however, a technology that is easily controllable. It may be used as an assessment of deterministic designs; it may be used to establish realistic criteria for deterministic designs; or it may be implemented as a preferred design approach. If used as the preferred design approach, probabilistic methods utilize a reliability target in lieu of factors-of-safety. Disclosure of risk characteristics alone should interest the designer in applying the technology.

Probabilistic design is an integrated process as shown schematically in Figure 6.5.1(a). The approach is to define/develop the functional relationships of the operations within the boxes, then build the relationships between them. This interconnects the entire process. In this way, when a factor in one operation changes, its effect can be determined on the others. The end result is the effect on failure probability.

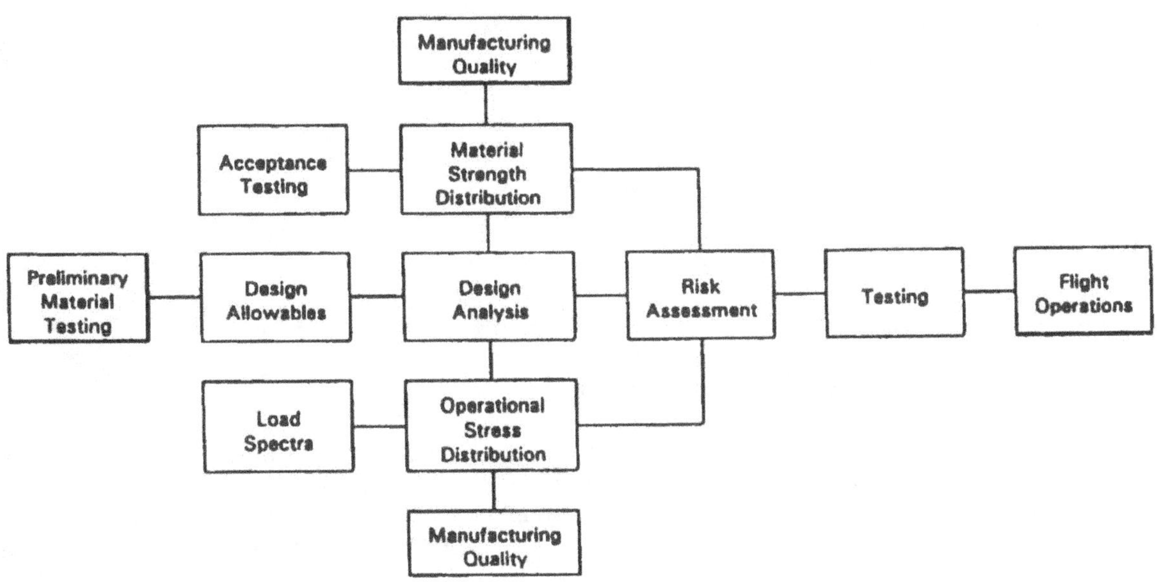

FIGURE 6.5.1(a) *Full probabilistic process.*

A flowchart of a Probabilistic Design model is shown in Figure 6.5.1(b). This model consists of four major activities; namely, the design process, material production, manufacturing, and operations. Output from the design process is the expected operating stress distribution resulting from the flight spectra. The remaining three activities provide the material strength distribution, determined through Monte Carlo simulation of random variables representing random variation of incoming material strength, manufacturing defects, and operational factors. Probability of failure occurs when the stress exceeds the strength. This is calculated by a double integral of the stress and the strength probability density functions to determine the probability that "stress exceeds strength".

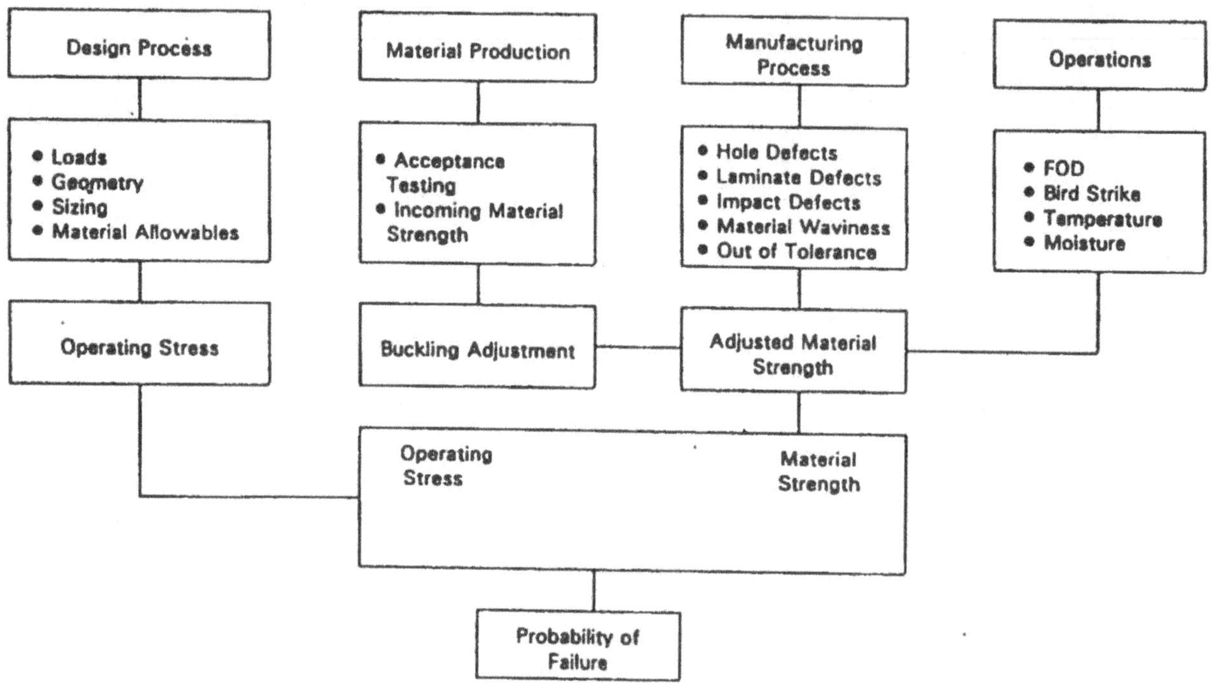

FIGURE 6.5.1(b) *Probabilistic design.*

6.5.2 Deterministic vs. Probabilistic Design Approach

Component dimensions, environmental factors, material properties, and external loads are design variables. They may be characterized with statistical modes. The deterministic approach seeks out and defines a worst case or an extreme value to meet in the design. The probabilistic approach utilizes the statistical characterization and attempts to provide a desired reliability in the design. The deterministic approach introduces conservatism by specifying a factor of safety to cover unknowns. The factor of safety is traditionally 1.5. The probabilistic approach depends on the statistical characterization of a variable to determine its magnitude and frequency. The amount of data (how well the variable is defined) influences its extreme values.

Application of a factor-of-safety to cover unknowns has a history of success. The danger in this approach is that the factor of safety may be too large, or in some cases, too small. Because it has worked in the past is no guarantee that it will suffice in the future. The whole approach of worst case extremes can lead to compounding and inefficiency. To select a factor-of-safety solely on the basis of "it worked in the past" should be examined.

Advanced composite materials were introduced in the early 1960's and since that time have undergone significant development. Some obstacles appeared insurmountable, including susceptibility of material strength degradation to elevated temperature, absorbed moisture, impact damage, and hidden flaws or damage. The approach to accommodate these material strength reduction factors has been to develop worst case manufacturing and operational scenarios and assume their existence for the life of the part. These factors, which are in reality variables, are thereby treated as constants.

Composite part design is governed by compounded conservatism illustrated by the following criteria:

- Worst case loading x safety factor (1.5)
- Worst case temperature
- Worst case moisture
- Worst case damage, undetected
- Material allowables derived from conservative statistical criteria

The effect of combining these conservative structural criteria is to produce inefficient products. Probabilistic methods offer an alternative to compound conservation. They quantify the degree of safety and permit the designer to discover the risk drivers.

6.5.3 Probabilistic Design Methodology

The basic probabilistic design mathematical theory, shown below, accounts for the probability distributions of both material strength and operating stress. Because failure is a local phenomenon, division of a component into N numbers of nodes is done to represent all the locations at which failure is possible to occur. In general, the distributions are assumed to be identical at all the nodes. Step 6 assumes that material strengths at the nodes are independent from each other.

Step No.

1. Establish allowable failure rate.

2. Establish the number of nodes where failure is possible.

3. Determine probability distribution for loads.

$$P(X_S < x_S) = F(x_S)$$

4. Determine the operating stress probability density function $f_S(x)$.

5. Determine probability distribution for strength.

$$P(Y_M < y_M) = F(y_M)$$

6. Calculate failure probability P.

$$P_f = \int_x f_s(z)[1-F_M(x))^N]dx$$

An alternative probabilistic design approach has been discussed in References 6.5.3(a) and 6.5.3(b). The fundamental elements of this approach are:

1. Identify all possible uncertain variables at all scales of composite structures. This includes variables at constituents scale, at all stages of fabrication process and assembly, and applied loads.

2. Assign a probabilistic distribution function for each variable.

3. Process all the random variables through an analyzer which consists of micro- and macro-composite mechanics and laminate theories, structural mechanics and probability theories.

4. Extract useful information from the output of the analyzer and check against defined probabilistic design criteria.

The IPACS (Integrated Probabilistic Assessment of Composite Structures) computer code developed at NASA Lewis integrated the above elements for probabilistic design of composite structures. A schematic of the computer code is shown in Figure 6.5.3.

6.5.4 Data Requirements

In order to conduct a probabilistic design exercise, the following parameters must be characterized as random variables:

1. Material mechanical properties

2. External loads anticipated during the life of the article

3. Manufacturing processes and their effect on material strength

4. Environmental effect on strength

5. Environmental history during operational usage

6. Flaw and/or damage locations, severity, probability of occurrence and effect on strength

7. Predictive Accuracy

Quality of incoming composite material is crucial to final product quality. To assure incoming material meets specifications, testing procedures and measurement value limits must be established to sufficiently discriminate between inferior and desired material. These criteria must be agreed upon by producer and consumer. Each wants to minimize their risk. The producer's risk is the probability of rejecting "good material" and the consumer's risk is the probability of accepting "inferior material".

6.5.5 Summary

Adopting specific structural criteria should not be done without a reason. The current criteria has its origins in metals technology. The goal of probabilistic design is to make reliability the foundation of composite structural criteria. It *will not* replace most structural mechanics functions.

Probabilistic Design is a powerful supplement/alternative to today's approach for composite design. It requires the development of sophisticated techniques in probability and characterization of statistical data for engineering variables. It is gaining momentum as more people become aware of its presence and benefits.

As the demand grows for more accurate, sophisticated designs, the requirement for probabilistic design methodology will become more and more accepted. The incorporation of Probabilistic Design, while quite challenging, offers significant payoffs not available with conventional technology.

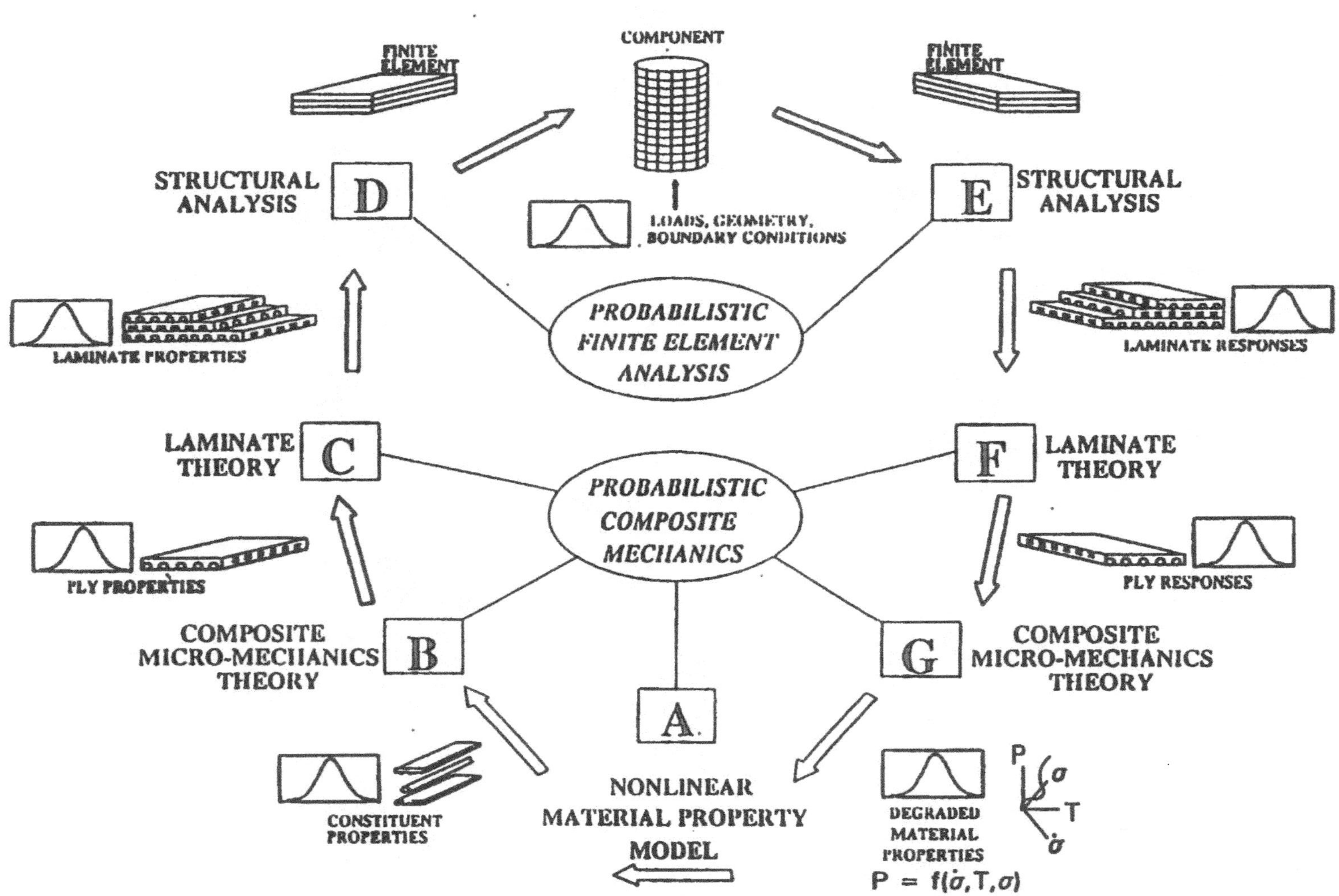

FIGURE 6.5.3 Schematic of the computer code IPACS.

REFERENCES

6.1(a) Doty, L.A., "Reliability for the Technologies," Industrial Press, New York, New York, 1985, pp. 1-9.

6.1(b) Bazovsky, I., "Reliability Theory and Practice," Prentice-Hall, Inc., Englewood Cliffs, New Jersey, 1961, pp. 1-16.

6.1(c) Whitehead, R.S., Kan, H.P., Cordero, R., and Saether, E.S., "Certification Testing Methodology for Composite Structures," Vols. I and II, Final Report, Contract No. N62269-84-C-0243, October 1986.

6.2.1 Muller, G.E. and Schmid, C.J., "Factor of Safety - USAF Design Practice," Final Report, AFFDL-TR-78-8, April 1978.

6.5.3(a) Chamis, C.C. and Murthy, P.L.N., "Probabilistic Composite Analysis." First NASA Advanced Composites Technology Conference, Part 2, NASA CP3104-PT-2, 1991, pp. 891-900.

6.5.3(b) Chamis, C.C., Shiao, M.C., and Kan, H.P., "Probabilistic Design and Assessment of Aircraft Composite Structures." Fourth NASA/DoD Advanced Composites Technology Conference, June 7-11, 1993, Salt Lake City, Utah.

This page intentionally left blank

CHAPTER 7 THICK SECTION COMPOSITES

Page

7.1 INTRODUCTION

Thick-section composites are ones where the effect of geometry (thickness-to-span ratio), material constituents (matrix and fiber stiffness/strength properties), lamination scheme, processing, and service loading exhibit three-dimensional states of stress. For instance, all loadings induce multiaxial stresses into individual plies of composite materials that are made of multi-directional ply laminates (either woven or nonwoven), even though the overall loadings may only be uniaxial. When transverse (through-thickness) stresses and strains occur to a significant degree, they must be accounted for in analysis, design and testing. A significant degree is achieved when these effects contribute to failure (e.g., delamination), excessive deflection or vibration. Frequently, these stresses and strains induce failures that cannot be accurately predicted by conventional two-dimensional analyses for thin laminates. These two-dimensional analyses are usually based on material response data obtained from traditional shear and uniaxial tensile/compressive testing techniques. In thick section composites, where any one of six stress components may significantly contribute to failure, a failure criteria must distinguish between different types of failure modes by associating the contribution of each three-dimensional stress component to a unique mode of failure, be it fiber, matrix or interface dominated. An appropriate failure criteria for thick section composites must consider the following laminate failure modes:

Fiber Dominated	Matrix Dominated	Interface Dominated
. Fiber pull-out	. Transverse cracking	. Interface disbonding
. Fiber tensile failure	. Interlaminar cracking	. Interface delamination
. Fiber micro-buckling	. Intralaminar cracking	. Compressive delamination
. Fiber shear failure	. Edge delamination	

For example, thick-section composites made of high stiffness and strength fiber-reinforced plies often exhibit significant transverse shear and transverse normal deformations (the type of three-dimensional stress contributions that are negligibly small in thin laminates). The thickness effect can also be influenced by short wavelength loadings and, in dynamics, high frequency vibrations. These three-dimensional effects are considerably more pronounced in composites than in homogeneous isotropic materials due to their inherently high material compliances in the transverse direction relative to the axial fiber direction. Moreover, composite laminates exhibit much lower strength in the transverse direction, and at ply interfaces, making them particularly susceptible to matrix cracking and delamination.

Thick section composites can also be defined from the standpoint of fabrication effects associated with a large number of plies. Process induced stresses can be significant and, therefore, warrant special attention. Fabrication effects of special concern include residual stresses, wrinkling, micro-cracking, exotherm, volatile removal, compaction, machining, and mechanical joining and/or adhesive bonding. To minimize these effects, special resins, processing, tooling, and cure cycles may be necessary.

In thick laminates, typically two competing objectives are desired, namely, minimization of process induced residual stresses and maximization of production rates (i.e., minimization of the processing time required to achieve complete cure). Fast cure cycle times, involving steep heating and cooling rates, will generally lead to high process induced residual stresses. On the other hand, slowly bringing all part thicknesses up to complete cure simultaneously will minimize, if not eliminate, all process induced residual stresses. This, however, is accomplished at the expense of extended cure cycle times. It is also important to note that process induced residual stresses may in fact be intentionally introduced to cancel, or otherwise mitigate, large superimposed in-service stresses.

In thick laminate design, cure simulation plays a very important role in developing a deeper understanding of the cure kinetics and the degree of cure at any point in the time domain. Such simulation is also able to predict processing stresses even during the cure cycle. This can be an important tool for prediction and preventing in-process part fabrication failures where both stresses and associated strengths are low.

The structural analyst needs to know the multiaxial strength and deformation characteristics for efficient thick composite material design. The full potential of thick composites cannot be realized until the material response under multiaxial service loadings can be established. Technical progress in the design, analysis and associated material testing of thick composites remain much less developed than the generally accepted methodology associated with thin composite material characterizations and applications.

The step-by-step method for analysis of thick section composites is illustrated by the flow chart in Figure 7.1.

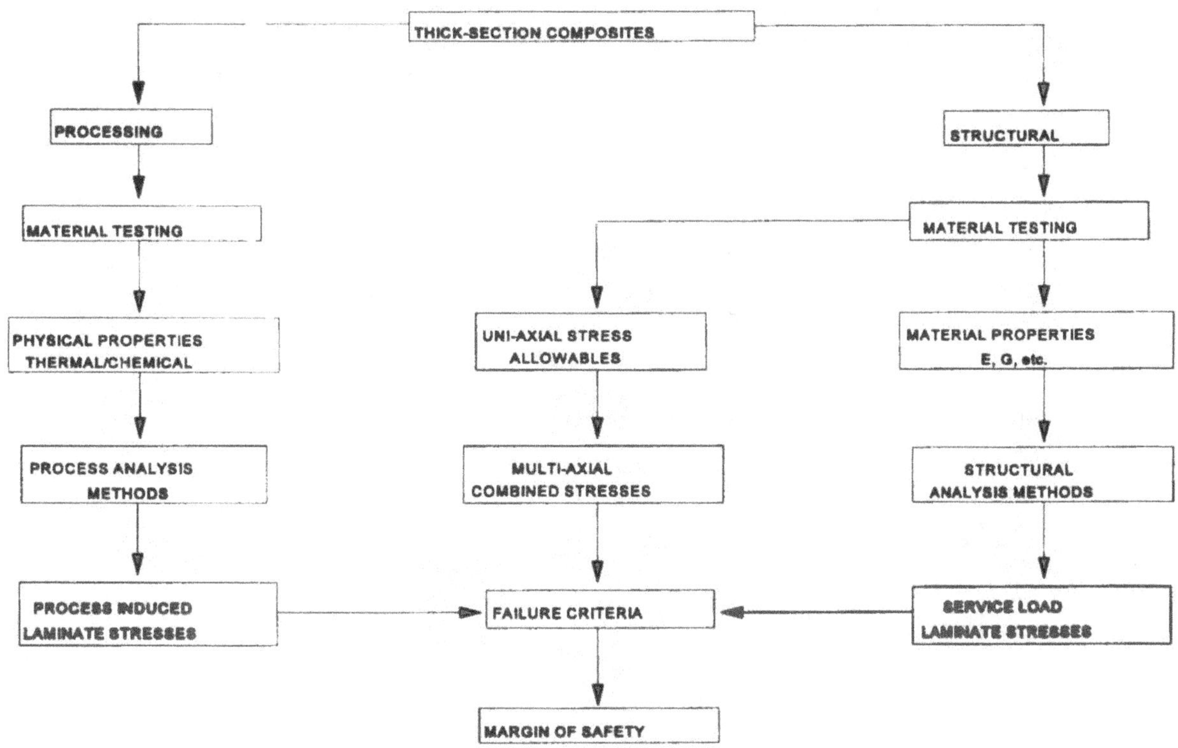

FIGURE 7.1 *Flowchart illustrating thick-section composites analysis method.*

7.2 MECHANICAL PROPERTIES REQUIRED FOR THICK SECTION COMPOSITE THREE-DIMENSIONAL ANALYSIS

The purpose of this section is to define the three-dimensional (3-D) orthotropic stiffness properties necessary to conduct a 3-D point stress analysis, and the failure strength and strain allowables required to calculate a margin of safety. This section will:

a) Define the stiffness properties currently used to conduct a conventional two-dimensional (2-D) analysis (Volume 1, Section 6.7).

b) Define the additional stiffness properties needed to conduct a three-dimensional (3-D) stress analysis.

c) Define the testing required to experimentally determine the 3-D stiffness properties and the failure strengths and strains for uniaxial loading (Section 7.2.3.1) and multiaxial loading (Section 7.2.3.2)

d) Discuss the methodology for predicting laminate stiffness properties through the thickness using the 3-D lamina properties (Section 7.2.4).

The symbols and nomenclature used in the handbook (Volume 3, Section 1.3.1) apply to 2-D and 3-D composites and utilize 1, 2, 3 for lamina axes and x, y, z for an oriented laminate axis directions.

7.2.1 2-D composite analysis

The two-dimensional composite analysis procedures (Volume 3, Section 4.3.1) apply when the through the thickness stresses are not significant. For unidirectional laminates that have low stresses in the thickness or 3-direction ($\sigma_3 = \tau_{23} = \tau_{13}$), plane stress), the stress-strain relationship (Reference 7.2.1) is,

$$\{\epsilon_{ij}\} = [S_{ij}]\{\sigma_{ij}\} \qquad 7.2.1(a)$$

$$\begin{Bmatrix} \epsilon_1 \\ \epsilon_2 \\ \gamma_{12} \end{Bmatrix} = \begin{bmatrix} S_{11} & S_{12} & 0 \\ S_{12} & S_{22} & 0 \\ 0 & 0 & S_{66} \end{bmatrix} \begin{Bmatrix} \sigma_1 \\ \sigma_2 \\ \tau_{12} \end{Bmatrix} \qquad 7.2.1(b)$$

In terms of the engineering elastic constants obtained by simple tests

$$\begin{Bmatrix} \epsilon_1 \\ \epsilon_2 \\ \gamma_{12} \end{Bmatrix} = \begin{bmatrix} \frac{1}{E_1} & -\frac{\nu_{12}}{E_1} & 0 \\ -\frac{\nu_{21}}{E_2} & \frac{1}{E_2} & 0 \\ 0 & 0 & \frac{1}{G_{12}} \end{bmatrix} \begin{Bmatrix} \sigma_1 \\ \sigma_2 \\ \tau_{12} \end{Bmatrix} \qquad 7.2.1(c)$$

The reciprocity relationships for stiffness is

$$\frac{\nu_{12}}{E_1} = \frac{\nu_{21}}{E_2} \qquad 7.2.1(d)$$

For the plane stress two-dimensional analysis, the four independent elastic material properties are:

$$E_1,\ E_2,\ G_{12},\ \nu_{12}$$

In-plane failure stress and strain values can be obtained from the same test used for determining the stiffness as discussed in Section 7.2.3.1.

7.2.2 3-D composite analysis

When the stresses and strains in the thickness direction are significant, (applied values are approaching their allowables) the problem requires a three-dimensional orthotropic stress analysis. A 3-D analysis is frequently necessary as the section thickness of a composite increases or when thin sections have out-of-plane loading (bending moment, lateral pressures, etc.) which results in, for example, interlaminar tensile stresses in a corner radius or interlaminar shear stresses in a beam or plate.

7.2.2.1 Unidirectional lamina 3-D properties

For the orthotropic unidirectional lamina there are nine independent constants as shown by the following stress-strain relationship (Reference 7.2.1):

$$\begin{Bmatrix} \epsilon_1 \\ \epsilon_2 \\ \epsilon_3 \\ \gamma_{23} \\ \gamma_{31} \\ \gamma_{12} \end{Bmatrix} = \begin{bmatrix} S_{11} & S_{12} & S_{13} & 0 & 0 & 0 \\ S_{12} & S_{22} & S_{23} & 0 & 0 & 0 \\ S_{13} & S_{23} & S_{33} & 0 & 0 & 0 \\ 0 & 0 & 0 & S_{44} & 0 & 0 \\ 0 & 0 & 0 & 0 & S_{55} & 0 \\ 0 & 0 & 0 & 0 & 0 & S_{66} \end{bmatrix} \begin{Bmatrix} \sigma_1 \\ \sigma_2 \\ \sigma_3 \\ \tau_{23} \\ \tau_{31} \\ \tau_{12} \end{Bmatrix} \quad 7.2.2.1(a)$$

or in terms of the engineering constants,

$$\begin{Bmatrix} \epsilon_1 \\ \epsilon_2 \\ \epsilon_3 \\ \gamma_{23} \\ \gamma_{31} \\ \gamma_{12} \end{Bmatrix} = \begin{bmatrix} \frac{1}{E_1} & -\frac{\nu_{21}}{E_2} & -\frac{\nu_{31}}{E_3} & 0 & 0 & 0 \\ -\frac{\nu_{12}}{E_1} & \frac{1}{E_2} & -\frac{\nu_{32}}{E_3} & 0 & 0 & 0 \\ -\frac{\nu_{13}}{E_1} & -\frac{\nu_{23}}{E_2} & \frac{1}{E_3} & 0 & 0 & 0 \\ 0 & 0 & 0 & \frac{1}{G_{23}} & 0 & 0 \\ 0 & 0 & 0 & 0 & \frac{1}{G_{31}} & 0 \\ 0 & 0 & 0 & 0 & 0 & \frac{1}{G_{12}} \end{bmatrix} \begin{Bmatrix} \sigma_1 \\ \sigma_2 \\ \sigma_3 \\ \tau_{23} \\ \tau_{31} \\ \tau_{12} \end{Bmatrix} \quad 7.2.2.1(b)$$

There are three reciprocal relationships that must be satisfied for an orthotropic material. They are

$$\frac{\nu_{12}}{E_1} = \frac{\nu_{21}}{E_2}, \qquad \frac{\nu_{13}}{E_1} = \frac{\nu_{31}}{E_3}, \qquad \frac{\nu_{23}}{E_2} = \frac{\nu_{32}}{E_3} \quad 7.2.2.1(c)$$

There are nine independent elastic material properties required for an orthotropic lamina

$$E_1,\ E_2,\ E_3,\ G_{12},\ G_{13},\ G_{23},\ \nu_{12},\ \nu_{13},\ \nu_{23}$$

When materials have a different stiffness in tension from in compression, it is common practice to use an average value when the difference is small. If the stiffness difference is significant, use the stiffness (tensile or compressive) that is representative of the application loading.

7.2.2.2 Oriented orthotropic laminate 3-D properties

The compliance matrix and associated nine elastic constants required to conduct a 3-D analysis are defined in this section and are for a oriented balanced and symmetric laminate loaded in the x, y, or z direction. Most practical composite laminate lay-ups generally are balanced and symmetric to prevent thermal warpage during processing. If the laminate is unbalanced and unsymmetric, or loaded "off-axis" to the principal orthogonal directions, then the matrix is fully populated with the Chentsov's coefficients ($\mu_{ij,kl}$) and coefficients of mutual influence ($\eta_{ij,i}$, $\eta_{i,ij}$) (see References 7.2.1, 7.2.2.2).

The compliance matrix for the balanced and symmetric laminate loaded in the x, y, or z direction is

$$\begin{Bmatrix} \epsilon_x \\ \epsilon_y \\ \epsilon_z \\ \gamma_{yz} \\ \gamma_{zx} \\ \gamma_{xy} \end{Bmatrix} = \begin{bmatrix} \bar{S}_{11} & \bar{S}_{12} & \bar{S}_{13} & 0 & 0 & 0 \\ \bar{S}_{12} & \bar{S}_{22} & \bar{S}_{23} & 0 & 0 & 0 \\ \bar{S}_{13} & \bar{S}_{23} & \bar{S}_{33} & 0 & 0 & 0 \\ 0 & 0 & 0 & \bar{S}_{44} & 0 & 0 \\ 0 & 0 & 0 & 0 & \bar{S}_{55} & 0 \\ 0 & 0 & 0 & 0 & 0 & \bar{S}_{66} \end{bmatrix} \begin{Bmatrix} \sigma_x \\ \sigma_y \\ \sigma_z \\ \tau_{yz} \\ \tau_{zx} \\ \tau_{xy} \end{Bmatrix} \qquad \text{7.2.2.2(a)}$$

In terms of the effective engineering elastic constants this relationship is,

$$\begin{Bmatrix} \epsilon_x \\ \epsilon_y \\ \epsilon_z \\ \gamma_{yz} \\ \gamma_{zx} \\ \gamma_{xy} \end{Bmatrix} = \begin{bmatrix} \frac{1}{E_x} & -\frac{\nu_{yx}}{E_y} & -\frac{\nu_{zx}}{E_z} & 0 & 0 & 0 \\ -\frac{\nu_{xy}}{E_x} & \frac{1}{E_y} & -\frac{\nu_{zy}}{E_z} & 0 & 0 & 0 \\ -\frac{\nu_{xz}}{E_x} & -\frac{\nu_{yz}}{E_y} & \frac{1}{E_z} & 0 & 0 & 0 \\ 0 & 0 & 0 & \frac{1}{G_{yz}} & 0 & 0 \\ 0 & 0 & 0 & 0 & \frac{1}{G_{zx}} & 0 \\ 0 & 0 & 0 & 0 & 0 & \frac{1}{G_{xy}} \end{bmatrix} \begin{Bmatrix} \sigma_x \\ \sigma_y \\ \sigma_z \\ \tau_{yz} \\ \tau_{zx} \\ \tau_{xy} \end{Bmatrix} \qquad \text{7.2.2.2(b)}$$

There are three reciprocal relationships that must be satisfied by the effective laminate stiffnesses. They are,

$$\frac{\nu_{xy}}{E_x} = \frac{\nu_{yx}}{E_y}, \qquad \frac{\nu_{xz}}{E_x} = \frac{\nu_{zx}}{E_z}, \qquad \frac{\nu_{yz}}{E_y} = \frac{\nu_{zy}}{E_z} \qquad \text{7.2.2.2(c)}$$

There are nine independent effective elastic material constants required for analysis of the oriented laminate,

$$E_x,\ E_y,\ E_z,\ G_{xy},\ G_{xz},\ G_{yz},\ \nu_{xy},\ \nu_{xz},\ \nu_{yz}$$

7.2.3 Experimental material property determination

The current and most commonly used approach for failure analysis of 2-D composites is to experimentally determine the strength and stiffness values for the unidirectional lamina from simple uniaxial tests and use a failure criterion to account for the various load direction interactions to calculate the margin of safety. These uniaxial tests are defined in Section 7.2.3.1 for 2-D and 3-D composites. Another approach is to conduct multiaxial tests that provide loading in the proper proportions to simulate the actual load applications. The multiaxial testing and methodology are discussed in Section 7.2.3.2.

There are considerable challenges associated with both uniaxial and multiaxial, mechanical testing of thick section composite materials. A partial list of experimental testing considerations is presented below:

- Test system and load introduction
- Gripping system and fixturing
- Computer control and interface
- Adequate displacement control over specimen centroid location
- Specimen design and optimization
- Unknown states of stress within thick composites
- Multiaxial extensometry and other measurement devices and techniques
- Inclusion and treatment of environmental effects
- Data acquisition and analysis
- Multiaxial yield and failure criteria
- Size effect and scaling law
- Edge effects treatment
- Static and dynamic testing, including fatigue and impact loadings
- Sensitivity to stress concentrations
- NDE of damage

7.2.3.1 Uniaxial tests

The type of common tests conducted on the unidirectional laminate to obtain the conventional 2-D in-plane tensile, compressive, and shear stiffness, as well as failure strength and strains are summarized in Figures 7.2.3.1(a) through 7.2.3.1(c). These tests are also discussed in detail in Volume 1, Section 6.7. The additional unidirectional laminate design property tests needed when a 3-D (thick section) analysis is required are summarized in Figure 7.2.3.1(d) and described in detail in Figures 7.2.3.1(e) and 7.2.3.1(f). Test methods available to obtain these properties are summarized in Table 7.2.3.1(a). Further test method development is needed for tension and compression testing in the 3 or through-thickness direction.

For oriented laminates, the additional design properties tests needed in addition to the 2-D tests for a 3-D analysis are summarized in Figure 7.2.3.1(g). The 3-D through the thickness stiffnesses can also be predicted from the unidirectional lamina stiffnesses by the methods discussed in Section 7.2.4 (Theoretical Property Determination). Table 7.2.3.1(b) summarizes the test methods available for determining 3-D properties for an oriented laminate. Furthermore, test method development is also needed for tension and compression testing in the z-thickness direction similar to the need for unidirectional laminate testing.

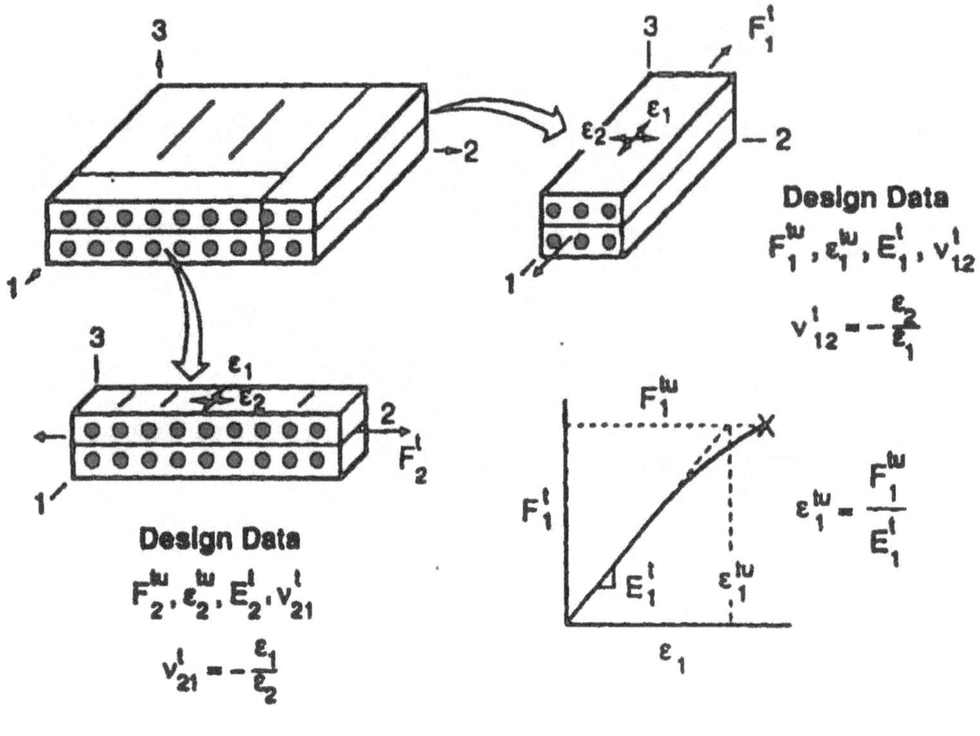

FIGURE 7.2.3.1(a) *Unidirectional laminate in-plane tensile design properties.*

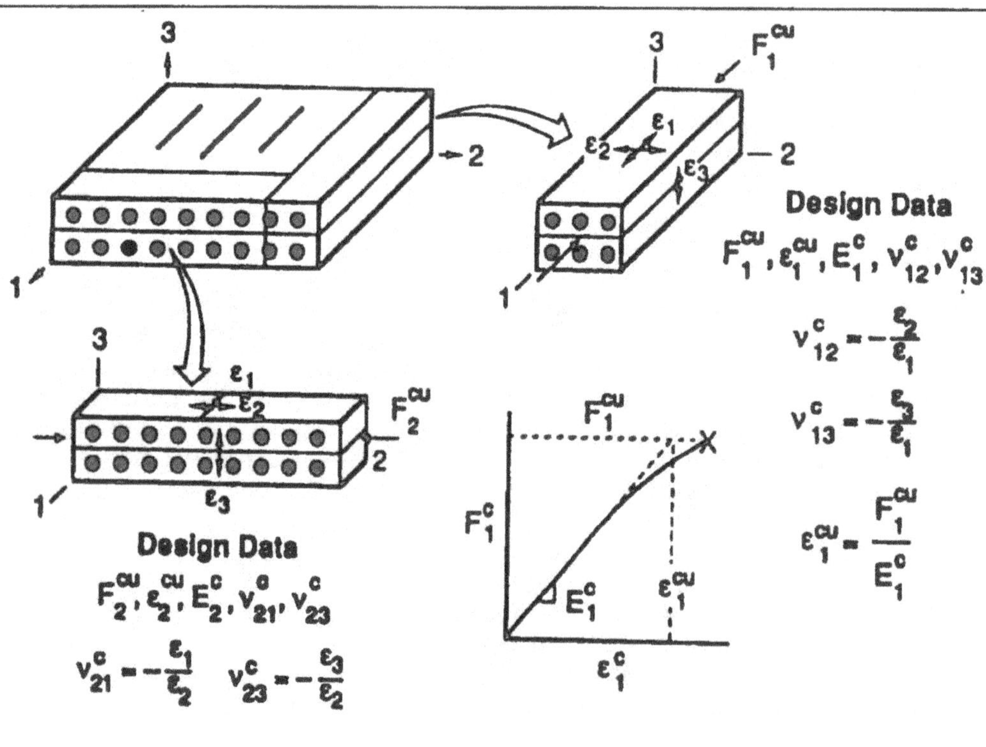

FIGURE 7.2.3.1(b) *Unidirectional laminate in-plane compressive design properties.*

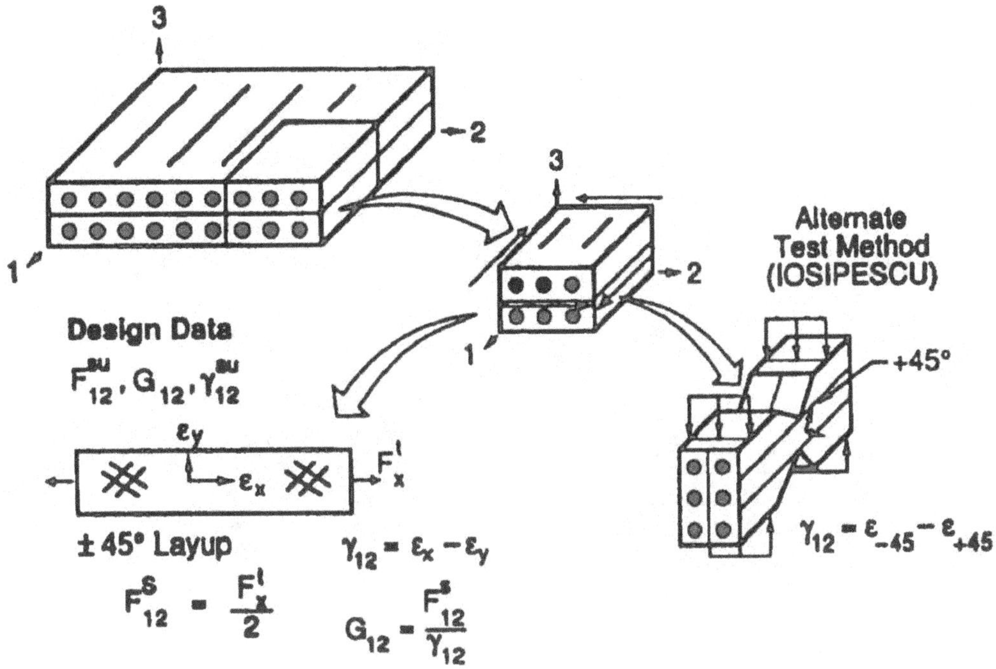

FIGURE 7.2.3.1(c) *Unidirectional laminate in-plane shear design properties.*

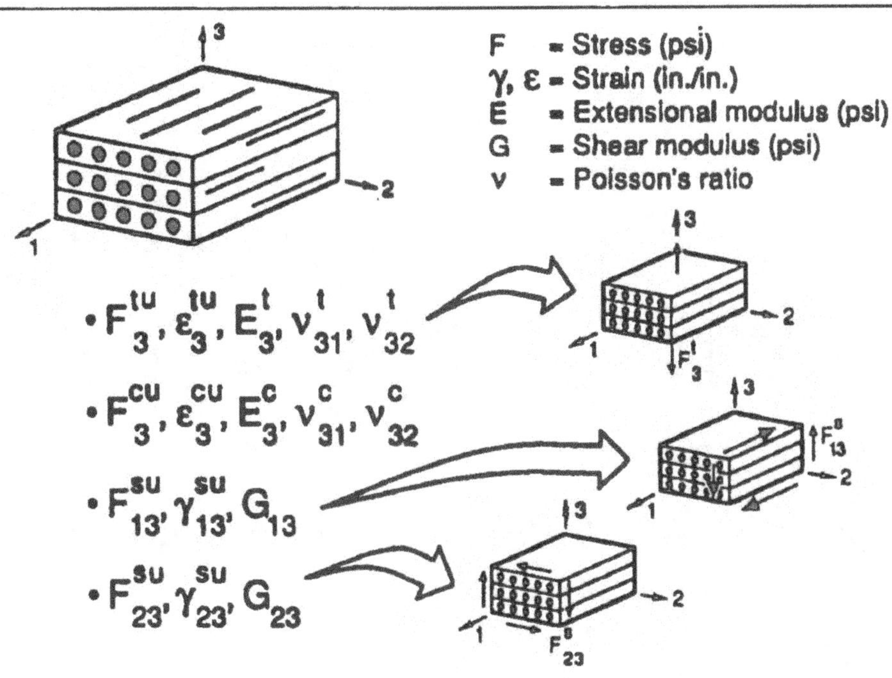

FIGURE 7.2.3.1(d) *Unidirectional laminate thickness direction design properties.*

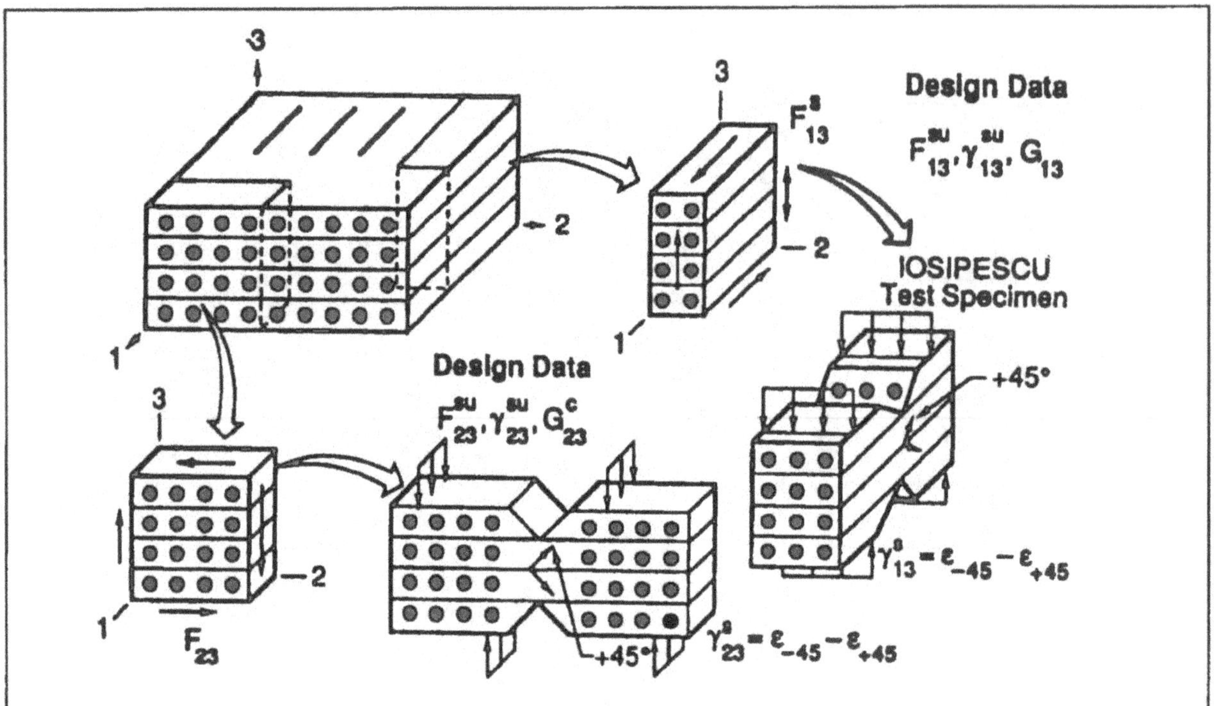

FIGURE 7.2.3.1(e) *Unidirectional laminate design properties for shear thickness direction.*

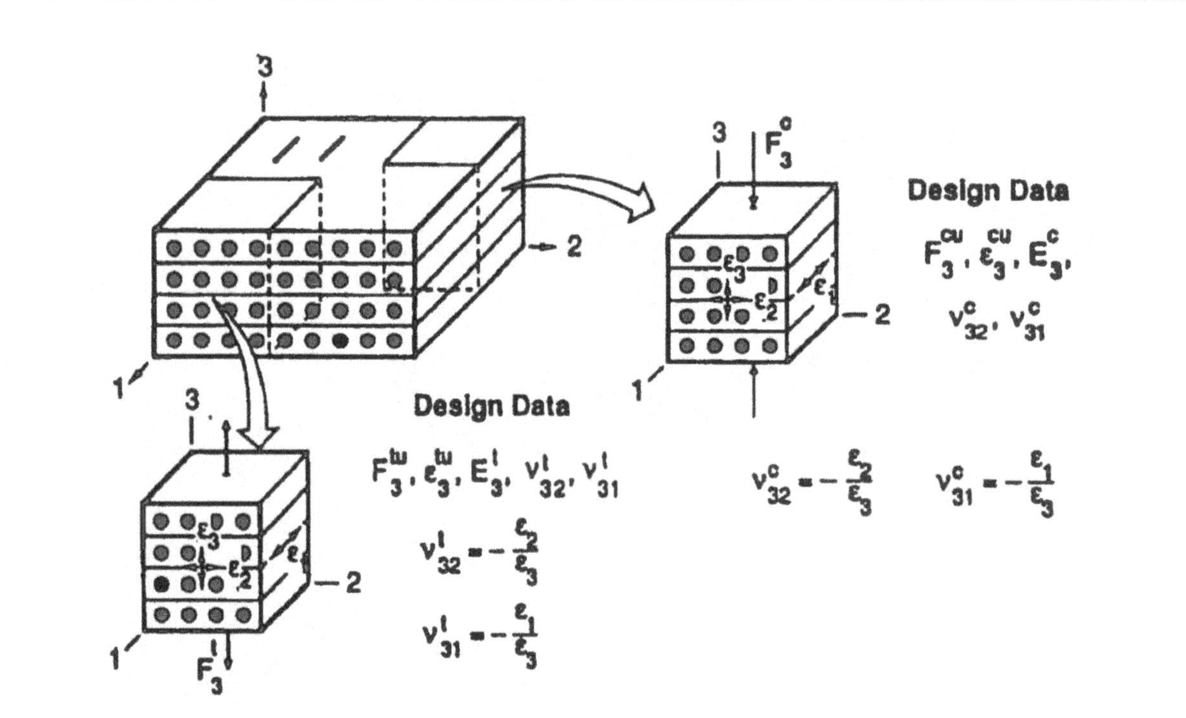

FIGURE 7.2.3.1(f) *Unidirectional laminate tensile and compressive design properties in thickness direction.*

TABLE 7.2.3.1(a) *Test methods available for determining 3-D lamina properties.*

Loading	Inplane Property	Test Method
1-Ten 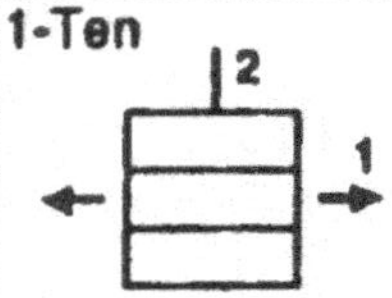	F_1^{tu}, ε_1^{tu}, E_1^t, ν_{12}^t	ASTM D3039 SACMA SRM-4
1-Comp 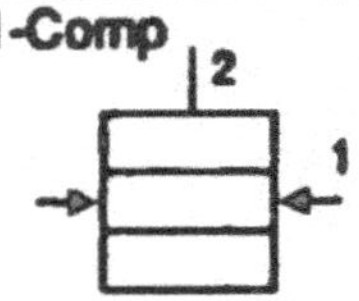	F_1^{cu}, ε_1^{cu}, E_1^c, ν_{12}^c, ν_{13}^c	ASTM D3410 SACMA SRM-1 ALLIANT TECHSYSTEMS DTRC ARL
2-Ten 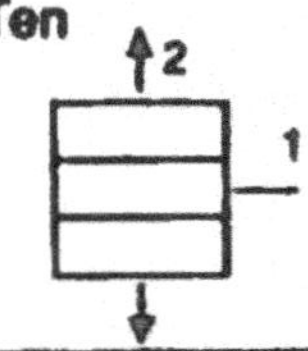	F_2^{tu}, ε_2^{tu}, E_2^t, ν_{21}^t	ASTM D3039 SACMA SRM-4
2-Comp 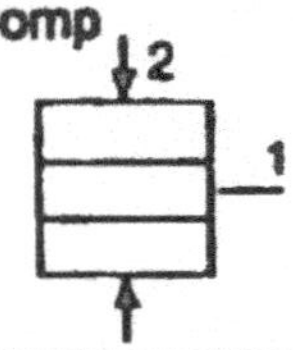	F_2^{cu}, ε_2^{cu}, E_2^c, ν_{21}^c, ν_{23}^c	ASTM D3410 SACMA SRM-1 ALLIANT TECHSYSTEMS DTRC ARL
12-Shear 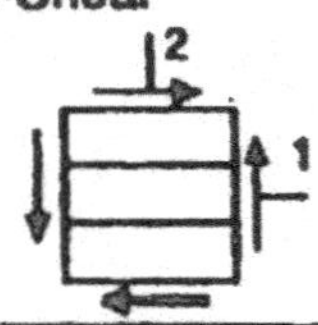	F_{12}^{su}, γ_{12}^{su}, G_{12}	ASTM D3518 SACMA SRM-7

Loading	Out-of-plane Property	Test Method
3-Ten 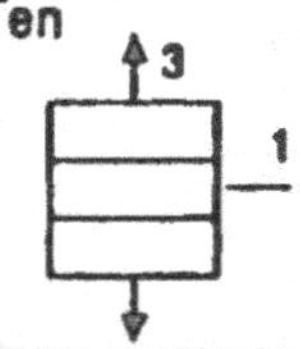	F_3^{tu}, ε_3^{tu}, E_3^t, ν_{31}^t, ν_{32}^t	To Be Developed
3-Comp 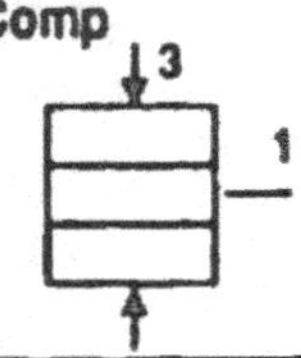	F_3^{cu}, ε_3^{cu}, E_3^c, ν_{31}^c, ν_{32}^c	To Be Developed
13-Shear 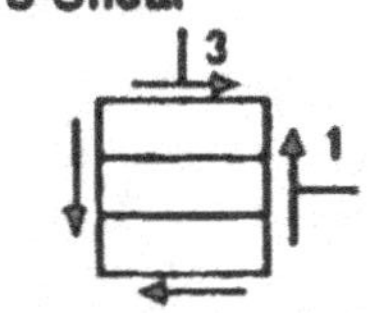	F_{13}^{su}, γ_{13}^{su}, G_{13}^t	ASTM D2344 SACMA SRM-8 IOSIPESCU
23-Shear 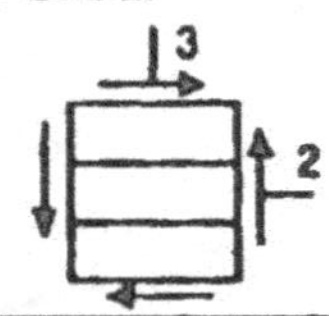	F_{23}^{su}, γ_{23}^{su}, G_{23}	IOSIPESCU

Notes:

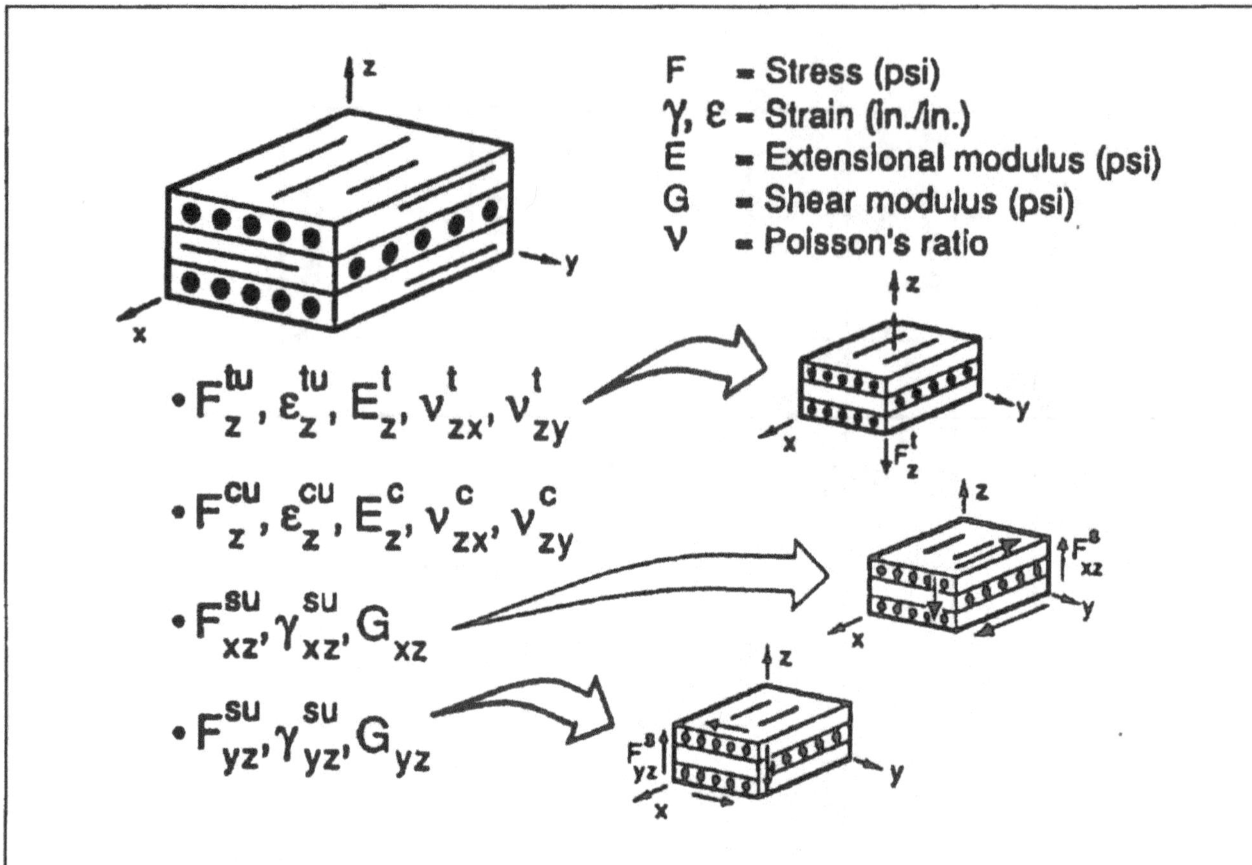

FIGURE 7.2.3.1(g) *Oriented laminate thickness direction design properties.*

TABLE 7.2.3.1(b) *Test methods available for determining 3-D lamina properties.*

Loading	Inplane Property	Test Method
x-Ten	F_x^{tu}, ε_x^{tu}, E_x^t, ν_{xy}^t	ASTM D3039? SACMA SRM-4?
x-Comp	F_x^{cu}, ε_x^{cu}, E_x^c, ν_{xy}^c, ν_{xz}^c	ASTM D3410? SACMA SRM-1? ALLIANT TECHSYSTEMS DTRC ARL
y-Ten	F_y^{tu}, ε_y^{tu}, E_y^t, ν_{yx}^t	ASTM D3039? SACMA SRM-1?
y-Comp	F_y^{cu}, ε_y^{cu}, E_y^c, ν_{yx}^c, ν_{yz}^c	ASTM D3410? SACMA SRM-1? ALLIANT TECHSYSTEMS DTRC ARL
xy-Shear	F_{xy}^{su}, γ_{xy}^{su}, G_{xy}	ASTM D4255? IOSIPESCU

Loading	Out-of-Plane Property	Test Method
z-Ten	F_z^{tu}, ε_z^{tu}, E_z^t, ν_{zx}^t, ν_{zy}^t	To Be Developed
z-Comp	F_z^{cu}, ε_z^{cu}, E_z^c, ν_{zx}^c, ν_{zy}^c	To Be Developed
xz-Shear	F_{xz}^{su}, γ_{xz}^{su}, G_{xz}	ASTM D2344? SACMA SRM-8? IOSIPESCU
yz-Shear	F_{yz}^{su}, γ_{yz}^{su}, G_{yz}	IOSIPESCU

Notes:
? - Applicability depends on laminate layup configuration.

An example of representative thick-section composite properties for an intermediate modulus carbon/epoxy material system are presented in Tables 7.2.3.1(c) and (d) for the unidirectional lamina and [0 /90] oriented laminate. The lamina properties were taken from Reference 7.2.3.1(a) and the [0/90] data were obtained by a Hercules test program from an 80-ply (t=0.59 in., 15mm) fiber-placed, autoclave-cured laminate (Reference 7.2.3.1(b)).

Tables 7.2.3.1(a) and (b) identify three uniaxial compression test methods for testing composites greater than 0.250 inches (6.35 mm) in thickness. Both the David Taylor Research Center (DTRC) and the Alliant Techsystems testing fixtures, which are shown in Figures 7.2.3.1(h) and 7.2.3.1(i), respectively (see References 7.2.3.1(a) and 7.2.3.1(c), respectively), were developed for uniaxial compression testing of thick prismatic columnar shaped composite material specimens. The US Army Research Laboratory - Materials Directorate (ARL) (Reference 7.2.3.1(d)) test method utilizes a cubic specimen loaded directly between two steel platens with no associated fixturing. The development of compression data relative to the different material orientations identified in Tables 7.2.3.1(a) and (b) is accomplished through independent, successive uniaxial load applications. Successive uniaxial compressive tests, that consist of one-directional load applications per material orientation, can be undertaken with conventional, medium-to-high capacity load frames. With proper care and specimen fixturing, these tests may also be used for determining unidirectional compressive material strengths and failure characteristics.

The primary feature that both the DTRC and the Alliant Techsystems test fixtures provide is that they have been developed for maintaining proper gripping and alignment of the test specimens as well as providing constraints to minimize any potential specimen end brooming (specimen splitting) under compressive load applications. Any potential onset of apparent, specimen end splitting and fixture-induced test specimen material cracking, may cause significant material strength reductions. Special tabbing as well as associated specimen-tabbing connection detail may be required for some uniaxial compression testing of thick composites.

7.2.3.2 Multiaxial tests

The purpose of this section is to provide information regarding multiaxial material testing methods. Some of these techniques, such as the two-dimensional methods (biaxial load applications), may be used for testing both thick and thin section composite materials. *However, the three-dimensional tests are primarily aimed at evaluating thick-section composite specimen material properties.* The importance of multiaxial testing becomes apparent when considering the need to evaluate the response of lamina and laminates to complex three-dimensional loads that result from service conditions. Multiaxial testing can help identify actual material strengths and failure mechanisms under properly proportioned loadings that simulate actual service conditions. Furthermore, multiaxial testing is recommended since the ability to predict the response of composites to multiaxial loadings has not been validated.

Currently, there is only limited experimental testing capability available to undertake all of the necessary work that is required to obtain a multidirectional material response data base. The testing procedures for thick composites are somewhat difficult to execute, have not yet been fully verified, and as such represent a major part of current and future research in themselves. However, the recently developed multiaxial testing techniques have been shown to be necessary in the determination of basic thick composite material parameters and actual material responses. These tests are also important in that the test results support the development of general and reliable three-dimensional numerical modeling, design, and analysis capabilities (i.e., finite element, boundary element, etc.) and failure theories for thick section composites in structural applications.

TABLE 7.2.3.1(c) *Intermediate modulus carbon/epoxy lamina typical 3-D properties.*

Loading	Inplane Property	RT/ Dry
1-Ten 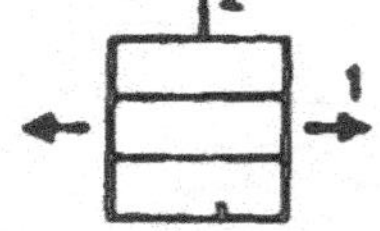	F_1^{tu} ε_1^{tu} E_1^t ν_{12}^t	250. ksi (1720 MPa) 15200 με 16.5 Msi (114 GPa) 0.33
1-Comp 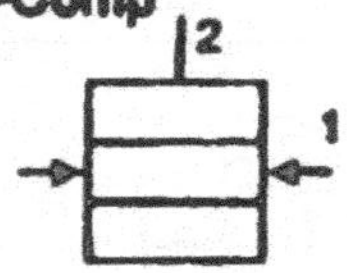	F_1^{cu} ε_1^{cu} E_1^c ν_{12}^c ν_{13}^c	170. ksi (1170 MPa) 10300 με 16.5 Msi (114 GPa)
2-Ten 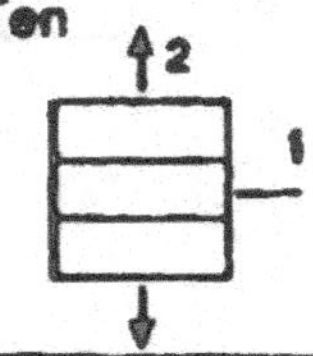	F_2^{tu} ε_2^{tu} E_2^t ν_{21}^t	8.00 ksi (55.2 MPa) 5700 με 1.40 Msi (9.65 GPa)
2-Comp 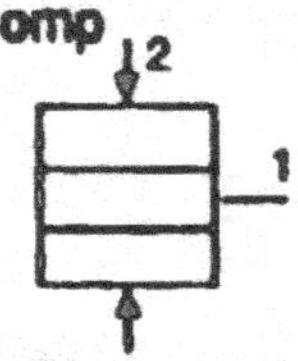	F_2^{cu} ε_2^{cu} E_2^c ν_{21}^c ν_{23}^c	30.0 ksi (207 MPa) 21500 με 1.40 Msi (9.65 GPa)
12-Shear 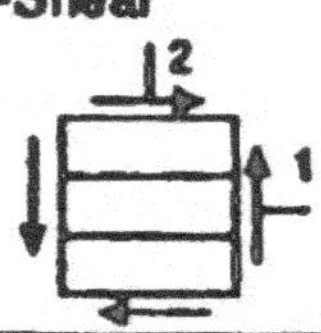	F_{12}^{su} γ_{12}^{su} G_{12}	15.0 ksi (103 MPa) 17000 με 0.87 Msi (6.0 GPa)

Loading	Out-of-plane Property	RT/ Dry
3-Ten 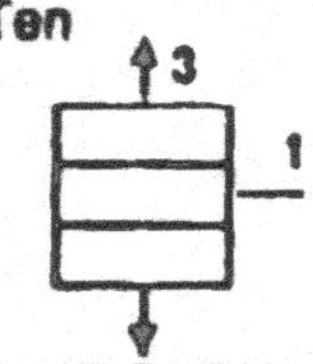	F_3^{tu} ε_3^{tu} E_3^t ν_{31}^t ν_{32}^t	8.00 ksi (55.2 (MPa) 5700 με 1.40 Msi (9.65 GPa)
3-Comp 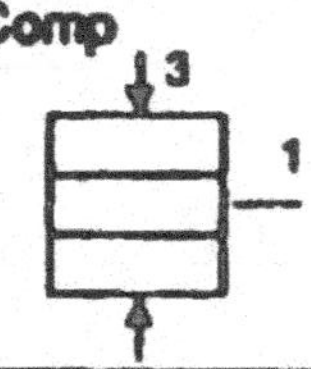	F_3^{cu} ε_3^{cu} E_3^c ν_{31}^c ν_{32}^c	30.0 ksi (207 MPa) 21500 με 1.40 Msi (9.65 GPa)
13-Shear 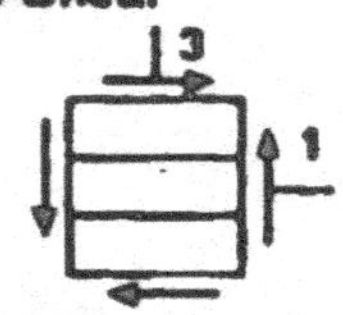	F_{13}^{su} γ_{13}^{su} G_{13}^t	12.0 ksi (82.7 MPa) 4000 με 0.87 Msi (6.0 GPa)
23-Shear 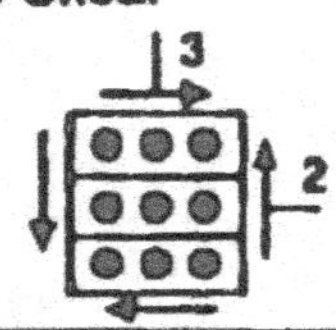	F_{23}^{su} γ_{23}^{su} G_{23}	12.0 ksi (82.7 MPa) 22000 με 0.55 Msi (3.8 GPa)

Notes:
1. Transverse isotropy assumed in 2-3 plane for stiffness, strength, strain.
2. Failure strain = strength/modulus.
3. 60% fiber volume.

TABLE 7.2.3.1(d) *Intermediate modulus carbon/epoxy [0_3, 90] laminate typical 3-D properties.*

Loading	Inplane Property	RT/ Dry	Loading	Out-of-Plane Property	RT/ Dry
x-Ten	F_x^{tu} ε_x^{tu} E_x^t ν_{xy}^t	140. ksi (965 MPa) 9330 με 15.0 Msi (103 GPa) 0.10	z-Ten	F_z^{tu} ε_z^{tu} E_z^t ν_{zx}^t ν_{zy}^t	3.40 ksi (23.4 MPa) 3040 με 1.12 Msi (7.72 GPa)
x-Comp	F_x^{cu} ε_x^{cu} E_x^c ν_{xy}^c ν_{xz}^c	111 ksi (765 MPa) 8600 με 12.9 Msi (88.9 GPa) 0.12	z-Comp	F_z^{cu} ε_z^{cu} E_z^c ν_{zx}^c ν_{zy}^c	60.0 ksi (414 MPa) 3660 με 1.64 Msi (11.3 GPa)
y-Ten	F_y^{tu} ε_y^{tu} E_y^t ν_{yx}^t	35.0 ksi (241 MPa) 6210 με 5.64 Msi (38.9 GPa) 0.03	xz-Shear	F_{xz}^{su} γ_{xz}^{su} G_{xz}	4.06 ksi (28.0 MPa) 7700 με 0.53 Msi (3.7 GPa)
y-Comp	F_y^{cu} ε_y^{cu} E_y^c ν_{yx}^c ν_{yz}^c	72.9 ksi (503 MPa) 12900 με 5.66 Msi (39.0 GPa) 0.029	yz-Shear	F_{yz}^{su} γ_{yz}^{su} G_{yz}	6.15 ksi (42.4 MPa) 9300 με 0.66 Msi (4.6 GPa)
xy-Shear	F_{xy}^{su} γ_{xy}^{su} G_{xy}	15.3 ksi (105 MPa) 22000 με 0.70 Msi (4.8 GPa)			

Notes:

1. [0_3,90] laminate (t =.59 in.) average properties from reference 4, fiber volume = 61.4%, void = 0.04%.
2. Fiber placed, autoclave cured, flat panel data, 5-7 specimen average.
3. Failure strain = strength/modulus.

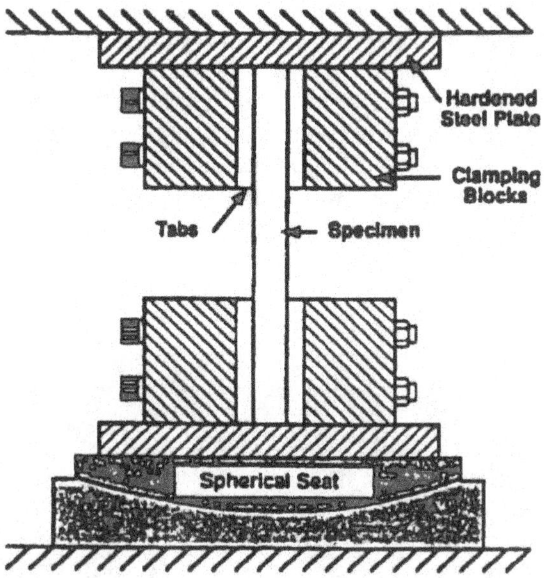

FIGURE 7.2.3.1(h) *Uniaxial thick-section compression test fixture - David Taylor Research Center (DTRC).*

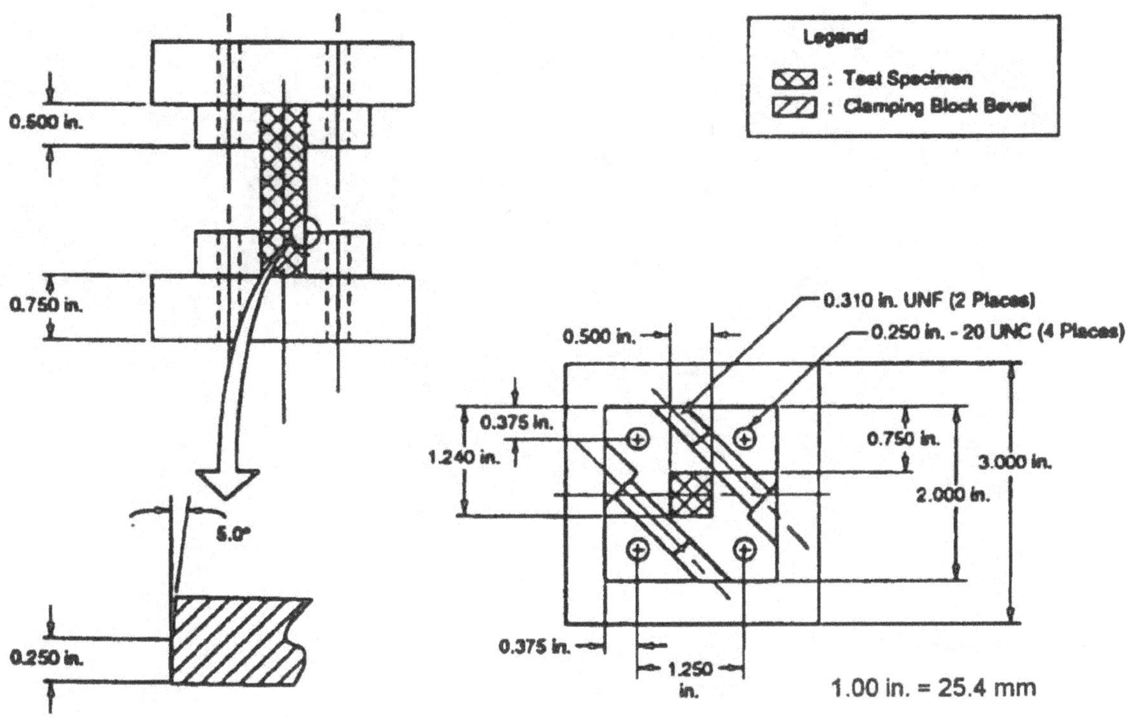

FIGURE 7.2.3.1(i) *Uniaxial thick-section compression test fixture - Alliant Techsystems Inc.*

There are in current use two distinctively different multiaxial composite material testing techniques, associated mechanical testing load frames, and specimen fixturing arrangements. One method utilizes testing machines that apply loads/displacements along primary, mutually orthogonal coordinate axes to lineal test specimens. This broad class of machines consists of planar biaxial machines (Figure 7.2.3.2(a)), and true triaxial test frames (Figure 7.2.3.2(b)). The second method employs a class of machines that apply loads/displacements on tubular test specimens. The biaxial machines consist of a basic uniaxial - universal testing machine that has the additional capability to also apply a torque about the primary axis of the cylindrical specimen. The corresponding triaxial machine (Figure 7.2.3.2(c)) is similar to the biaxial test frame except that it has the added capacity to also apply either an internally or externally induced pressure differential across the wall of the cylindrical test specimen.

FIGURE 7.2.3.2(a) *MTS biaxial tension/compression testing system.*

FIGURE 7.2.3.2(b) *Alliant Techsystems - University of Wyoming triaxial tension/compression testing system.*

FIGURE 7.2.3.2(c) *Three-dimensional axial/torsion* pressure testing systems.

7.2.3.2.1 Lineal test specimens/techniques

This material testing method utilizes the lineal test specimens, shown in Figure 7.2.3.2.1, such as cruciform or plate configured specimens for biaxial testing, and cubes or parallelopipeds for three-dimensional load applications.

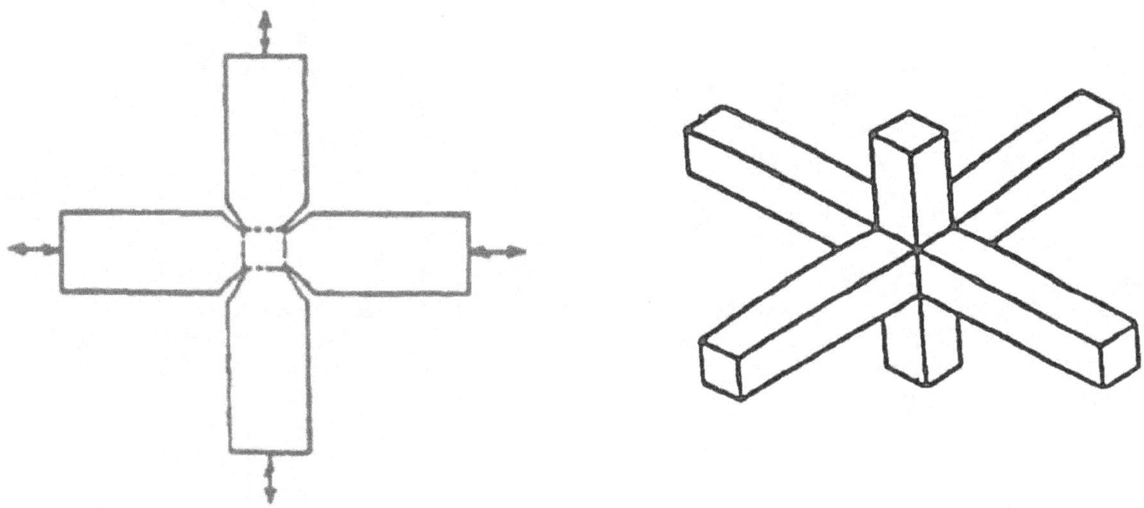

FIGURE 7.2.3.2.1 *Lineal biaxial test specimen geometries.*

Simultaneous, multiaxial tension/compression testing may be undertaken by applying loads along the principal, mutually orthogonal axes of the 2-D and 3-D specimens. Multiaxial testing is necessary for determining actual material strength/failure envelopes as well as for identifying failure mechanisms. This data is required in developing true multidirectional material constitutive equations and appropriate failure criteria.

There are available commercially fabricated, true biaxial machines for testing cruciform or plate configured material test specimens. These machines are typically of the servohydraulic-actuated type. There also exist special, non-commercially built, screw driven biaxial load frames. Both of these biaxial machine types are capable of simultaneously inducing tensile and/or compressive loads along two orthogonal axes. Thus, these load frames can be used to develop any general biaxial normal stress field within the test region of the material specimen. Special specimen fixturing, such as brush/comb flexible tabbing, has been developed and may be required to permit unrestricted in-plane movement and transverse constraints in order to minimize out-of-plane bending in biaxial tension/compression testing. This flexible specimen tabbing operates in a similar fashion as the brush/bearing platens typically used in compression testing of concrete.

True triaxial machines have also become available. These load frames have the capability of testing cubic anisotropic material specimens. These multidirectional, material testing machines may be either servo-hydraulically actuated or screw driven. Both of these types of three-dimensional machines may be used to apply any general combination of three-dimensional normal stress states to tabbed cubic test specimens. To the best of our knowledge, no comprehensive three-dimensional composite material testing has yet been undertaken with this equipment since special test specimen fixturing for these machines is currently under development and calibration of the load frames is in progress.

Both biaxial and triaxial machines require control systems that essentially maintain the test specimen centroid in a stationary position. This computer software - load frame control is a necessary feature in that it is recommended that the specimen not be subjected to unwanted eccentric loading conditions. Lack of proper test frame displacement or load control may produce erroneous test measurements, inappropriate material failure mechanisms, as well as failures occurring outside the instrumented gage areas. In summary, the proper utilization of these one-, two-. and three-dimensional load frames requires special test specimen holding fixtures, well-designed specimen geometries, and effective tabbing and/or specimen end constraining methods. Extreme care has to be exercised in designing test specimens, fixtures, tabbing, and load application methods in order to avoid developing undesirable edge or end effects along with stress concentrations. The three-dimensional test method, described above, is often referred to as a true triaxial method since the cubical test specimen geometry permits complete freedom as to the fiber lay-up orientations in relation to the load application axes.

7.2.3.2.2 Cylindrical test specimens/techniques

To date, the most frequently used multiaxial, two- and three-dimensional, composite material testing method utilizes cylindrical test specimens shown in Figure 7.2.3.2.2. Predominantly, these test specimens are thin-walled tubes. There are well over a hundred commercially built biaxial machines which can apply an axial load (tension or compression) in conjunction with a torsional twisting loading about the longitudinal axis of cylindrical test specimens.

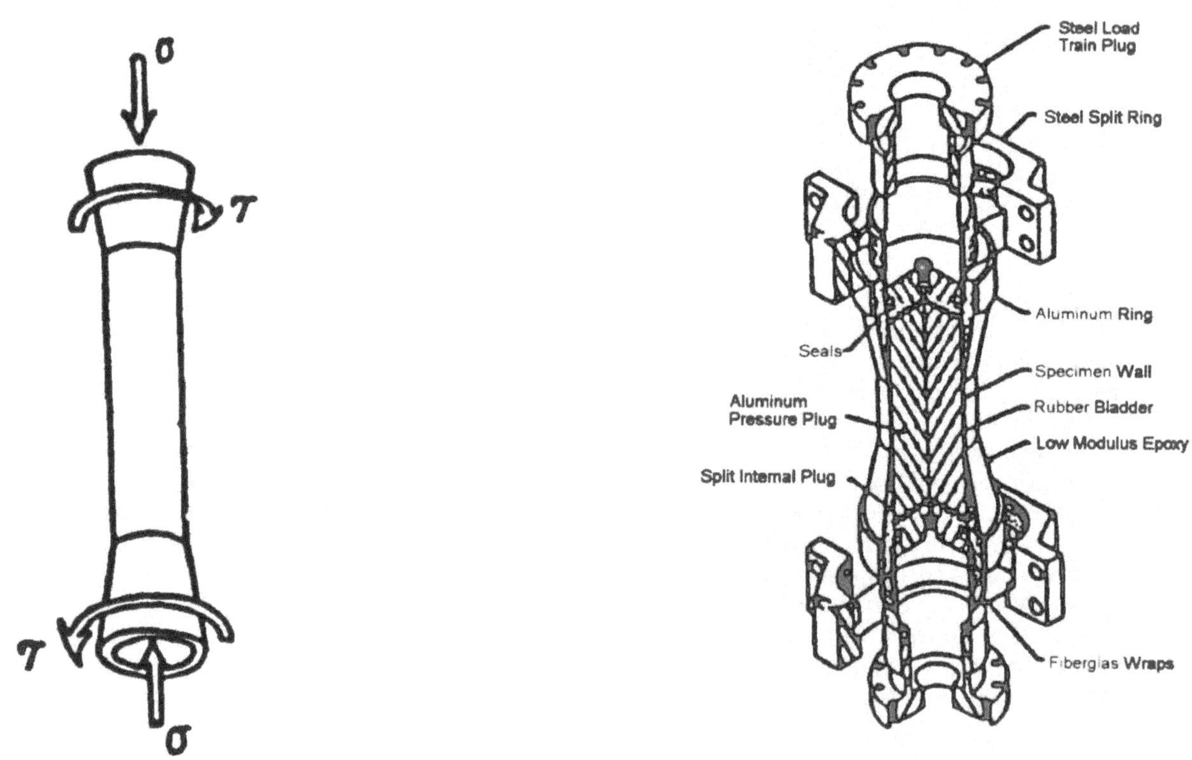

FIGURE 7.2.3.2.2 *Biaxial and triaxial thin tubular test specimen geometries, University of Utah triaxial specimen fixturing.*

The triaxial machines are similar to the 2-D test frames in their axial and torsional loads application utility. However, these load frames also have the additional capability of applying either an internal or external hydraulic pressurization across the wall thickness of hollow cylindrical specimens. By the very nature of hoop construction lay-ups of composite material cylinders, it would appear that this testing technique is very well suited for investigating material parameters and failure mechanisms of filament-wound test specimens. Typically, these cylindrical specimens do not exhibit edge effects in the gage section due to the geometric hoop continuity of cylindrical specimens. However, end effects such as brooming and shearing may be a problem and require careful structural design and analyses of the connection detail and specimen configuration. The potential occurrence of structural instability such as buckling of the cylindrical test specimens, that are subjected to either an individual or a combination of axial, torsional, and pressurization loadings is a major consideration with this testing method. The development of any structural buckling of the test cylinders would mask material strength measurements. It should also be noted that this multiaxial testing technique has been used primarily for investigating only thin-walled tubular specimens.

7.2.4 Theoretical property determination

In considering the use of theoretical procedures to determine the mechanical properties of composite materials the most fundamental level that can be addressed is that of the individual constituents, or the micromechanics level. A theoretical development of composite micromechanics is summarized in Volume 3 Section 4.2.2.1 of this Handbook and in Section 4 of Reference 7.2.4. 3-D laminate properties can be determined from constituent data using micromechanical analyses and these references should be consulted for additional information and references on this topic.

Since properties at the lamina or laminate level are typically used for the analysis of a composite structure, only these properties will be discussed in this section and all analyses considered are linear elastic.

7.2.4.1 3D lamina property determination

In Section 7.2, the nine independent elastic material properties required for a 3D lamina based analysis were listed as:

$$E_1,\ E_2,\ E_3\ ,G_{12},\ G_{13},\ G_{23},\ \nu_{12},\ \nu_{13},\ \nu_{23} \qquad \text{7.2.4.1 (a)}$$

Of these properties E_1, E_2, G_{12} and ν_{12} can be readily generated by conventional experimental methods. Methods for determining out-of-plane properties are discussed in Section 7.2.3. In the absence of experimental data for these properties, the assumption of transverse isotropy in the 2-3 plane is often reasonable. The validity of this assumption has been demonstrated by the experimental data available in References 7.2.4.1(a) - (c). The assumption of transverse isotropy implies

$$E_3 = E_2,\ G_{13} = G_{12},\ \nu_{13} = \nu_{12},\ \text{and}\ G_{23} = \frac{E_2}{2(1+\nu_{23})} \qquad \text{7.2.4.1 (b)}$$

Even with this simplifying assumption ν_{23} must be measured or estimated for full knowledge of the nine independent elastic material properties.

Values for ν_{23} that have been experimentally determined have been reported in References 7.2.4.1(a) - (c). A value for ν_{23} determined in compression for T300/5208 is reported in Ref. 7.2.4.1(a). A value for ν_{23} determined in tension and compression for T300/5208 is reported in Reference 7.2.4.1(b). Values for ν_{23} determined in compression for AS4/3501-6 and S2/3501-6 are reported in 7.2.4.1(c) and can be found in Table 7.2.4.1.

TABLE 7.2.4.1 *3-D elastic constants for carbon and S2 glass reinforced epoxy (Reference 7.2.4.1(c)), E and G in Msi (GPa).*

	AS4/3501-6 59.5% FVF	S2/3501-6 56.5% FVF
E_1 [1]	16.48 (113.6) (3.7)[2]	7.15 (49.3) (4.0)
E_2 [1]	1.40 (9.65) (3.6)	2.13 (14.7) (2.2)
E_3	1.40[3] (9.65)	2.13[3] (14.7)
ν_{12} [1]	0.334 (3.0)	0.296 (4.1)
ν_{13} [1]	0.328 (1.2)	0.306 (2.8)
ν_{23} [1]	0.540 (1.6)	0.499 (1.4)
G_{12}	0.87[4] (6.0)	0.98[4] (6.8)
G_{13}	0.87[5] (6.0)	0.98[4] (6.8)
G_{23}	0.45[6] (3.1)	0.71[4] (4.9)

1 E_1, E_2, ν_{12}, ν_{13}, and ν_{23} determined from thick, flat, compression test specimens
2 coefficient of variation (%)
3 E_3 assumed equal to E_2
4 G_{12} determined from $[\pm45]_{2s}$ tension test
5 G_{13} assumed equal to G_{12}
6 G_{23} from assumption of transverse isotropy

The need for all nine independent elastic constants does not imply that a 3-D analysis will be sensitive to the choice of the through-thickness material properties just discussed. For instance, a choice of ν_{23} of 0.50 versus 0.40 (a 20% difference) may only result in a 2% difference in the stress or strain results from a laminate or structural analysis. This sensitivity of a particular analysis to a particular material property should be evaluated through a parametric study if the value of the property is uncertain.

Likewise, the use of a linear analysis when certain material properties are extremely nonlinear (i.e., in-plane and through-thickness shear modulus) may not affect laminate or structural analysis and this too should be considered in 3-D analysis.

7.2.4.2 3-D laminate property determination

As for the case of a 3-D lamina properties, Section 7.2 lists the nine independent elastic material properties required for a 3-D laminate based analysis as:

$$E_x,\ E_y,\ E_z,\ G_{xy},\ G_{xz},\ G_{yz},\ \nu_{xy},\ \nu_{xz},\ \nu_{yz} \qquad 7.4.2.2\ (a)$$

E_x, E_y, G_{xy}, ν_{xy} can be readily determined by conventional experimental or theoretical methods. Theoretically they can be determined using classical lamination theory as presented in Volume 3 Section 4.3.2 of this Handbook. The determination of the remaining out-of-plane laminate properties present a much greater challenge than for the in-plane properties. Little experimental data exists for out-of-plane laminate properties and the test methodologies used to generate them can be described as very specific to the programs they have been used for, such as those in References 7.2.4.1(a) - (c).

A number of methods have been developed to theoretically predict the out-of-plane properties based on in-plane lamina properties (References 7.2.4.2(a) - (h)). These methods basically replace a layered inhomogeneous media of orthotropic layers with a homogeneous anisotropic media. This replacement is termed "smearing" and the resulting effective material properties are referred to as "smeared properties". These smeared anisotropic properties are commonly used in the analysis of composite structures. If average, global stress states or average displacements are sufficient for the analysis being conducted then an analysis with smeared properties is all that would be needed. If local stress states are needed then other analyses techniques must be employed such as a "global-local" technique. In this approach smeared anisotropic properties are used to determine global stress states, then this information is used to interrogate stress states in specific regions of concern on a ply-by-ply basis, therefore avoiding the costly use of a ply-by-ply analysis for an entire structure made of a thick-section composite material. The use of this global-local analysis technique is commonly referred to as the most rational way to approach the problem of design and analysis for thick composite materials.

The solution methods available to generate smeared anisotropic 3-D properties range from approximate formulations (Reference 7.2.4.2(a)) to exact formulations not including bending-extensional coupling (Reference 7.2.4.2(c)). The exact solutions by Pagano (Reference 7.2.4.2(c)) and Sun (Reference 7.2.4.2(b)) lend themselves to simple programming on personal computers. In fact Trethewey et al. (Reference 7.2.4.2(d)) and Peros (Reference 7.2.4.2(e)) have encoded the Pagano solution while Sun has encoded his own solution for a personal computer.

Tables 7.2.4.2(a) and (b) contain 3D laminate elastic constants for six laminate configurations and two materials as determined by laminate plate theory (LPT), and by the Pagano, Sun, and Roy solutions (Reference 7.2.4.2(g), (h)). Table 7.2.4.1 lists the lamina input properties used in each of the analyses. The three exact solutions (LPT, Pagano, Sun) yield identical results for both in-plane and through thickness properties for all of the cases presented. The results from the approximate solution by Roy differ from the others in the z-direction properties up to 12% in some cases.

Data verifying the results of these analyses are limited due to the difficulty in generating 3D experimental data. Data that does exist is documented in References 7.2.4.1(a) - (c), 7.2.4.2(h) and (i). Table 7.2.4.2(c) contains a comparison of theoretical predictions using the linear elastic theory by Pagano and experimental data from Reference 7.2.4.1(c) and Reference 7.2.4.2(i).

TABLE 7.2.4.2(a) *3-D effective properties of various AS4/3501-6 laminates, continued on next page.*

Laminate Properties for AS4/3501-6, E and G in Msi												
	$[0_2/90]_s$				$[0/90]_{2s}$				$[0/90/\pm45]_s$			
	LPT	Pagano	Sun	Roy	LPT	Pagano	Sun	Roy	LPT	Pagano	Sun	Roy
E_x	11.5	11.5	11.5	11.5	9.01	9.00	9.00	9.00	6.68	6.68	6.68	6.67
E_y	6.48	6.47	6.47	6.47	9.01	9.00	9.00	9.00	6.68	6.68	6.68	6.68
E_z	--	1.80	1.80	1.65	--	1.82	1.82	1.60	--	1.82	1.82	1.61
ν_{xy}	0.073	0.073	0.074	0.072	0.052	0.052	0.053	0.052	0.297	0.297	0.298	0.296
ν_{xz}	--	0.488	0.489	0.402	--	0.506	0.507	0.438	--	0.375	0.376	0.318
ν_{yz}	--	0.519	0.520	0.465	--	0.506	0.508	0.427	--	0.375	0.376	0.317
G_{xy}	0.870	0.870	0.870	0.870	0.870	0.870	0.870	0.870	2.58	2.57	2.57	2.57
G_{xz}	--	0.664	0.664	0.780	--	0.593	0.593	0.612	--	0.593	0.593	0.627
G_{yz}	--	0.536	0.536	0.503	--	0.593	0.593	0.573	--	0.593	0.593	0.519

Laminate Properties for AS4/3501-6, E and G in Msi												
	$[\pm30]_{2s}$				$[\pm45]_{2s}$				$[\pm60]_{2s}$			
	LPT	Pagano	Sun	Roy	LPT	Pagano	Sun	Roy	LPT	Pagano	Sun	Roy
E_x	6.84	6.84	6.84	6.84	2.94	2.94	2.94	2.94	1.77	1.77	1.77	1.77
E_y	1.77	1.77	1.77	1.77	2.94	2.94	2.94	2.94	6.84	6.84	6.83	6.85
E_z	--	1.66	1.66	1.50	--	1.82	1.82	1.71	--	1.66	1.66	1.74
ν_{xy}	1.14	1.41	1.41	1.13	0.691	0.691	0.691	0.689	0.295	0.295	0.295	0.294
ν_{xz}	--	-0.095	-0.094	-0.197	--	0.165	0.165	0.211	--	0.390	0.390	0.434
ν_{yz}	--	0.390	0.390	0.434	--	0.165	0.165	0.211	--	-0.095	-0.095	-0.197
G_{xy}	3.43	3.43	3.42	3.42	4.28	4.28	4.27	4.27	3.43	3.43	3.42	3.42
G_{xz}	--	0.705	0.705	0.708	--	0.593	0.593	0.596	--	0.512	0.512	0.515
G_{yz}	--	0.512	0.512	0.515	--	0.593	0.593	0.596	--	0.705	0.705	0.708

TABLE 7.2.4.2(a) *3-D effective properties of various AS4/3501-6 laminates, concluded.*

Laminate Properties for AS4/3501-6, E and G in GPa												
	$[0_2/90]_s$				$[0/90]_{2s}$				$[0/90/\pm45]_s$			
	LPT	Pagano	Sun	Roy	LPT	Pagano	Sun	Roy	LPT	Pagano	Sun	Roy
E_x	79.3	79.3	79.3	79.3	62.1	62.1	62.1	62.1	46.1	46.1	46.1	46.0
E_y	44.7	44.6	44.6	44.6	62.1	62.1	62.1	62.1	46.1	46.1	46.1	46.1
E_z	--	12.4	12.4	11.4	--	12.5	12.5	11.0	--	12.5	12.5	11.1
ν_{xy}	0.073	0.073	0.074	0.072	0.052	0.052	0.053	0.052	0.297	0.297	0.298	0.296
ν_{xz}	--	0.488	0.489	0.402	--	0.506	0.507	0.438	--	0.375	0.376	0.318
ν_{yz}	--	0.519	0.520	0.465	--	0.506	0.508	0.427	--	0.375	0.376	0.317
G_{xy}	6.00	6.00	6.00	6.00	6.00	6.00	6.00	6.00	17.8	17.7	17.7	17.7
G_{xz}	--	4.58	4.58	5.38	--	4.09	4.09	4.22	--	4.09	4.09	4.32
G_{yz}	--	3.70	3.70	3.47	--	4.09	4.09	3.95	--	4.09	4.09	3.58

Laminate Properties for AS4/3501-6, E and G in GPa												
	$[\pm30]_{2s}$				$[\pm45]_{2s}$				$[\pm60]_{2s}$			
	LPT	Pagano	Sun	Roy	LPT	Pagano	Sun	Roy	LPT	Pagano	Sun	Roy
E_x	47.2	47.2	47.2	47.2	20.3	20.3	20.3	20.3	12.2	12.2	12.2	12.2
E_y	12.2	12.2	12.2	12.2	20.3	20.3	20.3	20.3	47.2	47.2	47.1	47.2
E_z	--	11.4	11.4	10.3	--	12.5	12.5	11.8	--	11.4	11.4	12.0
ν_{xy}	1.14	1.41	1.41	1.13	0.691	0.691	0.691	0.689	0.295	0.295	0.295	0.294
ν_{xz}	--	-0.095	-0.094	-0.197	--	0.165	0.165	0.211	--	0.390	0.390	0.434
ν_{yz}	--	0.390	0.390	0.434	--	0.165	0.165	0.211	--	-0.095	-0.095	-0.197
G_{xy}	23.6	23.6	23.6	23.6	29.5	29.5	29.4	29.4	23.6	23.6	23.6	23.6
G_{xz}	--	4.86	4.86	4.88	--	4.09	4.09	4.11	--	3.53	3.53	3.55
G_{yz}	--	3.53	3.53	3.55	--	4.09	4.09	4.11	--	4.86	4.86	4.88

TABLE 7.2.4.2(b) *3-D effective properties of various S2/3501-6 laminates,* *continued on next page.*

Laminate Properties for S2/3501-6, E and G in Msi												
	$[0_2/90]_s$				$[0/90]_{2s}$				$[0/90/\pm45]_s$			
	LPT	Pagano	Sun	Roy	LPT	Pagano	Sun	Roy	LPT	Pagano	Sun	Roy
E_x	5.52	5.52	5.52	5.52	4.68	4.68	4.68	4.68	3.89	3.89	3.89	3.89
E_y	3.83	3.83	3.83	3.83	4.68	4.68	4.68	4.68	3.89	3.89	3.89	3.89
E_z	--	2.38	2.38	2.30	--	2.40	2.40	2.29	--	2.40	2.40	2.32
ν_{xy}	0.166	0.166	0.166	0.165	0.136	0.136	0.136	0.135	0.281	0.281	0.281	0.280
ν_{xz}	--	0.405	0.405	0.359	--	0.435	0.435	0.393	--	0.362	0.362	0.329
ν_{yz}	--	0.459	0.459	0.427	--	0.435	0.435	0.392	--	0.362	0.362	0.329
G_{xy}	0.980	0.980	0.980	0.980	0.980	0.980	0.980	0.980	1.52	1.52	1.52	1.52
G_{xz}	--	0.870	0.870	0.918	--	0.823	0.823	0.830	--	0.823	0.823	0.838
G_{yz}	--	0.782	0.782	0.754	--	0.823	0.823	0.811	--	0.823	0.823	0.781

Laminate Properties for AS4/3501-6, E and G in Msi												
	$[\pm30]_{2s}$				$[\pm45]_{2s}$				$[\pm60]_{2s}$			
	LPT	Pagano	Sun	Roy	LPT	Pagano	Sun	Roy	LPT	Pagano	Sun	Roy
E_x	4.45	4.45	4.45	4.45	2.88	2.88	2.88	2.88	2.26	2.26	2.26	2.26
E_y	2.26	2.26	2.26	2.26	2.88	2.88	2.88	2.88	4.45	4.45	4.45	4.45
E_z	--	2.30	2.30	2.16	--	2.40	2.40	2.33	--	2.30	2.30	2.44
ν_{xy}	0.546	0.546	0.546	0.545	0.468	0.468	0.468	0.467	0.278	0.278	0.278	0.277
ν_{xz}	--	0.200	0.200	0.136	--	0.267	0.267	0.284	--	0.387	0.387	0.406
ν_{yz}	--	0.387	0.387	0.406	--	0.267	0.267	0.284	--	0.200	0.200	0.136
G_{xy}	1.79	1.79	1.79	1.79	2.06	2.06	2.06	2.06	1.79	1.79	1.79	1.79
G_{xz}	--	0.895	0.895	0.895	--	0.823	0.823	0.823	--	0.762	0.762	0.763
G_{yz}	--	0.762	0.762	0.763	--	0.823	0.823	0.823	--	0.985	0.985	0.895

TABLE 7.2.4.2(b) *3-D effective properties of various S2/3501-6 laminates, concluded.*

Laminate Properties for S2/3501-6, E and G in GPa												
	$[0_2/90]_s$				$[0/90]_{2s}$				$[0/90/\pm45]_s$			
	LPT	Pagano	Sun	Roy	LPT	Pagano	Sun	Roy	LPT	Pagano	Sun	Roy
E_x	38.1	38.1	38.1	38.1	32.3	32.3	32.3	32.3	26.8	26.8	26.8	26.8
E_y	26.4	26.4	26.4	26.4	32.3	32.3	32.3	32.3	26.8	26.8	26.8	26.8
E_z	--	16.4	16.4	15.9	--	16.5	16.5	15.8	--	16.5	16.5	16.0
ν_{xy}	0.166	0.166	0.166	0.165	0.136	0.136	0.136	0.135	0.281	0.281	0.281	0.280
ν_{xz}	--	0.405	0.405	0.359	--	0.435	0.435	0.393	--	0.362	0.362	0.329
ν_{yz}	--	0.459	0.459	0.427	--	0.435	0.435	0.392	--	0.362	0.362	0.329
G_{xy}	6.76	6.76	6.76	6.76	6.76	6.76	6.76	6.76	10.5	10.5	10.5	10.5
G_{xz}	--	6.00	6.00	6.33	--	5.67	5.67	5.72	--	5.67	5.67	5.78
G_{yz}	--	5.39	5.39	5.20	--	5.67	5.67	5.59	--	5.67	5.67	5.04

Laminate Properties for AS4/3501-6, E and G in GPa												
	$[\pm30]_{2s}$				$[\pm45]_{2s}$				$[\pm60]_{2s}$			
	LPT	Pagano	Sun	Roy	LPT	Pagano	Sun	Roy	LPT	Pagano	Sun	Roy
E_x	30.7	30.7	30.7	30.7	19.9	19.9	19.9	19.9	15.6	15.6	15.6	15.6
E_y	15.6	15.6	15.6	15.6	19.9	19.9	19.9	19.9	30.7	30.7	30.7	30.7
E_z	--	15.9	15.9	14.9	--	16.5	16.5	16.1	--	15.9	15.9	16.8
ν_{xy}	0.546	0.546	0.546	0.545	0.468	0.468	0.468	0.467	0.278	0.278	0.278	0.277
ν_{xz}	--	0.200	0.200	0.136	--	0.267	0.267	0.284	--	0.387	0.387	0.406
ν_{yz}	--	0.387	0.387	0.406	--	0.267	0.267	0.284	--	0.200	0.200	0.136
G_{xy}	12.3	12.3	12.3	12.3	14.2	14.2	14.2	14.2	12.3	12.3	12.3	12.3
G_{xz}	--	6.17	6.17	6.17	--	5.67	5.67	5.67	--	5.25	5.25	5.26
G_{yz}	--	5.25	5.25	5.26	--	5.67	5.67	5.67	--	6.79	6.79	6.17

TABLE 7.2.4.2(c) *Comparison of theoretical and experimental laminate results* $[0_2/90]_{ns}$ *from Reference 7.2.4.1(c),* $[0_3/90]_{ns}$ *from Reference 7.2.4.2(i), E and G in Msi (GPa).*

	AS4/3501-6 $[0_2/90]_{ns}$		S2 glass/3501-6 $[0_2/90]_{ns}$		AS4/3501-6 $[0_3/90]_{ns}$	
	Theoretical	Experimental	Theoretical	Experimental	Theoretical	Experimental
E_x	11.53 (79.5)	11.63^A (80.2) $[4.0]^B[32]$	5.52 (38.1)	5.82 (40.1) [6.9][32]	12.80 (88.3)	12.90^A (88.9)
E_y	6.47 (44.6)		3.83 (26.4)		5.27 (36.3)	5.66^A (39.0)
E_z	1.80 (12.4)		2.38 (16.4)		1.63 (11.2)	1.64^A (11.3)
ν_{xy}	0.073	0.069^A $[6.7][7]^C$	0.166	0.166 [4.3][7]	0.090	0.120^A
ν_{xz}	0.488	0.469^A [3.0][14]	0.405	0.363 [2.7][14]	0.440	---
ν_{yz}	0.519		0.459		0.452	---
G_{xy}	0.87 (6.0)		0.98 (6.8)		0.87 (6.0)	0.70^D (4.8)
G_{xz}	0.73 (5.0)		0.78 (5.4)		0.72 (5.0)	0.53^D (3.7)
G_{yz}	0.63 (4.3)		0.64 (4.4)		0.54 (3.7)	0.66^D (4.6)

A data from thick, flat, compression test specimens
B coefficient of variation (%)
C number of data points in average
D data from Iosipescu shear test specimens

7.2.5 Test specimen design considerations

This section is reserved for future work.

7.3 STRUCTURAL ANALYSIS METHODS FOR THICK-SECTION COMPOSITES

This section is reserved for future work.

7.4 PHYSICAL PROPERTY ANALYSIS REQUIRED FOR THICK-SECTION COMPOSITE THREE-DIMENSIONAL ANALYSIS

This section is reserved for future work.

7.5 PROCESS ANALYSIS METHODS FOR THICK-SECTION COMPOSITES

This section is reserved for future work.

7.6 FAILURE CRITERIA

This section is reserved for future work.

7.7 FACTORS INFLUENCING THICK-SECTION ALLOWABLES (i.e., SAFETY MARGINS)

This section is reserved for future work.

7.8 THICK LAMINATE DEMONSTRATION PROBLEM

This section is reserved for future work.

This page intentionally left blank

REFERENCES

7.2.1 Jones, R.M., *Mechanics of Composite Materials*, 1975 Edition, Hemisphere Publishing Corporation.

7.2.2.2 Lekhnitskii, S.G., *Elasticity of an Anisotropic Body*, p. 30.

7.2.3.1(a) Camponeschi, E.T., Jr., *Compression Response of Thick-Section Composite Materials*, DTRC-SME-90/90, August 1990.

7.2.3.1(b) Abdallah, M.G., et al., *A New Test Method for External Hydrostatic Compressive Loading of Composites in Ring Specimens*, Fourth Annual Thick Composites in Compression Workshop, Knoxville, TN, June 27-28, 1990.

7.2.3.1(c) Bode, J.H., *A Uniaxial Compression Test Fixture for Testing Thick-Section Composites*, Fourth Annual Thick Composites in Compression Workshop, Knoxville, TN, June 27-28, 1990.

7.2.3.1(d) Goeke, E.C., "Comparison of Compression Test Methods for "Thick" Composites," *Composite Materials; Testing and Design (Eleventh Volume), ASTM STP 1206*, ed. E.T. Camponeschi, American Society for Testing and Materials, 1993.

7.2.4. *Engineering Materials Handbook, Vol. 1, Composites*, ASM International, 1987.

7.2.4.1(a) Knight, M., "Three-Dimensional Elastic Moduli of Graphite/Epoxy Composites," *Journal of Composite Materials*, Vol. 16, 1982, pp. 153-159.

7.2.4.1(b) Sandorf, P.E., "Transverse Shear Stiffness of T300/5208 Graphite-Epoxy in Simple Bending," Lockheed-California Co. Report No. LR 29763, Burbank, CA, Nov. 30, 1981.

7.2.4.1(c) Camponeschi, E.T., Jr., "Compression Response of Thick-Section Composite Materials," David Taylor Research Center Report No. DTRC SME-90-60, Oct. 1990.

7.2.4.2(a) Christensen, R.M. and Zywicz, E., "A Three-Dimensional Constitutive Theory for Fiber Composite Laminated Media," *Journal of Applied Mechanics*, Jan. 1990.

7.2.4.2(b) Sun, C.T. and Li, S., "Three-Dimensional Effective Elastic Constants for Thick Laminates," *Journal of Composite Materials*, Vol. 22, No. 7, July, 1988.

7.2.4.2(c) Pagano, N.J., "Exact Moduli of Anisotropic Laminates," *Mechanics of Composite Materials*, ed. G. Sendeckyj, Academic Press, 1984, pp. 23-44.

7.2.4.2(d) Trethewey, B.R., Jr., Wilkins, D.J., and Gillespie, J.W., Jr., "Three-Dimensional Elastic Properties of Laminate Composites," CCM Report 89-04, University of Delaware Center for Composite Materials, 1989.

7.2.4.2(e) Peros, V., "Thick-Walled Composite Material Pressure Hulls: Three-Dimensional Laminate Analysis Considerations," University of Delaware Masters Thesis, Dec. 1987.

7.2.4.2(f) Herakovich, C.T., "Composite Laminates With Negative Through-The-Thickness Poisson's Ratios," *Journal of Composite Materials*, Vol. 18, Sept., 1984.

7.2.4.2(g) Roy, A.K. and Tsai, S.W., "Three-Dimensional Effective Moduli of Orthotropic and Symmetric Laminates," to appear in the *Journal of Applied Mechanics*, Trans. of ASME, 1991.

7.2.4.2(h) Roy, A.K. and Kim, R.Y., "Effective Interlaminar Normal Stiffness and Strength of Orthotropic Laminates," Proceeding of the 45th Meeting of the Mechanical Failures Prevention Group, Vibration Institute, Willowbrook, IL, 1991, pp. 165-173.

7.2.4.2(i) Abdallah, M., Williams, T.O., and Muller, C.S., "Experimental Mechanics of Thick Laminates: Flat Laminate Mechanical Property Characterization," Hercules, Inc. IR&D Progress Report No. DDR 153253, Misc: 2/2-3249, June, 1990.

This page intentionally left blank

CHAPTER 8 SUPPORTABILITY

Page

8.1 INTRODUCTION

Composite structures offer significant benefits over metallic components in terms of lighter weight, and fatigue and corrosion resistance. These performance benefits are often compromised by increased acquisition and support costs. While repair techniques for composites are often different from those for metals, bonded repair techniques exhibit similar traits between both material types. Many of the traditional bonding concerns (i.e., surface preparation, uniform heating/curling, etc.) discussed in this section are applicable to any bonded repair application. This section will provide design teams with information needed to develop high performance composite structural components that can be supported by field-level technicians using a minimum of facilities and equipment.

Supportability is an integral part of the design process that ensures supportability requirements are incorporated in the design and logistics resources are defined to support the system during its operating or useful life. Support resource requirements include the skills, tools, equipment, facilities, spares, techniques, documentation, data, materials, and analysis required to ensure that a composite component maintains structural integrity over its intended lifetime. When the load carrying capability of an aircraft is compromised, (i.e., loss of design function), the damaged structure must be restored quickly and at low cost. Customer requirements can dictate maintenance philosophy, materials availability, and repair capabilities that a design team must incorporate throughout the design process.

Since the operating and support cost of a vehicle continues to escalate throughout its life, it becomes imperative to select and optimize those designs that maximize supportability. Life cycle cost, being comprised of research and development, acquisition, operational and support, and disposal costs, is often a crucial customer requirement for any new weapon system or commercial transport. Often, design changes that enhance producibility, improve vehicle availability, and reduce operational and support costs, far outweigh the short-term increases in acquisition costs. Lost airline profits and reduced wartime readiness are a direct result of designs that did not incorporate supportability early in the design process. Telltale indicators of non-supportable designs include expensive spares, excessive repair times, and unneeded inspections.

Aircraft users are often constrained to perform maintenance during aircraft turnaround, after each days usage, and during scheduled maintenance. Repair time limitations can range from several minutes to several days. In each case users of aircraft containing composite components require durable structures that, when damaged, can be repaired within the available support infrastructure including skills, materials, equipment, and technical data.

Composite designs are usually tailored to maximize performance by defining application dependent materials, ply orientation, stiffening concepts, and attachment mechanisms. High performance designs are often less supportable due to increased strain levels, fewer redundant load paths, and a mix of highly tailored materials and geometries. Product design teams should focus on a variety of features that improve supportability including compatibility of available repair materials with those used on the parent structure, available equipment and skill, improving subsystem accessibility, and extended shelf-life composite repair materials. Structural elements and materials should be selected that are impervious to inherent and induced damage especially delaminations and hail damage. Each supportability enhancement feature results from the designer having an explicit knowledge of the aircraft's operational and maintenance environment and associated requirements and characteristics. Other design considerations also have an impact on supportability including durability, reliability, damage tolerance, and survivability. A supportable design integrates all the requirements, criteria, and features necessary to provide highly valued products in terms of performance, affordability and availability.

This section is designed to assist integrated product teams in the development of supportable products through five basic sections: 1) Introduction - which provides an overview of the Supportability chapter; 2) Design for Supportability - which provides the designer with design criteria, guidelines and checklists to ensure a supportable design; 3) Support Implementation - which defines and demonstrates those key elements of supportability that must be performed to insure mission success; 4) Logistics Requirements - which establishes the support resources needed to maintain the backbone of the support structure; and 5) Terminology - which provides a common forum for the terms used in supporting composite structures. Each section provides the designer and aircraft user with the supportability data and lessons learned that will reduce cost of ownership and improve aircraft availability. Other sections throughout MIL-HDBK-17 discuss the details needed to design supportable components. Sections contained in Volume 1 include material and structural testing, material types and properties, and joint types; in Volume 3 include materials and processes, quality, design, joints, reliability, and lessons learned needed to supplement those decisions that influence supportability.

8.2 DESIGN FOR SUPPORTABILITY

Guidelines are provided to support the design of a new structure. Items for consideration are as outlined.

8.2.1 Inspectability

Designing of structures should include for the aspect of inspectability for potential damage. The methods chosen for inspection may be restricted due to the size and configuration of the component, limitations of the instrument for detecting certain types of damage, and operating characteristics of the instruments (environmental, health hazards, etc.).

8.2.1.1 Methods of inspection

A variety of nondestructive inspection techniques (NDI) are utilized as inspection tools for process-related and determining service-related defects in composite structures. However as in metallic structures, no single nondestructive inspection method can locate and isolate all defects. The designer should assess based on the structure and its configuration which methods of inspection are practical and usable to detect damage. See Volume I, Chapter 6.2 for description of inspection methods.

8.2.2 Material selection

8.2.2.1 Introduction

Chapter 2 offers an in-depth review of advanced composite materials. Each one of the composite materials described in Chapter 2 can offer benefits over metallic materials to the designer in terms of performance and costs. However, these benefits will be erased if, when designing a component, the design is focused only on the mechanical and thermal performance of the component and does not take into consideration where the part will be used and how it will be repaired if it is damaged. The goal of the designer must be a part that will be both damage tolerant and damage resistant as well as easy to maintain and repair. This selection is offered as a guideline for the designer when selecting a material system.

8.2.2.2 Resins and fibers

When selecting a resin, it is important to look at where the resin system will be used, how the resin system has to be processed, what is its shelf life and storage requirements and is it compatible with surrounding materials. Table 8.2.2.2 describes the common resin types, their process conditions and their advantages and disadvantages in terms of repairability. An in-depth review of these materials can be found in Section 2.2.

TABLE 8.2.2.2 *Supportability concerns with resin types.*

Resin Type	Cure Temp. Ranges	Pressure Ranges	Processing Options	Advantages	Disadvantages
Epoxy Non-Toughened	RT to 350°F (180°C)	Vacuum to 100 psi (690 kPa)	Autoclave, press, vacuum bag, resin transfer molding	Low level of volatiles, low temp processing, vacuum bagable	Storage requirements
Epoxy Non-Toughened	RT to 350°F (180°C)	Vacuum to 100 psi (690 kPa)	Autoclave, press, vacuum bag and resin transfer molding	Low level of volatiles, low temp processing, vacuum bagable	Storage requirements
Polyester	RT to 350°F (180°C)	Vacuum Bag to 100 psi (690 kPa)	Same as epoxies	Ease of processing, quick cure with elevated temp., low cost	Poor elevated temp performance, health (Styrene)
Phenolic	250 to 350°F (120 to 180°C) with post cure	Vacuum Bag to 100 psi (690 kPa); lower pressure gives high void content	Autoclave, press molding		Water off gassing, high temp cure/post cure, high void content
Bismaleimides (BMI)	350F (180°C) with 400 to 500°F (200 to 260°C) post cure required	45 to 100 psi (310 to 690 kPa)	Autoclave, press molding, RTM	Lower pressure processing than polyimides	High temperature processing
Polyimides	350 to 700°F (180 to 370°C) post cure required	85 to 200+ psi (590 to 1400+ kPa)	Autoclave and press molding		Cost, availability of adhesives, high pressure
Structural Thermoplastic	500°F+ (260°C+)	Vacuum bag to 200 psi (1400 kPa)	Autoclave and press molding	Reformable	High temperature processing

Refer to Section 2.3 for available fibers for composite structures.

In terms of supportability, the minimum number of resin systems and material specifications should be chosen. This will reduce the logistic problems of storage, shelf life limitations and inventory control.

8.2.2.3 Product forms

A detailed description of available composite product forms can be found in Section 2.3.

The goal when repairing a composite part is to return it to its original performance capability while incurring the least cost and weight gain. Therefore, the ease of repairing different product forms should be taken into consideration when selecting the material system. Figure 8.2.2.3 shows the relative ease of repairing various product forms.

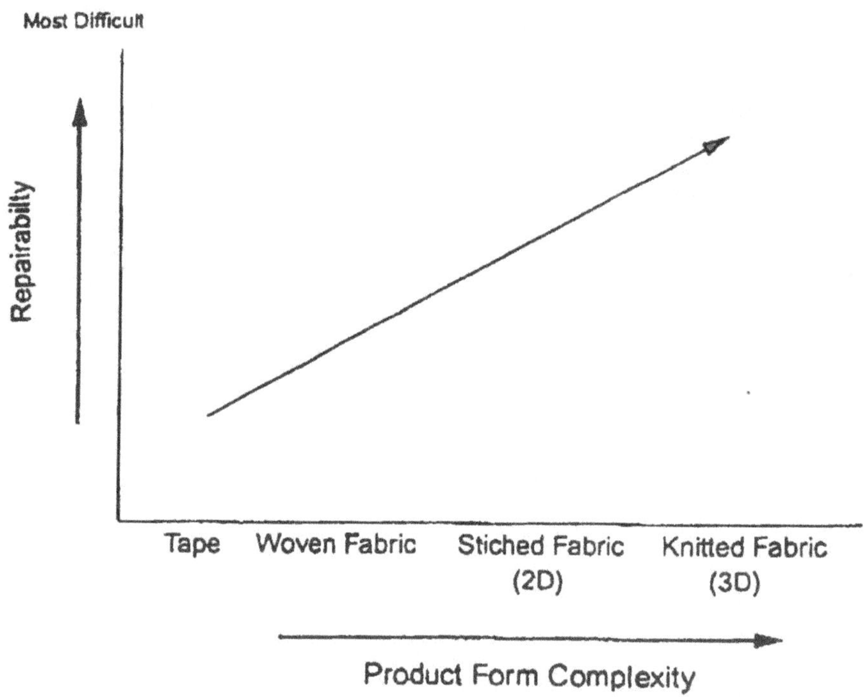

FIGURE 8.2.2.3 *Difficulty of repairing product forms.*

8.2.2.4 Supportability issues

Table 8.2.2.4 offers a list of Material Support issues for your consideration.

8.2.2.5 Environmental concerns

Health and safety: There are recognized hazards that go with advanced composite materials. If you know about these hazards, you can protect yourself and others from exposure to them. It is important that you read and understand the Material Safety Data Sheets (MSDS) and handle all chemicals, resins and fibers correctly. Refer to SACMA publication "Safe Handling of Advanced Composite Materials" for additional information (Reference 8.2.2.5).

TABLE 8.2.2.4 *Material support issues.*

Issue	Support Impact
Autoclave only cure	1. Equipment availability in the field and at small repair facilities 2. Part has to be removed for repair
Press curing	1. Equipment availability 2. Part has to be removed for repair
High temperature cure	1. Damage to surrounding structure in repair on aircraft 2. Protective equipment needed to handle high temperatures
Freezer storage required	1. Equipment availability

Disposal of scrap and waste: When selecting materials, consideration must be given to the disposal of scrap and waste. Disposal of scrap and waste should be specified under federal, state and local laws.

8.2.3 Damage tolerance and durability

See Section 4.11.1 for damage tolerance criteria.

In normal operating conditions, components can be expected to be subjected to potential damage from sources such as maintenance personnel, tools, runway debris, service equipment, hail, lightning, etc. During initial manufacturing and assembly, these components may be subject to the same or similar conditions. The given structure must be able to endure a reasonable level of such incidents without costly rework or downtime.

Minimum established levels of damage resistance are based on the type of structure and type of impact and the level of impact energy. In addition to strength and stiffness, the structure is zoned based on regions which have high or low susceptibility to damage and its required damage resistance. In defining the requirements, the type of structure; primary or secondary structure, construction method; sandwich or solid laminate, and whether its a removable or non-removable structure are pertinent.

As a general rule of thumb for design purposes, damage resistance is improved by utilizing thicker laminates and for sandwich applications, use of denser core materials and core materials with a minimum thickness - provides a degree of protection of the inner face. The selection of reinforcement fibers can also have a improved affect; use of high modulus, high strain fibers. Additionally, the selection of toughened matrix materials can greatly enhance damage resistance.

Other items include laminate layup; use of a layer of fabric as the exterior ply over tape to resist scratches, abrasion, softening of impact and reduction of fiber breakout. Laminate edge placement should not be positioned as exposed directly into the air stream - possibly subjected to delaminations. Utilize non-erosive edge protection, replaceable sacrificial materials or locate the forward edge below the level of the aft edge of the next panel forward. Resistance to damage requires structural robustness.

Areas prone to high energy lightning strike should utilize replaceable conductive materials, provide protection at tips and trailing edge surfaces and make all conductive path attachments easily accessible.

8.2.3.1 In-service damage detectability

See Section 4.11.1.2.

8.2.3.2 Design strain allowables

See Section 4 for criteria for design strain allowables.

8.2.3.3 Durability

See Volume 3, Section 4.11.2.

8.2.3.4 Margin of safety

The designer must take into account the margin of safety when designing a structure. Designs with a minimum or zero margin of safety should not be allowed. Margins of safety are based on such parameters as material properties, the effect of structural performance degradation due to exposure to moisture, light, thermal degradation and other operational characteristics. In turn, these margins of safety when coupled with a damaged structure must be sufficiently adequate to perform the mission requirements without failure. Margins need to be assessed for processing changes to affect a repair, change of materials, i.e., matrix materials and degradation due to undetected damage of a structure with environmental conditioning.

Current design practice is to use a fixed design allowable for compressively loaded laminated components and a fixed value for tension loaded laminated components. A location that is more susceptible to damage during service should be designed using lower strain levels. And locations where such threats are absent may be designed to larger strain levels, resulting in lighter, more efficient structures. The margin of safety should be based on the type of repair in a given location for a given type of structure.

8.2.4 Environmental Compliance

8.2.5 Reliability and maintainability

The maintainability of a structure is achieved by developing schemes for methods of inspection and maintenance procedures during the design phase. The designer with the overall knowledge of the performance and operational characteristics of the structure should access, based on the construction method, and configuration, material selection, etc., whether the structure is maintainable. Such factors in assessment would include development of cradle-to-grave inspection methodology, techniques, protection schemes and defined inspection intervals for maintenance.

8.2.6 Interchangeability and replaceability

8.2.6.1 Interchangeability

Interchangeability of high-unit cost, frequently-damaged or heavily loaded components that cannot be repaired should be a factor in the design of a structure. When damaged, a component is either repaired in place or removed, repaired and reinstalled. "Spares" or interchangeable components have not been traditionally utilized due to the lack of mandatory enforcement of MIL-I-8500, the use of interchangeable components (Reference 8.2.6.1).

Typically, components are trimmed and match drilled upon final assembly; essentially a custom, drilled installation. The designer in accessing this condition should tailor the structure so that it is easily removable. The process used for trimming and drilling of processed components should be standardized with the use of drill fixtures. The overall design should account for tolerancing factors which would prevent the use of interchangeable components. Interchangeability means no final trimming or match drilling of components.

8.2.6.2 Replaceability

An important aspect for the designer to utilize is the concept of removeability, the ability to remove a component or module from a structure for repair or replacement without damage to the existing and replacing structures.

The design would require the use of fastener systems that do not initiate damage upon removal and replacement. A bonded system may require the investigation of bonding agents that can be heated for removal and rebonding purposes.

8.2.7 Accessibility

Accessibility is an important factor when designing structures for repair. Sufficient access should always be provided to properly inspect, prepare the damage structure, fit and install the repair parts and use repair tools and bonding equipment. Limited access may dictate the repair approach, i.e., use of precured patches, use of mechanical fasteners in lieu of cocuring, etc. If feasible, two-sided access is preferred.

8.2.8 Repairability

The repairability of a structure is influenced by several design factors:

1. **Type of Structure** - The type of structure can be generally defined as primary, secondary or tertiary. Based on the structural loading condition, the type of repairs, and overall repair limitation can be assessed.

2. **Design** - The design of the overall structure, affects the repairability. Several of the state-of-the-art automated processing methods do not lend themselves to producing repairable structures but are primarily utilized to reduce production costs and maintain consistency in product properties. Repairability needs to be incorporated rather than simply removing and replacing. Methods of attachment are affected by design which could prevent repair of a structure.

3. **Operating Parameters** - Other parameters that can affect the repairability of a structure are its operating characteristics. In achieving such working parameters by the use of materials and construction methods, the repairability of a structure should be considered.

Structures that provide ballistic protection, armor, containment from radiation bombardment, chemical hazards or shielding requires special consideration to attempt to provide for repairability. In many cases, the protection is compromised or deteriorated beyond a repairable condition.

Other parameters would include for multiple repairs, flutter and mass balance. The design should allow to reestablish mass balance or allow for the increase of weight due to the repair.

8.3 SUPPORT IMPLEMENTATION

A repair has the objective of restoring a damaged structure to an acceptable capability in terms of strength, stiffness, functional performance, safety, cosmetic appearance or service life. Ideally, the repair will return the structure to original capability and appearance.

The design assessment of a repair for a given loading condition involves the selection of a repair concept, the choice of the appropriate repair materials and processes, then specifying the detailed configuration and size of the repair. Most repairs are basically designed as a joint to transfer load into and out of a patch. To ensure that the repair configuration will have adequate strength and stiffness, the repair joint must be analyzed to predict its strength.

The selection of the type of load-transfer joint to be used for a patch/strap is a tradeoff between simplicity, strength and stiffness. The easier configurations are generally not as strong as the more difficult ones. It is critical that the materials and process information is available prior to the system being put into place.

8.3.1 Inspection

8.3.2 Assessment

8.3.3 Repair

8.3.4 Repair design criteria

8.3.4.1 Thermal zoning

8.3.4.2 Thermal aging due to multiple/subsequent cure cycles - repair related

8.3.4.3 Corrosion

The use of metallic hardware, e.g., fasteners, attaching hardware or materials, e.g., metal-filled adhesives, metallic core when contacted with graphite materials has resulted in galvanic corrosion. The repair materials shall be chosen so that they are galvanically compatible with themselves and the parent structure. Barrier plies and sealants are utilized to "break" the direct contact between these materials and minimizes corrosion. In addition, barrier ply materials should be included in the repair scheme.

8.3.4.4 Mean-time-to-repair

8.3.4.5 Weight/mass balance

8.3.4.6 Conductivity restoration

Composite structures are generally poor conductors of electricity. Electrically conductive materials may be incorporated in a structure to provide ground planes for antennas, discharge paths for lightning strikes and similar purposes. The materials used include tapes, fabrics, foils, mesh materials and coatings. When conductive treatments are damaged, they must be restored, either by replacing the original material or by using a substitute material to bridge the disrupted path. Complicated designs should not be considered; if utilized, have acceptable and workable procedures for repair.

8.3.4.7 Stress/strength criteria

8.3.4.8 Compatibility with existing and surrounding structure

8.3.4.9 Allowables

8.3.4.10 Durability

Durability is another consideration in the design of repair joints. It is desirable that the repair joints withstand the static and fatigue stresses of the operating environment for the life of the structure without suffering environmental degradation. Some of the ways in which durability can be achieved are:

1. Avoidance of thermal mismatches between adherends which can produce residual bond stresses.
2. Avoidance of structural eccentricities which can cause stress risers and lead to repair joint deterioration and failure.
3. Attention to fastener fit and spacing to properly distribute fastener bearing loads.
4. Application of knockdown factors to account for environmental effects.
5. Adequate sealing to prevent moisture absorption and entrapment.
6. Repair should not degrade design fatigue life.
7. Bond durability is achieved by proper preparation of surfaces and cleanliness precautions are applied.

8.3.5 Replace

8.3.6 Disposal

8.4 LOGISTICS REQUIREMENTS

8.4.1 Training

8.4.2 Spares

8.4.3 Facilities

8.4.4 Technical data

8.4.5 Support equipment

8.5 TERMINOLOGY

REFERENCES

8.2.2.5 "Safe Handling of Advanced Composite Materials," Suppliers of Advanced Composite Materials Association, Arlington, VA.

8.2.6.1 MIL-I-8500, "Interchangeability and Replaceability of Component Parts for Aerospace Vehicles".

This page intentionally left blank

CHAPTER 9 LESSONS LEARNED

9.1 INTRODUCTION

The focus of much of what is in this handbook concentrates on establishing proper techniques for development and utilization of composite material property data. The motivation prompting specific choices is not always evident. This chapter provides a depository of knowledge gained from a number of involved contractors, agencies, and businesses for the purpose of disseminating lessons learned to potential users who might otherwise repeat past mistakes. Many of the contractors involved in developing the lessons learned are aerospace oriented. Thus, the lessons learned may have a decidedly aerospace viewpoint.

The chapter starts with a discussion of some of the characteristics of composite materials that makes them different from metals. These characteristics are the primary cause for establishing the methods and techniques contained in the handbook.

Specific lessons learned are defined in later sections. They contain the specific "rule of thumb" and the reason for its creation or the possible consequence if it is not followed. The lessons learned are organized into six different categories for convenience.

9.2 UNIQUE ISSUES FOR COMPOSITES

Composites are different from metals in several ways. These include their largely elastic response, their ability to be tailored in strength and stiffness, their damage tolerance characteristics, and their sensitivity to environmental factors. These differences force a different approach to analysis and design, processing, fabrication and assembly, quality control, testing, and certification.

9.2.1 Elastic properties

The elastic properties of a material are a measure of its stiffness. This property is necessary to determine the deformations that are produced by loads. In composites, the stiffness is dominated by the fibers; the role of the matrix is to prevent lateral deflections of the fibers and to provide a mechanism for shearing load from one fiber to another. Continuous fiber composites are transversely isotropic and in a two-dimensional stress state require four elastic properties to characterize the material:

Modulus of elasticity parallel to the fiber, E_1
Modulus of elasticity perpendicular to the fiber, E_2
Shear modulus, G_{12}
Major Poisson's ratio, ν_{12}

In general, material characterization may require additional properties not defined above. A thorough discussion of this subject is given in Section 4.3.1. Only two elastic properties are required for isotropic materials, the modulus of elasticity and Poisson's ratio.

The stress-strain response of commonly used fiber-dominated orientations of composite materials is almost linear to failure although some glasses and ceramics have nonlinear or bilinear behavior. This is contrasted to metals that exhibit nonlinear response above the proportional limit and eventual plastic deformation above the yield point. Many composites exhibit very little, if any, yielding in fiber dominated behavior. Toughened materials and thermoplastics can show considerable yielding, particularly in matrix dominated directions. This factor requires composites to be given special consideration in structural details where there are stress risers (holes, cutouts, notches, radii, tapers, etc.). These types of stress risers in metal are not a major concern for static strength analysis (they do play a big role in durability and damage tolerance analysis, however). In composites they must be considered in static strength analysis. In general, if these stress risers are properly considered in design/analysis of laminated parts, fatigue loadings will not be critical.

Another unique characteristic of composite material elastic response is its orthotropy. When metals are extended in one direction, they contract in the perpendicular direction in an amount equal to the Poisson's ratio times the longitudinal strain. This is true regardless of which direction is extended. In composites, an extension in the longitudinal (1 or x) direction produces a contraction in the transverse direction (2 or y) equal to the "major" Poisson's ratio, ν_{xy}, times the longitudinal extension. If this is reversed, an extension in the transverse direction produces a much lower contraction in the longitudinal direction. In fiber dominated laminates, Poisson's ratio can vary from <0.1 to >0.5.

The most unusual characteristic of composites is the response produced when the lay-up is unbalanced and/or unsymmetric. Such a laminate exhibits anisotropic warping characteristics. In this condition an extension in one direction can produce an in-plane shear deformation. It can also cause an out-of-plane bending or torsional response. All these effects are sometimes observed in one laminate. This type of response is generally undesirable because of warping or built-in stresses that occur. Hence, most laminate configurations are balanced and symmetric.

Classical lamination theory is used to combine the individual lamina properties to predict the linear elastic behavior of arbitrary laminates. Lamination theory requires the definition of lamina elastic properties, their orientation within the laminate, and their stacking position. The process assumes plane sections remain plane and enforces equilibrium. Lamination theory will solve for the loads/stresses/strains for each lamina within the laminate at a given location for a given set of applied loads. This combined with appropriate failure theory will predict the strength of the laminate (empirically modified input ply properties are often necessary).

9.2.2 Tailored properties and out-of-plane loads

The properties of a composite laminate depend on the orientation of the individual plies. This provides the engineer with the ability to tailor a laminate to fit a particular requirement. For high axial loads predominantly in one direction, the laminate should have a majority of its plies oriented parallel to that loading direction. If the laminate is loaded mostly in shear, there should be a high percent of ±45° pairs. For loads in multi-directions, the laminate should be quasi-isotropic. An all 0° laminate represents the maximum strength and stiffness that can be attained in any given direction, but is impractical for most applications since the transverse properties are so weak that machining and handling can cause damage. Fiber-dominated, balanced and symmetric, laminate designs that have a minimum of 10% of the plies in each of the 0°, +45°, -45°, and 90° directions are most commonly used.

Tailoring also means an engineer is not able to cite a strength or stiffness value for a composite laminate until he knows the laminate's ply percentages in each direction. Carpet plots of various properties vs. the percent of plies in each direction are commonly used for balanced and symmetric laminates. An example for stiffness is shown in Figure 9.2.2. Similar plots for strength can also be developed.

Out-of-plane loads can also be troublesome for composites. These loads cause interlaminar shear and tension in the laminate. Interlaminar shear stress can cause failure of the matrix or the fiber-matrix interphase region. Interlaminar shear and tension stresses can delaminate or disbond a laminate. Such loading should be avoided if possible. Design situations that tend to create interlaminar shear loading include high out-of-plane loads (such as fuel pressure), buckling, abrupt changes in cross-section (such as stiffener terminations), ply drop-offs, and in some cases laminate ply orientations that cause unbalanced or unsymmetric lay-ups. Interlaminar stresses will arise at any free edge. Interlaminar stresses will arise between plies of dissimilar orientation wherever there is a gradient in the components of in-plane stress.

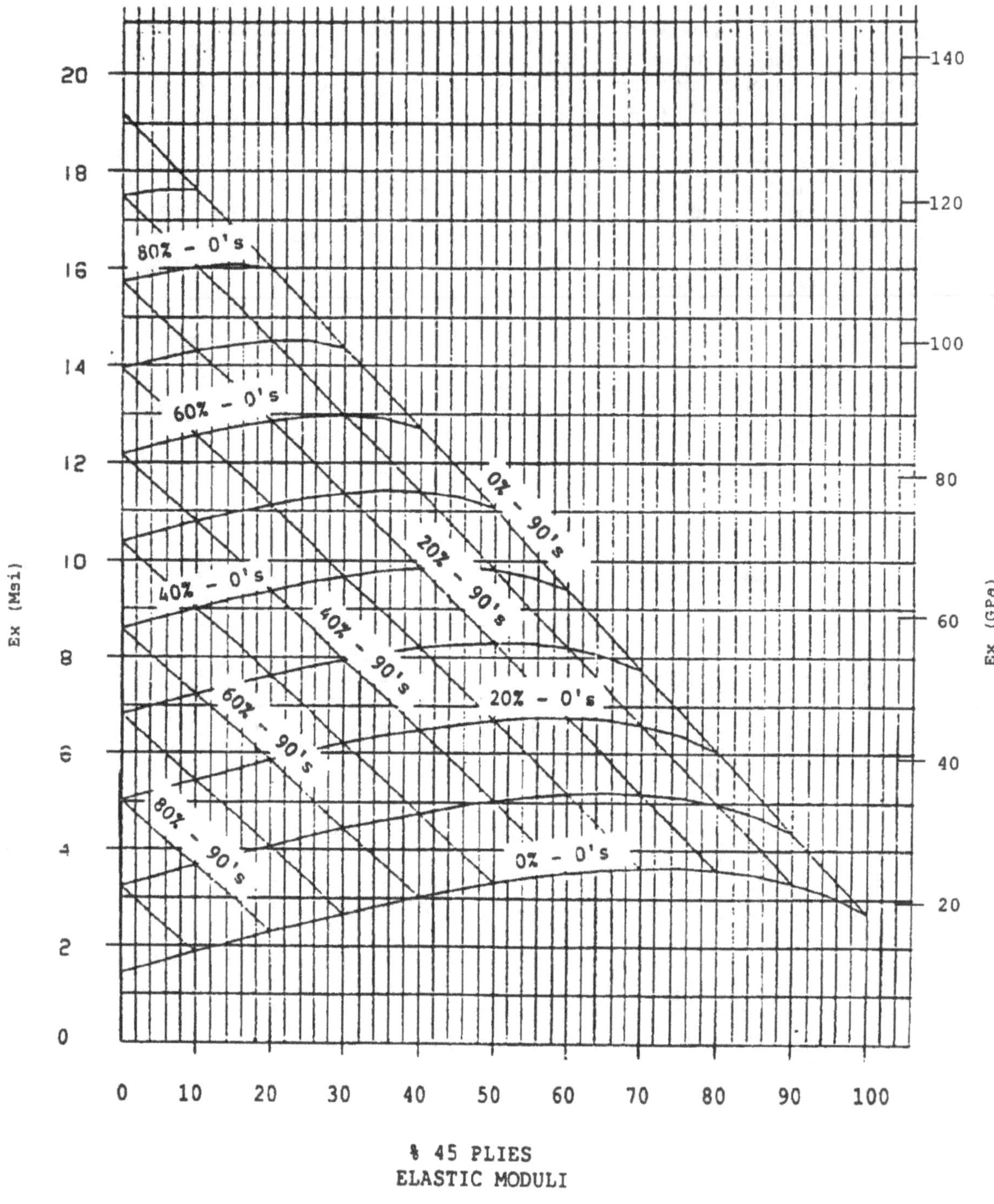

FIGURE 9.2.2 *Sample carpet plot.*

9.2.3 Damage tolerance

Damage tolerance is the measure of the structure's ability to sustain a level of damage or presence of a defect and be able to perform its operating functions. The concern is with the damaged structure having adequate residual strength and stiffness to continue in service safely: 1) until the damage can be detected by scheduled maintenance inspection and repaired, or 2) if the damage is undetected, for the remainder of the aircraft's life. Thus, safety is the primary goal of damage tolerance. Both static load and durability related damage tolerance must be interrogated experimentally because there are few, if any, accurate analytical methods.

There are basically two types of damage that are categorized by their occurrence during the fabrication and use of the part, i.e., damage occurring during manufacturing or damage occurring in service. It is hoped that the occurrence of the majority of manufacturing associated damage, if beyond specification limits, will be detected by routine quality inspection. Nevertheless, some "rogue" defects or damage beyond specification limits may go undetected. Consequently, their occurrence must be assumed in the design procedure and subsequent testing (static and fatigue) performed to verify the structural integrity.

Service damage concerns are similar to those for manufacturing. Types of service damage include edge and surface gouges and cuts or foreign object collision and blunt object impact damage caused by dropped tools or contact with service equipment. A level of non-detectable damage should be established and verified by test that will not endanger the normal operation of the aircraft structure for two lifetimes. A certain level (maximum allowed) damage that can be found by inspection should be defined such that the vehicle can operate for a specified number of hours before repair or replacement at loads not exceeding design limit. This damage should also be tested (statically and in fatigue) to verify the structural integrity.

Delaminations can also be critical defects. However, unless they are very large, historically more than 2 inches (50 mm) in diameter, the problem is mostly with thin laminates. Effects of manufacturing defects such as porosity and flawed fastener holes that are slightly in excess of the maximum allowable are usually less severe. They are generally accounted for by the use of design allowable properties that have been obtained by testing specimens with stress concentrations, e.g., notches. Most commonly these are specimens with a centered hole. Open holes are typically used for compression specimens while either open or filled holes (holes with an installed fastener) are used for tension testing. (Open holes are more critical than filled holes for compression. Filled holes may be more critical in tension, especially for laminates with ply orientations with a predominate number of plies in the load direction.) Consequently, the design allowables thus produced may be used to account for a nominal design stress concentration caused by an installed or missing fastener, at least to a 0.25 inch (6.4 mm) diameter, as well as accounting for many other manufacturing defects. This is sometimes called the "rogue flaw" approach to laminate design, see Reference 9.2.3.

9.2.4 Durability

Durability of a structure is its ability to maintain strength and stiffness throughout the service life of the structure. A structure must have adequate durability when subjected to the expected service loads and environment spectra to prevent excessive maintenance, repair, or modification costs over the service life. Thus, durability is primarily an economic consideration.

Metallic structure can be very sensitive to durability issues; major factors limiting life are corrosion and fatigue. Metal fatigue is dictated by the number of load cycles required to start a crack (crack initiation) and the number of load cycles for the crack to grow to its critical length, reaching catastrophic failure (crack growth). Crack/damage growth rate is very dependent on the concentration of stress around the crack.

In composites, it has been demonstrated that one of the most common damage growth mechanisms is intercracking (delamination). This makes composites most sensitive to compression-dominated fatigue loading. A second common fatigue failure mode is fastener hole wear caused by high bearing stresses. In this failure mode the hole gradually elongates. The most serious damage to composite parts is low velocity impact damage which can reduce static strength, fatigue strength, or residual strength after fatigue. Again, testing is a must!

The strain level of composites in most actual vehicle applications to date has been held to relatively low values. Composites under in-plane loads have relatively flat stress-life (S-N) curves with high fatigue thresholds (endurance limits). These two factors combined have resulted in insensitivity to fatigue for most load cases. However, the greater variability found with composites requires an engineer to still characterize the composite's fatigue life to failure to correctly characterize its fatigue scatter.

9.2.5 Environmental sensitivity

When a composite with a polymeric matrix is placed in a wet environment, the matrix will absorb moisture. The moisture absorption of most fibers used in practice is negligible; however, aramid fibers (e.g., Kevlar) absorb significant amounts of moisture when exposed to high humidity. The absorption of moisture at the interface of glass/quartz fibers is a well-known degrading phenomena.

When a composite has been exposed to moisture and sufficient time has elapsed, the moisture concentration throughout the matrix will be uniform. A typical equilibrium moisture content for severe humidity exposure of common epoxy composites is 1.1 to 1.3 percent weight gain. The principal strength degrading effect is related to a change in the glass transition temperature of the matrix material. As moisture is absorbed, the temperature at which the matrix changes from a glassy state to a viscous state decreases. Thus, the strength properties decrease with increasing moisture content. Current data indicate this process is reversible. When the moisture content is decreased, the glass transition temperature increases and the original strength properties return. With glass/quartz fibers there is additional degradation at the interface with the matrix. For aramid fibers there is additional degradation at the interface with the matrix and, also, in the fibers.

The same considerations also apply for a temperature rise. The matrix, and therefore the lamina, loses strength and stiffness when the temperature rises. This effect is primarily important for the matrix-dominated properties. Temperature rise also worsens the fiber/matrix interface degradation for glass/quartz fibers and aramid fibers. The aramid fiber properties are also degraded by a rise in temperature.

The approach for design purposes is to assume a worst case. If the material is assumed to be fully saturated and at the maximum temperature, material allowables can be derived for this extreme. This is a conservative approach, since typical service environments do not generate full saturation for most complex structures. Once the diffusivity of a composite material is known, the moisture content and through the thickness distribution can be accurately predicted by Fickian equations. This depends on an accurate characterization of the temperature-humidity service environment.

Thermal expansion characteristics of common composites, like carbon/epoxy, are quite different from metals. In the (0 or 1) longitudinal direction, the thermal expansion coefficient of carbon/epoxy is almost zero. Transverse to the fiber (90 or 2 direction), the thermal expansion is the same magnitude as aluminum. This property gives composites the ability to provide a dimensionally stable structure throughout a wide range of temperatures.

Another feature of composites that is related to environment is resistance to corrosion. Polymer matrix composites (with the exception of some carbon/bismaleimides) are immune to salt water and most chemical substances as far as corrosion sensitivity. One precaution in this regard is galvanic corrosion. Carbon fiber is cathodic (noble); aluminum and steel are anodic (least noble). Thus carbon in contact with aluminum or steel promotes galvanic action which results in corrosion of the metal. Corrosion barriers (such as fiberglass and sealants) are placed at interfaces between composites and metals to prevent metal corrosion. Another precaution regards the use of paint strippers around most polymers. Chemical paint strippers are very powerful and attack the matrix of composites very destructively. Thus, chemical paint stripping is forbidden on composite structure.

Other environmental effects worth noting include the effect of long term exposure to radiation. Ultraviolet rays from the sun can degrade epoxy resins. This is easily protected by a surface finish such as a coat of paint. Another factor is erosion or pitting caused by high speed impact with rain or dust particles. This is likely to occur on unprotected leading edges. There are surface finishes such as rain erosion coats and paints for preventing surface wear. Lightning strike is also a concern to composites. A direct strike can cause considerable damage to a laminate. Lightning strike protection in the form of conductive surfaces is applied in susceptible areas. In cases where substructure is also composite, the inside end of attachment bolts may need to be connected with each other and to ground by a conducting wire.

9.2.6 Joints

Successful joint design relies on knowledge of potential failure modes. Failure modes depend on joint geometry and laminate lay-up for one given material. The type of fastener used can also influence the occurrence of a particular failure mode. Different materials will give different failure modes.

Net-section tension/compression failures occur when the bolt diameter is a sufficiently large fraction of the strip width. For most successful designs, this fraction (D/W) is about one-quarter or more for near-isotropic lay-ups in carbon/epoxy systems that have a D/E of one-third or less.

Shear-out and shear-out delamination failures occur because the bolt is too close to the edge of the laminate. Such a failure can be triggered when there is only a partial net-section tension or bearing failure. D/t ratios should be 0.75 to 1.25.

In some instances the bolt head may be pulled through the laminate after the bolt is bent and deformed. This mode is frequently seen with countersunk fasteners and is highly dependent on the particular fastener used.

Bearing strength is a function of joint geometry, fastener and member stiffnesses. For a 0/±45/90 family of laminates with 20-40% of 0° plies and 40-60% of ±45° plies, plus a minimum (10%) of 90° plies, the bearing strength is relatively constant. Fastener characteristics such as clamp-up force and head configuration have a significant effect. However, for a specific laminate family, a specific fastener, and equal thickness laminate joining members, the parameter with the greatest influence is D/t.

Composite joints require smaller D/W and D/E ratios than do metals to get bearing failures.

Composite joint strength characteristics differ from metals because the strength is influenced by the bypass load going around the joint. This occurs when two or more fasteners are arranged in a line to transfer the load through a joint. Since not all of the load is reacted by one fastener, some of the load by-passes it. The by-pass effects become prominent once the ratio of by-pass to fastener bearing load exceeds 20%.

Titanium fasteners are the most common means of mechanical attachment in composites. This is because titanium is non-corrosive in the galvanic atmosphere created by the dissimilar materials. Titanium is closer to carbon on the cathodic scale.

9.2.7 Design

The design of composite structure is complicated by the fact that every ply must be defined. Drawings or design packages must describe the ply orientation, its position within the stack, and its boundaries. This is straightforward for a simple, constant thickness laminate. For complex parts with tapered thicknesses and ply build-ups around joints and cutouts, this can become extremely complex. The need to maintain relative balance and symmetry throughout the structure increases the difficulty.

Composites can not be designed without concurrence. Design details depend on tooling and processing as does assembly and inspection. Parts and processes are so interdependent it could be disastrous to attempt sequential design and manufacturing phasing.

Another factor approached differently in composite design is the accommodation of thickness tolerances at interfaces. If a composite part must fit into a space between two other parts or between a substructure and an outer mold line, the thickness requires special tolerances. The composite part thickness is controlled by the number of plies and the per-ply-thickness. Each ply has a range of possible thicknesses. When these are layed up to form the laminate they may not match the space available for assembly within other constraints. This discrepancy can be handled by using shims or by adding "sacrificial" plies to the laminate (for subsequent machining to a closer tolerance than is possible with nominal per-ply-thickness variations). The use of shims has design implications regarding load eccentricities. Another approach is to use closed die molding at the fit-up edges to mold to exact thickness needed.

The anisotropy of special laminates, while more complicated, enables a designer to tailor a structure for desired deflection characteristics. This has been applied to some extent for aeroelastic tailoring of wing skins.

Composites are most efficient when used in large, relatively uninterrupted structures. The cost is also related to the number of detail parts and the number of fasteners required. These two factors drive designs towards integration of features into large cocured structures. The nature of composites enables this possibility. Well designed, high quality tooling will reduce manufacturing and inspection cost and rejection rate and result in high quality parts.

9.2.8 Handling and storage

Epoxy resins are the most common form of matrix material used in composites. Epoxies are perishable. They must be stored below freezing temperature and even then have limited shelf life. Once the material is brought out of storage there is limited time it can be used to make parts (30 days is common). For very complex parts with many plies, the material's permissible out-time can be a controlling factor. If the material is not completely used, it may be returned to storage. An out-time record should be kept. In addition, freezer storage of these materials is usually limited by the vendor to 6 to 12 months. Overage material will produce laminates with a high level of porosity.

The perishability of the material also requires that it be shipped refrigerated from the supplier. Upon arrival at the contractor's facility, there must be provisions to prevent it being left on-dock for long periods of time.

Tack is another composite material characteristic that is unique. Tack is "stickiness" of the prepreg. It is both an aid and a hindrance. Tack is helpful to maintain location of a ply once it is placed in position. It also makes it difficult to adjust the location once the ply has been placed.

9.2.9 Processing and fabrication

Composite parts are fabricated by successive placement of plies one after the other. Parts are built-up rather than machined down. Many metal fabrication steps require successive removal of material starting from large ingots, plates, or forgings. Prepreg "tape" material typically comes in rolls of relatively thin strips (0.005-0.015 inches or 0.13 - 0.38 mm). These strips are a variety of widths: 3", 6", and 36". Prepreg "fabric" is usually thicker than tape (0.007-0.020 inches or 0.18 - 0.51 mm) and usually comes in 36-inch (0.9 m) wide rolls.

Fabrication of a detail part requires the material to be taken out of the freezer in a sealed bag and allowed to come to room temperature prior to any operations. Placement of the prepreg on the tool (if not automated) requires care. The plies must be aligned properly to the desired angle and stacked in the prescribed sequence. Prepreg plies come with a backing material to keep them from sticking together on the rolls. This backing material must be removed to prevent contamination of the laminate. Care must be exercised when handling the material to prevent splinters from piercing the hands.

Part lay-up (particularly when done by hand) can lead to air entrapment between plies. This creates difficulty when the part is cured because the air may not escape, causing porosity. Thus, thick parts are normally pre-compacted using a vacuum periodically during the lay-up.

Some prepreg materials contain an excess of resin. This excess is expected to be "bled" away during cure. Bleeder plies are placed under the vacuum bag to soak up the excess resin. However, most current prepreg materials are "net resin" so no bleeding is required.

Composite processing requires careful attention to tool design. The tools must sustain high pressures under elevated temperature conditions. The composite material has different expansion characteristics than most tooling materials, thus thermal stresses are created in the part and in the tool. Tool surfaces are treated with a release agent to facilitate removal of the part after cure. Tools must also be pressure tight because autoclave processing requires application of a vacuum on the laminate as well as positive autoclave pressure. Lastly, tool design must account for the rate of manufacture and the number of parts to be processed.

Prepreg material is not fully cured. Curing requires application of heat and pressure that is usually performed in the autoclave. Autoclaves typically apply 85 psi (590 kPa) pressure up to 350°F (180°C). They can go beyond these values if required for other materials (such as polyimides), but they must be qualified for higher extremes. Autoclave size may limit the size of a part to be designed and manufactured. Very large autoclaves are available, but they are expensive and costly to run. Common problems that occur in autoclave operations include blown vacuum bags, improper heat-up rates, and loss of pressure.

Once the part is cured it may still require drilling, trimming and machining. Drilling of composites requires very sharp bits, careful feed and speed, and support of the back face to prevent splintering. Water-jet cutters are very useful for trimming. Machining produces a fine dust that requires protection for the operator's safety.

9.2.10 Quality control

The quality control function for composite materials starts at a much earlier phase than for metals. There is much coordination and interaction occurring between the material supplier and the user before the material is ever shipped. These controls are defined by the material and process specifications and in some cases design allowables requirements. The supplier is often required to perform chemical and mechanical tests on the material prior to shipment. These involve the individual material constituents, the prepreg, and cured laminates.

Material processing and handling must be monitored throughout the various manufacturing phases. Receiving inspections are performed on the prepreg and cured laminates when the material first comes in. From this time on the material is tracked to account for its shelf life and out-time.

Quality control activities include verification of the ply lay-up angle, its position in the stack, the number of plies, and the proper trim. During lay-up it is necessary to ensure all potential contaminates and foreign materials are not allowed to invade the material.

The curing process is monitored to ensure proper conformance to time-temperature-pressure profiles. These records are maintained for complete traceability of the parts.

After the part is cured, there are a number of methods to verify its adequacy. One of the most common is Through-Transmission-Ultrasonics (TTU). Parts with high porosity or delaminations can not transmit sound as well as unflawed parts. Thus ultrasound transmission is attenuated in a flawed part. Other techniques used to verify part quality include traveler specimens, specimens cut from excess material on the part, tracer yarns within the laminate, and in some cases proof loading. Visual inspections, thickness measurements, and tap testing also serve to interrogate composite parts.

One of the most crucial aspects of quality control is information on the effect of defects. It is not enough to discover a flaw or suspected non-conformity. There must also be sufficient information to evaluate the impact of that rejection. The quality control function in its entirety includes the dispositioning of exposed non-conformances. Dispositioning includes acceptance as-is, repair or rework, and scrappage. If proper dispositioning is not possible because of a lack of knowledge about the effect of defects, an inordinate expense will be incurred scrapping or reworking affected parts.

9.3 LESSONS LEARNED

9.3.1 Design and analysis

	LESSON	REASON OR CONSEQUENCE
A-1.	"Concurrent Engineering", whereby a new product or system is developed jointly and concurrently by a team composed of designers, stress analysts, materials and processes, manufacturing, quality control, and support engineers, (reliability, maintainability, survivability), as well as cost estimators, has become the accepted design approach.	To improve the quality and performance and reduce the development and production costs of complex systems
A-2.	In general, design large cocured assemblies. Large assemblies must include consideration for handling and repair.	Lower cost due to reduced part count and assembly time. If the assembly requires overly complex tooling, the potential cost savings can be negated.
A-3.	Structural designs and the associated tooling should be able to accommodate design changes associated with the inevitable increases in design loads.	To avoid scab-on reinforcements and similar last minute disruptions.
A-4.	Not all parts are suited to composite construction. Material selection should be based on a thorough analysis that includes consideration of performance, cost, schedule, and risk.	The type of material greatly influences performance characteristics as well as producibility factors.
A-5.	Uni-woven and bi-directional woven fabric should be used only when justified by trade studies (reduced fabrication costs). If justified, woven fabric may be used for 45° or 0°/90° plies.	Fabric has reduced strength and stiffness properties and the prepreg material costs more than tape. Fabric may be necessary for complex shapes and some applications may require the use of fabric for its drapeability.
A-6.	Whenever possible, mating surfaces should be tool surfaces to help maintain dimensional control. If this is not possible, either liquid shims or, if the gap is large, a combination of precured and liquid shims should be used.	To avoid excessive out-of-plane loads that can be imposed if adjoining surfaces are forced into place. Large gaps may require testing.
A-7.	Part thickness tolerance varies directly with part thickness; thick parts require larger tolerance.	Thickness tolerance is a function of the number of plies and the associated per-ply-thickness variation.
A-8.	Carbon fibers must be isolated from aluminum or steel by using an adhesive a layer and/or a thin glass-fiber ply at faying surfaces.	Galvanic interaction between carbon and aluminum or steel will cause corrosion of the metal.

A-9.	The inspectability of structures, both during production and in-service, must be considered in the design. Larage defects or damage sizes must be assumed to exist when designing composite structures if reliable inspection procedures are not available.	There is a much better chance that problems will be found if a structure is easily inspected.
A-10.	In Finite Element Analysis (FEA) a fine mesh must be used in regions of high stress gradients, such as around cut-outs and at ply and stiffener drop-offs.	Improper definition or management of the stresses around discontinuities can cause premature failures.
A-11.	Eliminate or reduce stress risers whenever possible.	Composite (fiber-dominated) laminates are generally linear to failure. The material will not yield locally and redistribute stresses. Thus, stress risers reduce the static strength of the laminate.
A-12.	Avoid or minimize conditions which cause peel stresses such as excessive abrupt laminate terminations or cocured structures with significantly different flexural stiffnesses (i.e., $EI_1 >> EI_2$).	Peel stresses are out-of-plane to the laminate and hence, in its weakest direction.
A-13.	Buckling or wrinkling is permissible in thin composite laminates provided all other potential failure modes are properly accounted for. In general, avoid instability in thick laminates.	Significant weight savings are possible with postbuckled design.
A-14.	Locating 90° and ±45° plies toward the exterior surfaces improves the buckling allowables in many cases. Locate 45° plies toward the exterior surface of the laminate where local buckling is critical.	Increases the load carrying capability of the structure.
A-15.	When adding plies, maintain balance and symmetry. Add between continuous plies in the same direction. Exterior surface plies should be continuous.	Minimizes warping and interlaminar shear. Develops strength of plies. Continuous surface plies minimize damage to edge of ply and help to prevent delamination.
A-16.	Never terminate plies in fastener patterns.	Reduces profiling requirements on substructure. Prevents delamination caused by hold drilling. Improves bearing strength.

A-17. Stacking order of plies should be balanced and symmetrical about the laminate midplane. Any unavoidable unsymmetric or unbalanced plies should be placed near the laminate midplane.	Prevents warpage after cure. Reduces residual stresses. Eliminates "coupling" stresses.
A-18. Use fiber dominated laminate wherever possible. The [0°/±45°/90°] orientation is recommended for major load carrying structures. A minimum of 10% of the fibers should be oriented in each direction.	Fibers carry the load; the resin is relatively weak. This will minimize matrix and stiffness degradation.
A-19. When there are multiple load conditions, do not optimize the laminate for only the most severe load case.	Optimizing for a single load case can produce excessive resin or matrix stresses for the other load cases.
A-20. If the structure is mechanically fastened, an excess of 40% of the fibers oriented in any one direction is inadvisable.	Bearing strength of laminate is adversely affected.
A-21. Whenever possible maintain a dispersed stacking sequence and avoid grouping similar plies. If plies must be grouped, avoid grouping more than 4 plies of the same orientation together.	Increases strength and minimizes the tendency to delaminate. Creates a more homogeneous laminate. Minimizes interlaminar stresses. Minimizes matrix microcracking during and after service.
A-22. If possible, avoid grouping 90° plies. Separate 90° plies by a 0° or ±45° plies where 0° is direction of critical load.	Minimizes interlaminar shear and normal stresses. Minimizes multiple transverse fracture. Minimizes grouping of matrix critical plies.
A-23. Two conflicting requirements are involved in the pairing or separating of ±θ° plies (such as ±45°) in a laminate. Laminate architecture should minimize interlaminar shear between plies and reduce bending/twisting coupling.	Separating ±θ° plies reduces interlaminar shear stresses between plies. Grouping ±θ° plies together in the laminate reduces bending/twisting coupling.
A-24. Locate at least one pair of ±45° plies at each laminate surface. A single ply of fabric will suffice.	Minimizes splintering when drilling. Protects basic load carrying plies.
A-25. Avoid abrupt ply terminations. Try not to exceed dropping more than 2 plies per increment. The plies that are dropped should not be adjacent to each other in the laminate.	Ply drops create stress concentrations and load path eccentricities. Thickness transitions can cause wrinkling of fibers and possible delaminations under load. Dropping non-adjacent plies minimizes the joggle of other plies.

A-26.	Ply drop-offs should not exceed 0.010 inch (0.25mm) thick per drop with a minimum spacing of 0.20 inch (0.51 mm) in the major load direction. If possible, ply drop-offs should be symmetric about the laminate midplane with the shortest length ply nearest the exterior faces of the laminate. Shop tolerance for drop-offs should be 0.04 inch (1 mm).	Minimizes load introduction into the ply drop-off creating interlaminar shear stresses. Promotes a smooth contour. Minimizes stress concentration.
A-27.	Skin ply drop-offs should not occur across the width of spars, rib, or frame flange.	Provides a better load path and fit-up between parts.
A-28.	In areas of load introduction there should be equal numbers of +45° and -45° plies on each side of the mid-plane.	Balanced and symmetric pairs of ±45° plies are strongest for in-plane shear loads which are common at load introduction points.
A-29.	A continuous ply should not be butt-spliced transverse to the load direction.	Introduces a weak spot in the load path.
A-30.	A continuous ply may be butt-spliced parallel to the load direction if coincident splices are separated by at least four plies of any orientation.	Eliminates the possibility of a weak spot where plies are butted together.
A-31.	The butt joint of plies of the same orientation separated by less than four plies of any direction must be staggered by at least 0.6 inch (15 mm).	Minimizes the weak spot where plies are butted together.
A-32.	Overlaps of plies are not permitted. Gaps shall not exceed 0.08 inch (2 mm).	Plies will bridge a gap, but must joggle over an overlap.

9.3.1.1 Sandwich design

B-1.	Facesheets should be designed to minimize people induced damage during handling or maintenance of component.	Thin skin honeycomb structure is very susceptible to damage by harsh handling.
B-2.	When possible avoid laminate buildup on the core side of the laminate.	Minimizes machining of the core.
B-3.	Core edge chamfers should not exceed 20° (from the horizontal plane). Larger angles may require core stabilization. Flex core is more sensitive than rigid core.	Prevents core collapse during cure cycle.

B-4.	Use only non-metallic or corrosion resistant metal honeycomb core in composite sandwich assemblies.	Prevents core corrosion
B-5.	Choice of honeycomb core density should satisfy strength requirements for resisting the curing temperature and pressure during bonding or cocuring involving the core. 3.1 PCF (50 g/m^3) is a minimum for non-walking surfaces.	Prevents crushing of the core.
B-6.	For sandwich structure used as a walking surface, a core density of 6.1 PCF (98 g/m^3) is recommended.	3.1 PCF (50 g/m^3) core density will result in heel damage to the walking surface.
B-7.	Do not use honeycomb core cell size greater than 3/16 inch (4.8 mm) for cocuring sandwich assemblies (1/8 inch (3.2 mm) cell size preferred).	Prevents dimpling of face sheets.
B-8.	When core is required to be filled around bolt holes, etc., this should be done using an approved filler to a minimum of 2D from the bolt center.	Prevents core crushing and possible laminate damage when bolt is installed.
B-9.	Two extra layers of adhesive should be applied to the inner moldline at the core run out (edge chamfer). This should be applied a minimum of 0.6 in. (15 mm) from the intersection of the inner skin and edge band up the ramp and a minimum of 0.2 in. (5 mm) from that point into the edge band.	Curing pressures tend to cause the inner skin to "bridge" in this area creating a void in the adhesive (skin to core bond). 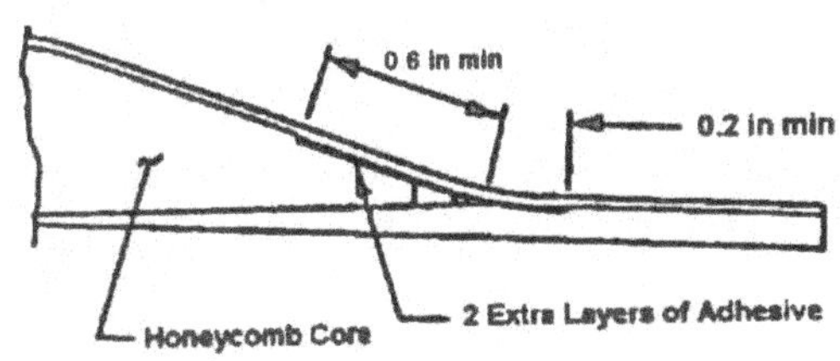
B-10.	The use of honeycomb sandwich construction must be carefully evaluated in terms of its intended use, environment, inspectability, repairability, and customer acceptance.	Thin skin honeycomb is susceptible to impact damage, water intrusion due to freeze/thaw cycles, and is difficult to repair.

9.3.1.2 Bolted joints

C-1.	Design the joints first and fill in the basic structure afterwards.	Optimizing the "basic" structure first compromises the joint design and results in low overall structural efficiency.

	Guideline	Rationale
C-2.	Joint analysis should include the effects of shimming to the limits permitted by drawings.	Shimming can reduce joint strength.
C-3.	Design joints to accommodate the next larger fastener size.	To accommodate routine MRB and repair activities.
C-4.	Bolted joint strength varies far less with percentage of 0° plies in fiber pattern than does unnotched laminate strength.	The stress concentration factor, K_t, is highly dependent on 0° plies.
C-5.	Optimum single-row joints have approximately three-fourths of the strength of optimum four-row joints.	Optimum single-row joints operate at higher bearing stress than the most critical row in an optimized multi-row joint.
C-6.	Common errors in composite bolted joints are to use too few bolts, space them too far apart, and to use too small a diameter.	Does not maximize the strength of the laminate.
C-7.	Rated shear strength of fasteners does not usually control the joint design.	Bolt diameter is usually governed by the need not to exceed the allowable bearing stress in the laminate.
C-8.	The peak hoop tension stress around bolt holes is roughly equal to the average bearing stress.	Keeping the laminate tension strength high requires keeping the bearing stress low.
C-9.	Maximum torque values should be controlled, particularly with large diameter fasteners.	Avoids crushing the composite.
C-10.	Bolt bending is much more significant in composites than for metals.	Composites tend to be thicker (for a given load) and more sensitive to non-uniform bearing stresses (because of brittle failure modes.
C-11.	Optimum w/d ratio for multi-row bolted joints varies along length of joint. w/d = 5 at first row to minimize load transfer, w/d = 3 at last row to maximize transfer, w/d = 4 for intermediate bolts.	Maximizes joint strength.
C-12.	Stainless steel fasteners in contact with carbon should be permanent and installed wet with sealant.	Prevents galvanic corrosion.
C-13.	Use a layer of fiberglass or Kevlar (0.005 inch (0.13 mm) minimum) or adhesive with serim on faying surfaces of carbon epoxy panels to aluminum.	Prevents corrosion of aluminum.

C-14. Bolt stresses need careful analysis, particularly for the effects of permissible manufacturing parameters, for example, hole perpendicularity (±10°), shimming, loose holes.	Bolt failures are increasingly becoming the "weak link" with current high strength composite materials.
C-15. Bolted joint data bases should include the full range of all permitted design features.	Establishes that failure modes remain consistent and that there are no detrimental interaction effects between design parameters.
C-16. The design data base should be sufficient to validate all analysis methods over the entire range permitted in design.	For proper verification of analytical accuracy.
C-17. Mechanical joint data bases should contain information pertaining to durability issues such as clamp-up, wear at interfaces, and hole elongation. Manufacturing permitted anomalies such as hole quality, edge finish, and fiber breakout also need to be evaluated.	Practical occurrences can affect strength and durability.
C-18. Use drilling procedures that prevent fiber break out on the back side of the component.	Improper back side support or drilling procedures can damage surface plies on the back side.
C-19. Splice plate stresses should be lower than the stresses in skins to prevent delaminations.	Splice plates see less clamp up than the skin sandwiched in between, because of bolt bending.
C-20. The best bolted joints can barely exceed half the strength of unnotched laminates.	The strength reduction is caused by stress concentrations around the hole for the fastener.
C-21. Laminate percentages for efficient load transfer: 0° = 30 - 50%; ±45° = 40 - 60%; 90° = minimum of 10%.	Best range for bearing and by-pass strength.
C-22. Countersink depths should not exceed 70% of laminate thickness.	Deep countersinks result in degraded bearing properties and increased hole wear.
C-23. Fastener edge distance and pitch: Use 3.0D edge distance in direction of major load; use 2.5D + 0.06 side distance. (D is diameter of fastener.)	Maximizes joint strength.

C-24. Gap between attached parts should not exceed 0.03 inch (0.8 mm) for non-structural shim.	Large gaps cause excessive bolt bending, non-uniform bearing stresses, and eccentric load path.
C-25. Any gap in excess of 0.005 inch should be shimmed.	Minimizes interlaminar stresses due to clamp-up.
C-26. Use "form-in-place" gaskets on carbon/epoxy doors over anodized aluminum substructure. Allow for a seal thickness of 0.010 ± 0.005 inch (-0.25 ± 0.13 mm) minmum.	Prevents corrosion of aluminum.
C-27. Use only titanium, A286, PH13-8 MO, monel or PH17-4 stainless steel fastener with carbon/epoxy.	Prevents galvanic corrosion.
C-28. Do not buck rivets in composite structure.	The bucking force can damage the laminate.
C-29. The use of interference fit fasteners should be checked before permitting their use in design.	Installation of interference-fit fasteners can damage laminates if a loose-fit sleeve is not installed first.
C-30. Fastener-to-hole size tolerance for primary structure joints must be assessed and controlled.	Tight fitting fastener promotes uniform bearing stress in a single fastener hole, and promotes proper load sharing in a multi-fastener joint.
C-31. Squeeze rivets can be used if washer is provided on tail side.	Washer helps protect the hole.
C-32. For blind attachments to composite substructure, use fastener with large blind side footprint of titanium or A286.	Prevents damage to composite sub-structure by locking collars of fasteners.
C-33. Tension head fasteners are preferred for most applications. Shear head fasteners may be used in special applications only with stress approval.	Shear head fasteners.
C-34. Avoid putting fastener threads in bearing against the laminate.	Fastener threads can gouge and damage the laminate.
C-35. Tapered splice plates should be used to tailor the load transfer, row by row, to minimize the bearing stress at the most critical row.	Multi-row bolted joints between uniformly thick members will have high peak bearing loads in outermost rows of fasteners.

9.3.1.3 Bonded joints

	Guideline	Rationale
D-1.	Use secondary adhesive bonding extensively for thin, lightly loaded, composite structures, restricting the use of mechanical fastening to thicker, more heavily loaded structures.	Reduces cost. Reduces the number of holes in composite components. Reduces weight by eliminating build-ups for fastener countersinking and bearing strength.
D-2.	Never design for an adhesive bond to be the weak link in a structure. The bonds should always be stronger than the members being joined.	Maximizes the strength of the structure. The bond could act as a weak-link fuse and unzip catastrophically from a local defect.
D-3.	Thick bonded structures need complex stepped-lap joints to develop adequate efficiency.	Large loads require many steps to transfer the load and assure that adhesive develops the strength of the adherends.
D-4.	Anticipate bolted repairs for thick structures by reducing strain levels.	Thick structures are impractical to repair by bonding, except for one-shot and throwaway structures.
D-5.	When there is no need for repair, as in missiles and unmanned aircraft, bonding permits extremely high structural efficiencies to be obtained, even on thick structures.	Load transfer is performed without drilling holes for fasteners.
D-6.	Proper surface preparation is a "must" - beware of "cleaning" solvents and peel plies. Mechanical abrasion is more reliable.	Maintaining joint strength in service is very dependent on the condition of the surfaces to be bonded.
D-7.	Laminates must be dried before performing bonded repairs.	Heat applied to the laminates during repair can cause any moisture present to vaporize and cause blisters.
D-8.	Adherend overlaps must not go below specified minimums.	Key to durability of bonded joints is that some of the adhesive must be lightly stressed to resist creep.
D-9.	Bonded overlaps are usually sized to survive hot/wet environmental conditions.	Elevated temperature and moisture degrade the strength and stiffness of the adhesive.
D-10.	Bonded joint strength can also be degraded by cold environment where adhesive is brittle.	The brittleness of the adhesive limits joint strength.

D-11.	Taper ends of bonded overlaps down to 0.020 inch (0.51 mm) thick with a 1--in-10 slope.	Minimizes induced peel stresses that would cause premature failures.
D-12.	Adhesives work best in shear, are poor in peel, but composites are even weaker in interlaminar tension.	Joint must be designed to minimize out-of-plane stresses.
D-13.	Design of simple, uniformly thick (for near quasi-isotropic carbon/epoxy) bonded splices is very simple. Use 30 t overlap in double shear, 80 t overlap for single-lap joints, 1-in-50 slope for scarf joint.	Provides a bonded joint with good strength capability.
D-14.	Design of stepped-lap joints for thick structure needs a nonlinear analysis program.	Complex stress states in stepped-lap joints. Nonlinear adhesive characteristics.
D-15.	Adhesives are well characterized by thick-adherend test coupon, generating complete nonlinear shear stress-strain curve.	This test provides ample data for analysis of joints critical in shear.
D-16.	For highly loaded bonded joints a co-cured, multiple step, double sided lap is preferred.	Very efficient joint design.
D-17.	Never design a bonded joint such that the adhesive is primarily loaded in either peel or cleavage.	Adhesive peel strength is very poor and unpredictable.
D-18.	Ductile adhesives are preferred over brittle ones.	Ductile adhesives are more forgiving.
D-19.	Film adhesives are preferred over paste adhesives for large area bonds.	Provides more uniform bond line, easier to contain when heated.
D-20.	Balanced adherend stiffnesses improve joint strength.	Reduces peel stresses.
D-21.	Minimize joint eccentricities.	Reduces peel stresses.
D-22.	Use adherends of similar coefficients of thermal expansion.	Reduces residual stresses.
D-23.	Insure the bonded joint configuration is 100% visually inspectable.	Improves reliability and confidence. Need to emphasize process control.

9.3.1.4 *Composite to metal splice joints*

	Rule	Reason
E-1.	Bonding composites to titanium is preferred; steel is acceptable; aluminum is not recommended.	Minimizes differences in thermal expansion coefficient.
E-2.	Bonded step joints preferred over scarf joints.	Better fit, higher strength.
E-3.	Where possible, 45° plies (primary load direction) should be placed adjacent to the bondline; 0° plies are also acceptable. 90° plies should never be placed adjacent to the bondline unless it is also the primary load direction.	Minimizes the distance between the bondline and the plies that carry the load. Prevents failure of surface ply by "rolling log" mechanism.
E-4.	For a stepped joint, the metal thickness at the end step should be 0.030 inch (0.76 mm) minimum and the step no longer than 0.375 in (9.5 mm).	Prevents metal failure of end step.
E-5.	If possible, have ±45° plies end on first and last step of bonded step joint.	Reduces peak interlaminar shear stresses at end steps.
E-6.	If possible, do not end more than two 0° plies (not more than 0.014 inch (0.36 mm) maximum thickness) on any one step surface. For 0° plies ending on last step (longest 0° ply) serrated edges have been shown to reduce stress concentration.	Reduces stress concentration at end of joint.
E-7.	45° or 90° plies should butt up against the first step of a step joint.	Reduces magnitude of interruption in load path.
E-8.	Tension and peel stresses should be avoided in adhesive bonded joints.	Minimum strength direction of adhesive.

9.3.1.5 *Composite to metal continuous joints*

	Rule	Reason
F-1.	Bonding composites to titanium preferred, steel is acceptable.	Minimizes differences in thermal expansion coefficient.
F-2.	No composite to aluminum structural adhesive bond except for corrosion resistant aluminum honeycomb core and lightly loaded secondary structure.	Minimizes interlaminar shear stress due to large difference in thermal expansion coefficient between composites and aluminum.

9.3.1.6 Composite to composite splice jJoints

G-1.	Scarfed joints are never preferred over stepped joints, except for repairs of thin structures.	Improves strength of joint.
G-2.	Cocured joints are preferred over pre-cured joints if there are fit-up problems.	Less sensitive to tolerance mismatches.
G-3.	For pre-cured parts, machined scarfs are preferred over layed up scarfs.	For improved fit.
G-4.	Use of cocured bonded subassemblies should be evaluated in terms of supportability.	Reduces ply count and assembly time, but increases rework cost.
G-5.	Bonded repairs are not acceptable for thick laminates.	Taper ratio requirement makes bonded repair impractical.

9.3.2 Materials and processes

	LESSON	REASON OR CONSEQUENCE
H-1.	Materials selection forms the foundation for structural and manufacturing development and supportability procedures.	The material selected influences critical issues, how parts are fabricated, inspected, and assembled, and how much previous data/learning is available.
H-2.	Material selection must be based on a thorough analysis and occur early in the process.	Various materials have various advantages. Specific applications should use materials that best fit the needs of the application.
H-3.	Imide-based polymer composites shall consider galvanic degradation.	Some of these materials have exhibited galvanic corrosion in the presence of salt water.
H-4.	Net-resin prepregs improve quality at reduced cost.	Minimizes (eliminates) bleeding of prepreg during cure.
H-5.	Composite material applications must have a margin between the wet T_g and the use temperature (usually 50°F).	To prevent the material from operating in an environment where its properties become greatly decreased and widely scattered.

H-6.	Specific issues impacting materials selection/use: Fluid/solvent degradation High residual thermal stresses Mechanical performance Out-time/tack time Effects of defects Sensitivity to processing variations OSHA/EPA requirements Cost (Procurement, Manufacturing, Quality) Environmental degradation Cocure compatibility with other composites and adhesives.	Ignoring key material features could result in an inferior product.

9.3.3 Fabrication and assembly

	LESSON	REASON OR CONSEQUENCE
I-1.	Highly integral cocured structures are weight and cost effective, however, they place a high burden on tooling design.	Integrally cured structure eliminates parts and fasteners. The tools to perform the fabrication are complex and greatly influence the quality of the part.
I-2.	Machining/drilling must be rigorously controlled; this includes feeds, speeds, lubrication, and tool replacement.	Backside breakout is a major nonconformance on all programs. Composite to metal drilling must avoid chip scoring. Highly directionally stacked laminates tend to gouge during drilling in the stacked areas.
I-3.	Waterjet trimming of cured laminates has been shown to be highly successful.	Produces a clean, smooth edge very rapidly.
I-4.	Sanding/trimming must consider out-of-plane damage. Tool rotation must be in the same plane as the laminate.	These operations tend to produce forces in the weakest direction of the laminate.
I-5.	Waterjet prepreg cutting can fray pre-preg edges. Frequent nozzle replacement may be necessary.	Produces acceptable cuts.
I-6.	Lay-up shop temperature/humidity directly impacts handleability.	Tack and drapeability are influenced by temperature and moisture in the prepreg.
I-7.	Unauthorized hand creams can lead to extensive porosity and contamination. The use of gloves can prevent this risk.	Some hand creams contain ingredients that are contaminates.

I-8.	Irons and hot air guns used for ply locating and compaction must be calibrated.	Avoids ply damage due to overheating.
I-9.	FOD control in the lay-up shop is absolutely necessary.	Can lead to foreign materials in the laminate.
I-10.	Hand drills can cause significant damage.	Feed and speed are less precise. Hole perpendicularity may be imperfect.
I-11.	Ply placement tolerances must be able to meet design requirements.	Strength/stiffness analysis is based on assumptions regarding angle of the plies and their location.
I-12.	Assembly jigs must provide the dimensional rigidity necessary to meet assembly tolerances.	Composites are less tolerant of pull-up stresses imposed by poor fit.
I-13.	Engineering drawings and specifications should be supplemented by fully illustrated planning documents or handbooks.	Drawings and specifications tend to be highly complex and detailed. They are not easy to follow on the factory floor.
I-14.	Consider two-step curing process in bonding and cocuring operations.	Alleviates problems such as core slippage and crushing, skin movement, and ply wrinkling.
I-15.	Fastener grip lengths should take into account actual thicknesses (including shims) at the fastener location.	A fastener with excessive grip length may not provide proper clamp-up. Too short a grip length may put threads in bearing or result in an improperly formed head.
I-16.	Tolerance requirements have a big impact on selecting manufacturing and tooling processes and therefore cost.	Different processes produce varying tolerance control.
I-17.	If possible, the mating surfaces should be tool surfaces.	Maintains the best possible dimensional control.
I-18.	The use of molded rubber and trapped rubber tools has had mixed success. Rubber can be used successfully in local areas as a pressure intensifier, such as inside radii on stiffeners of cocured structure.	Rubber tools are difficult to remove, tend to become entrapped. They do not wear well.

I-19.	Analyses can be done to predict distortion or "spring back" of a part after it is removed from a tool. The problem is usually solved by trial and error methods through tool modifications. The "spring back" problem is generally more pronounced on metal tools than on CFRP tools.	Residual or curing stresses build up in composite laminates formed to various shapes. When the structure is removed from the tool, the residual stresses tend to relieve themselves causing "spring back".
I-20.	Tool design, including tool material selection, must be an integral part of the overall design process.	Tool design is dependent on part size and configuration, production rate and quantity, and company experience.
I-21.	Aluminum tools have been used successfully on small parts but are avoided on large parts and female molds.	Thermal expansion mismatch.
I-22.	Invar is often used for production tooling.	Invar has good durability and low thermal expansion.
I-23.	Electroformed nickel also produces a durable, high quality tool, but is less frequently used.	More expensive.
I-24.	Steel or Invar tools are needed for curing high temperature resins such as polyimides and bismaleimides.	The thermal mismatch with other materials is magnified at the higher cure temperature of these resins.
I-25.	Air bubbles in a silicone rubber tool will cause "bumps" in the cured laminate.	The tool fails to provide support for the laminate and apply uniform pressure.
I-26.	Resin containment is essential to part thickness control.	Uncontained resin will cause resin rich and resin starved areas.

9.3.4 Quality control

	LESSON	REASON OR CONSEQUENCE
J-1.	Continuing process control and process monitoring are required during production.	Assures that neither the process nor the material is changing.
J-2.	Ultrasonic C-Scan is the most commonly used NDI technique. It may be supplemented by other techniques such as X-ray, shearography, and thermography.	Useful for detecting porosity, disbonds and delaminations.

	LESSON	REASON OR CONSEQUENCE
J-3.	Determine and understand the effect of defects on part performance.	Minimizes the cost of MRB activity.
J-4.	There is no substitute for destructive, tear-down inspections of complex parts under development.	Not all discrepancies can be detected by NDI methods.

9.3.5 Testing

	LESSON	REASON OR CONSEQUENCE
K-1.	The testing of joints and demonstration of damage tolerance should include sufficient detail to adequately evaluate structural details and size effects.	Small details and size effects can have a large influence on the response of composite structure. In general, damage tolerance of composites exhibits size effects. Bolted and bonded joints, if properly designed, do not.
K-2.	A well planned test program must include an accelerated approach for taking into account the effects of moisture, temperature, impact damage, etc.	Including moisture and elevated temperature on a real-time basis for full-scale testing is impractical for most components.
K-3.	A finite element analysis should be performed prior to conducting a full-scale test. The analysis must accurately simulate the test article and the boundary conditions of the test fixture and loads applied during the test.	For a more accurate assessment of the internal loads and failure prediction of the test article.
K-4.	Traceability of test coupons to batch, constituent material lots, autoclave run, panel, position in panel, and technicians is essential to data analysis.	If full traceability is not maintained and documented, the cause of outlier data points or unexpected failure modes may be difficult to identify. The result is that "bad" data, which might legitimately be discarded for cause, might be retained and add undeserved variability to the data set.
K-5	Adequate instrumentation is essential for all design/development or concept validation testing. Placement of strain gages, LVDT's, etc., should be based on analysis.	A good understanding of local failure modes and correlation of test results with analysis will aid the design process.

9.3.6 Certification

	LESSON	REASON OR CONSEQUENCE
L-1.	The "building block" approach is an excellent method for developing and validating the details of the design.	A wide variety of issues and details can be evaluated cost effectively. Hardware serves a dual purpose - engineering and manufacturing.
L-2.	Component qualification is complicated by the fact that critical design conditions include hot, wet environments. This is often accomplished by overloading a test article that is in ambient conditions, or by analysis of failure modes coupled with strain measurements related back to subcomponent hot, wet tests.	It is generally impractical to try to ingest moisture in full scale test articles and test them hot.

9.3.7 In-service and repair

	LESSON	REASON OR CONSEQUENCE
M-1.	In spite of concerns about the sensitivity of composites to damage, experience in service has been good. Navy aircraft have not experienced any delamination failures in service. Most damage has occurred during assembly or routine service performed on the aircraft.	Current design, fabrication, and certification procedures adequately prepare the structure to survive its intended environment.
M-2.	Composite components located in the vicinity of engine exhaust are subject to thermal damage. At present there are no acceptable NDI methods for detecting thermal damage of matrix materials.	Composite components exposed to engine exhaust or other heat sources s-hould be shielded or insulated to keep temperatures down to an acceptable level.
M-3.	Moisture ingestion is the biggest problem with honeycomb sandwich structure. The thin, stabilized skins that make honeycomb structurally efficient are also the reason they are damage prone. Panels get walked on and damaged.	Honeycomb design must be applied judiciously. Repair must account for the possibility of water in the core.
M-4.	Aircraft are commonly painted and repainted. Paint stripping has been done with solvents. Solvents can damage epoxy matrices.	Increased use of water-based paints and solvent-less stripping of paint is desirable.

M-5.	Records pertaining to MRB actions and in plant repairs of composite parts should be readily available to personnel responsible for in-service maintenance.	During routine maintenance checks, depot personnel sometimes find defects or discrepancies. In some cases they have been able to determine that the "defect" was in the part at delivery and considered acceptable.
M-6.	Supportability and repair must be responsive to service environment.	It is necessary to account for equipment, facilities, and personnel capabilities.

REFERENCES

9.2.3 Grimes, G.C., "Tape Composite Material Allowables Application in Airframe Design/Analysis," *Composites Engineering*, Vol. 3, Nos. 7-8, pp. 777-804, 1993.

This page intentionally left blank

INDEX

Volume 3

Custodians:
- Army - MR
- Navy - AS
- Air Force - 11
- DOT - FAA

Review activities:
- Army - AT, AV, MI
- Air Force - 15

Civil agencies:
- DOT - ACO
- NASA - NA

OASD section:
- SI - IQ

Preparing Activity:
- Army - MR
- Project No. CMPS - 0131

This page intentionally left blank

For Product Safety Concerns and Information please contact our EU
representative GPSR@taylorandfrancis.com
Taylor & Francis Verlag GmbH, Kaufingerstraße 24, 80331 München, Germany

www.ingramcontent.com/pod-product-compliance
Lightning Source LLC
LaVergne TN
LVHW080754300926
849162LV00008BA/1809
9781566768283